MCGRAW-HILL PUBLICATIONS
IN THE AGRICULTURAL SCIENCES

ADRIANCE AND BRISON, *Propagation of Horticultural Plants*

AHLGREN, *Forage Crops*

ANDERSON, *Diseases of Fruit Crops*

BROWN AND WARE, *Cotton*

CARROLL, KRIDER, AND ANDREWS, *Swine Production*

CHRISTOPHER, *Introductory Horticulture*

CRAFTS AND ROBBINS, *Weed Control*

CRUESS, *Commercial Fruit and Vegetable Products*

DICKSON, *Diseases of Field Crops*

ECKLES, COMBS, AND MACY, *Milk and Milk Products*

ELLIOTT, *Plant Breeding and Cytogenetics*

FERNALD AND SHEPARD, *Applied Entomology*

GARDNER, BRADFORD, AND HOOKER, *The Fundamentals of Fruit Production*

GUSTAFSON, *Conservation of the Soil*

GUSTAFSON, *Soils and Soil Management*

HAYES, IMMER, AND SMITH, *Methods of Plant Breeding*

HERRINGTON, *Milk and Milk Processing*

JENNY, *Factors and Soil Formation*

JULL, *Poultry Husbandry*

KOHNKE AND BERTRAND, *Soil Conservation*

LAURIE AND RIES, *Floriculture*

LEACH, *Insect Transmission of Plant Diseases*

MAYNARD AND LOOSLI, *Animal Nutrition*

METCALF, FLINT, AND METCALF, *Destructive and Useful Insects*

NEVENS, *Principles of Milk Production*

PATERSON, *Statistical Technique in Agricultural Research*

PETERS AND GRUMMER, *Livestock Production*

RATHER AND HARRISON, *Field Crops*

RICE, ANDREWS, WARWICK, AND LEGATES, *Breeding and Improvement of Farm Animals*

ROADHOUSE AND HENDERSON, *The Market-milk Industry*

STEINHAUS, *Principles of Insect Pathology*

THOMPSON, *Soils and Soil Fertility*

THOMPSON AND KELLY, *Vegetable Crops*

THORNE, *Principles of Nematology*

TRACY, ARMERDING, AND HANNAH, *Dairy Plant Management*

WALKER, *Diseases of Vegetable Crops*

WALKER, *Plant Pathology*

WILSON, *Grain Crops*

WOLFE AND KIPPS, *Production of Field Crops*

*Professor R. A. Brink was Consulting Editor of this series from 1948
until January 1, 1961
The late Leon J. Cole was Consulting Editor of this series from 1937 to 1948.
There are also the related series of McGraw-Hill Publications in the Botanical
Sciences, of which Edmund W. Sinnott is Consulting Editor, and in the Zoological
Sciences, of which Edgar J. Boell is Consulting Editor. Titles in the Agricultural
Sciences were published in these series in the period 1917 to 1937.*

Breeding and Improvement
of Farm Animals

SIXTH EDITION

VICTOR ARTHUR RICE

Deceased

FREDERICK NEWCOMB ANDREWS

*Professor of Animal Science, Vice President for Research
and Dean of the Graduate School, Purdue University*

EVERETT JAMES WARWICK

*Chief, Beef Cattle Research Branch
Animal Husbandry Research Division
United States Department of Agriculture*

JAMES EDWARD LEGATES

*William Neal Reynolds Professor of Animal Science
Head, Animal Breeding Section, Department of Animal Science
North Carolina State University*

MCGRAW-HILL BOOK COMPANY

New York, St. Louis, San Francisco, Toronto, London, Sydney

In the preface to the first edition, published in 1926, Professor Rice wrote, "This book has been written primarily for use as a textbook of animal breeding. It treats the subject of breeding practice from the scientific standpoint in so far as this is possible at the present time. Breeding is an art to be learned only by practice, but knowledge of principles supplies the only firm foundation for its practice. Superior animals will be more numerous when breeders know why as well as how."

Of the sixth edition it might be said that only the title and the basic philosophy, "Breeding is an art to be learned only by practice, but knowledge of principles supplies the only firm foundation for its practice," remain unchanged.

Professor Rice participated in the planning and general format of this edition and had completed a draft of the introductory chapter just prior to his death on June 25, 1964. His literary style was known to thousands of college students and his brilliant classroom techniques to a smaller but still substantial number of privileged listeners. As in the fifth edition, the authors decided to produce an entirely new manuscript and, after long and searching analysis, were convinced that the format should be substantially altered. The information on the basic principles of genetics and the basic principles of reproductive physiology, which was once the very heart of the book, has been greatly condensed and now serves only as a review for the advanced undergraduate student. For in the intervening years most universities have strengthened the curriculum by adding separate courses in genetics and the physiology of reproduction.

Uncounted generations of livestock breeders have made valuable contributions to the practical art of animal breeding by the persistent practice of concepts such as "mate the best with the best" and "like tends to beget like." During past centuries and especially in the recent decades, there have been marked changes in the total animal environment and also in human demand. The once desirable objectives of the selection of dairy cattle for milk of high butterfat content, of the lard-type hog, and of marked fat deposition in early-maturing meat animals are undergoing reevaluation. The contemporary animal producer needs to understand not only the basic principles of genetics and animal breeding but how to evaluate and apply them in a changing environment of human wants, economic pressures, and livestock feeding and management practices.

We, like all authors, are greatly indebted to earlier workers and writers in our field and to a host of colleagues and contemporaries. No list of acknowledgments could possibly be complete. We would, however, like to give special thanks to several who have given us ideas, constructive criticism, or materials not otherwise acknowledged which have found their way into this book: J. O. Almquist, Lemuel Goode, L. N. Hazel, C. M. Kincaid, S. M. Scheinberg, C. E. Shelby, O. W. Robison, T. G. Martin, E. U. Dilliard, and C. E. Terrill.

It is the authors' hope that this revised sixth edition will provide insight into some of the current theories and practices of animal breeding that we may serve our fellow men.

Frederick Newcomb Andrews
Everett James Warwick
James Edward Legates

CONTENTS

CHAPTER ONE

Introduction

Since his emergence on this planet, man's greatest problem has been that of securing enough food of a kind suitable to meet his nutritional needs. As has been so well said, "Without food, nothing else matters." In the United States of America we have an abundance of food, in many items more than we can consume, but Asia's teeming millions are constantly haunted by the fear of starvation—a fear that becomes an actuality for thousands each year. And the future looks even blacker.

Most of the Earth's people do not get enough food of any sort to meet their daily needs and far too little of the nutritious, palatable, protective foods such as milk, eggs, and meat that are required for normal growth, energy, and resistance to various diseases.

We have the knowledge to make the Earth's arable acres yield more abundantly and to increase the number of those acres, but ignorance, indolence, illiteracy, tradition, superstition, religious taboos, and, most of all, lack of capital prevent advanced agricultural technology from being put to work.

The above is written in terms of the 3 billion present inhabitants of Earth. But the world's reproductive rate is such that the 3 billion will become 6 billion in 40 years. There is no reason to expect it will not grow to 12 billion in the next 40 years. Perhaps the greatest paradox for man will lie in the fact that his conquest of death led to his destruction. Fewer babies die and people live to a greater average age. In short, man has lowered the death rate but as yet has done nothing effective about the world's birth rate. Imagine a house with front and back doors in which there is a certain comfortable number of people, babies coming in the front door and old people going out the door at the back. If as many come in the front as go out the back, everyone remains comfortable. But if we leave the front door (birth rate) wide open and make it harder to get out the back door (lower the death rate), sooner or later the house will become overcrowded. This analogy fits our present Earth. High birth rates and declining death rates, especially in parts of Asia, Africa, and Latin America, are compounding man's social difficulties, thwarting his attempts to find amicable solutions to world problems, making much "foreign aid" useless, since the population increase continues to outstrip whatever assistance the "aid" might bring. Demagoguery of all descriptions fattens on poverty, and because of man's runaway birth rate the Earth's poverty increases daily as it struggles unsuccessfully to accommodate a net increase of 135,000 people daily, or about 50 million each year. What the Earth will be like 100 years (or a thousand or a million) from now depends upon, more than anything else, how many babies are born and whether among parents those better endowed have as many or more offspring than the nincompoops. The world at large needs food in increased quantity and improved quality. It is in the latter category that animal husbandry can make its greatest contribution. All the energy reaching our planet Earth originates in the sun, and a goodly portion of it becomes stored through photosynthesis in forms which man cannot utilize directly as a source of food energy. Man therefore utilizes animals to convert a wide variety of materials into more palatable and more nutritious forms. In addition to converting waste into wealth, an agriculture based largely on animals also helps to conserve our topsoil and to return a large measure of necessary chemical elements to the soil.

There are essential differences among animals in the over-all efficiency with which they convert plant tissues into foodstuffs or other animal products. Improvement of animals through breeding and selection has but one goal—to improve continually their efficiency as converters and the quality of their end products. Considerable progress toward this goal has been made since animals were first domesticated. When presently known principles of breeding come into more general understanding and use, further improvements in livestock production and efficiency will result. Concurrent with hereditary improvement we need also to invoke proved methods of improving animal environment. Milk yield from all cows in the United States now averages about 6,000 lb per year. Although specific proof may be lacking, it seems probable that this figure is at least one or two thousand pounds below their average hereditary limit under good environmental conditions and far below what it might be under the best conditions. The basic job of the animal breeder is to learn all he can about the probable genetic make-up of his animals from their own individuality, their close relatives, and especially from their performance records, and then through systems of breeding and selection, to mold or guide these superior hereditary determiners into more favorable combinations.

MODERN IDEAS ON LIFE'S ORIGIN

The best information available seems to indicate that under suitable conditions various elements and compounds in the warm seas became organized into very simple forms of living material with the power of nourishing and reproducing themselves. The materials so formed were probably relatively stable, but with a tendency for polymerization as well as for various rearrangements of their atoms and molecules to produce somewhat different forms. From these earliest living forms there perhaps eventually came to be organizations of materials resembling the gene. You will learn considerable about genes in later chapters; suffice it here to say that genes are the ultimate, indivisible units of heredity—the causative agents responsible for the expression of any characteristic. They are the mechanism of life and heredity today, and it seems quite probable that they have played this role since life first appeared on our planet.

Existing singly at first, the genes perhaps combined in various patterns to form chromosomes or chromosomelike units. Later, cytoplasmic materials and a cell wall may have been added, and after a few hundred million years development may have reached the one-celled stage. Growth, aggregation, variation, differentiation, and specialization of function proceeded, producing finally the hundreds of thousands of plants and animals which at present inhabit the Earth.

We cannot here trace even briefly the various steps in the evolution of life. Suffice it to say that it seems to have proceeded from the gene, to single-celled animals, to many-celled animals involving even greater complexity in structure and function. When life became possible in the sea and later on the land, it radiated in all directions, continually producing new forms. Animal life progressed through the invertebrates (creatures with no backbone), the fishes, amphibia, reptiles, and birds, and finally came the origin and differentiation of the mammals.

The lowest mammalian forms are those like the spiny anteater and the duckbill, or platypus. In these forms the egg is still laid outside the body, as in lower forms, and the digestive tract and urogenital organs empty into a common vent, or cloaca, as in reptiles and birds. Hair has now replaced scales, and a diaphragm separates the abdominal from the thoracic cavity. The brain is not yet deeply wrinkled. The primitive mammary glands are scattered over the belly region of the female and secrete a fluid that is licked up by the young.

The second stage in mammalian development is represented by the marsupials, such as the kangaroo and the opossum. The young are discharged from the uterus in a very immature state and are transferred to a pouch on the mother's abdominal wall, where they complete their fetal growth. These species are viviparous, as were some reptiles and fish before them, and have separate openings for urogenital and alimentary tracts as well as a distinctly four-chambered heart.

The highest form of life process yet produced is that of the placental mammals. In these the fetus is retained for a longer time in the mother's womb and is nourished by means of the placenta. The brain is more convoluted, or creased. The mammae are well developed for suckling and are arranged over the belly region usually in two rows, although in the highest types they become restricted to the breast region. Representative mammalian forms are the ungulates, or hoofed animals, such as horses, cattle, sheep, and swine; the carnivores, such as cats, dogs, wolves, and tigers; the rodents, such as rats and squirrels; the anthropoids, such as apes, gibbons, gorillas, orangs, and chimpanzees; and, finally, man.

THE THEORY OF EVOLUTION

By definition, the theory of evolution is that the various types of contemporary plants and animals have arisen by descent with modification from other preexisting types. Many famous persons, beginning with early Hindu and Greek philosophers, have been responsible for man's slowly acquired knowledge of the evolution of organisms, but space permits the mention of only a few.

Buffon (1707–1788), a French naturalist, greatly enlarged the concept of mutability through direct environmental means, an idea which he developed in later life, having earlier subscribed to the idea of immutability. He was a precursor of Lamarck and perhaps also of Darwin in regard to pangenesis as well as the struggle for existence and the survival of the fittest. Coupled with his idea of change due to environmental influences was his further implied belief that these so-called *acquired* characters were heritable.

Erasmus Darwin (1731–1802), an English physician and poet and grandfather of Charles Darwin, is one of the most imposing figures in the field of human inquiry. He borrowed and enlarged on many old ideas concerning evolution and made distinctive new contributions. He was the first to stress the idea that evolution has been operating from the time of the first primordial life. To him new forms were but the flowering of potentialities originally deposited in the life stream by the Creator and called into being by the necessity for adaptation to the environment.

Lamarck (1744–1829), a French naturalist, can be rightfully called the father of the modern theory of evolution. He was the first to devise a classification scheme or phyletic tree to include all plants and animals, and he was the first to state that all animals formed a branching series of related forms, shading into one another by very gradual steps. His theory of evolution, propounded in 1809, had three main points:

1. The environment, directly in plants but indirectly in animals through the medium of the nervous system, causes changes in the organism.
2. The use or disuse of parts leads, respectively, to their further development or to their atrophy.
3. These so-called *acquired* characteristics are inherited.

Lamarck's belief that the environment is the principal cause of change is similar in its general outline to Buffon's earlier belief that something external to the organism brings about changes. Geoffroy Saint-Hilaire (1772–1844), a French zoologist, subscribed to Lamarck's idea that environmentally induced changes are heritable, but he believed that the changes were produced in the embryonic or germinal condition rather than in the adult. He thus, in a sense, anticipated Weismann's idea of germinal variation and causation as well as De Vries' idea of evolution by means of large jumps or mutations.

DARWIN

Charles Darwin (1809–1882), the best-known proponent of the theory of evolution, was an English naturalist. Darwin was not the father of the idea of a gradual progressive change. He might perhaps more correctly be called the attending physician who brought the concept safely into the world or, as Butler said, "Darwin's chief glory is not that he discovered evolution but that he made men believe in it, and what glory," he added, "could be greater than this?"

Darwin sailed on the H.M.S. *Beagle* as naturalist on a round-the-world trip lasting from 1831 to 1836. On this voyage he had the opportunity of studying the rich fauna and flora of South America and several island archipelagoes. He was struck by the manner in which animal types shaded into one another and by the distinctive forms found on separate islands. He had begun his journey with a belief in separate creation, but during the course of the voyage he became convinced of the mutability of species. On his return to England in 1837 he began to organize the known facts on plant and animal variation and to try to discover new ones. During the next year, while reading Malthus' "Essay on Population," the idea of the struggle for existence that is constantly going on in nature as an explanation of the great variety of plant and animal species came to him, as it had previously come to Wells, Matthew, and perhaps many others.

Darwin's theories involved four points:

1. Organisms vary, i.e., are not exactly like their parents.
2. These variations are or may be hereditary, i.e., passed along to descendants.
3. Due to the dynamics of reproduction in all species there is a continual struggle for existence in nature.
4. Those best adapted to survive in a given environment will be the most likely to survive and should therefore leave the most descendants.

Darwin thought that the working of the above-named four principles would account for the great variety of forms of plants and animals that now inhabit the Earth or have inhabited it in the past. He accepted evolution as a working hypothesis, accepted the obvious facts of variation and hereditary transmission of potentialities, and adduced a tremendous amount of data that seemed to support his thesis that natural selection or, to use Spencer's term, "the survival of the fittest," is the leading actor in this drama of survival.

EVIDENCE FOR EVOLUTION

Space does not permit a lengthy discussion of the various lines of evidence that seem to indicate the validity of the general theory of evolution. In outline, they are as follows:

1. *Classification, or taxonomy.* The fact that the 800,000 described species of animals can be arranged into 11 phyla and the 250,000 known species of plants into 4 phyla, the whole resembling a branching tree.
2. *Comparative embryology.* The fact that all embryos start as single-celled zygotes, progress along similar lines of development, and that those of the higher species pass hurriedly through stages which are the end stage of organisms lower down the evolutionary scale.
3. *Comparative anatomy.* The fact that the brain, heart, and other organs show a progressive development from lower to higher species; that all mammals have the same bony framework—the bones in the flipper of a whale, the wing of a bird, the front leg of a horse, and the arm of a man being similar in general form and articulation.

4. *Vestigial structures.* The fact that man during his embryonic life develops a set of gill clefts and arches, a tail, and a fairly heavy covering of hair, all of which disappear before birth, and that adult man has ear and tail muscles, an appendix, and many other vestigial structures which are well-developed and functional in lower forms.

5. *Paleontology.* The fact that fossil remains of animals show a progressive development from lower to higher in ascending strata of the Earth's crust.

6. *Zoogeography.* The fact that species grow increasingly divergent in form in widening circles from their point of origin.

7. *Blood tests.* The fact that the blood of closely related animals shows small incompatibilities as measured by amount of cellular agglutination, whereas that of distantly related ones shows great incompatibilities.

8. *Observation and experiment.* The fact that all of our breeds of farm animals have been developed by a process of selection during about the past 200 years and that many new breeds are in the process of formation.

DOMESTICATION OF ANIMALS

The domestication of animals and plants provided the foundation on which civilization could be built. Without a productive agriculture it seems reasonable to conclude that social and scientific progress would have been limited, if not entirely prevented. Just when the domestication of animals began is not known; it certainly was long before there was any thought of, or means for, recording historical data. Domestication very likely began at the end of the Old Stone Age and received decided impetus during the New Stone Age. Each domestic animal has had a rather restricted and humble origin, yet through constant selection for one or another character all the multifarious forms have evolved— some useful, some ornamental, some having both qualities, and some neither.

THE DOG

Dogs are represented in Egyptian monuments as far back as 3400 B.C., and various breeds of dogs existed during the height of the Roman civilization. Their skeletal remains are found as far back as the New Stone Age and possibly as far back as the Old Stone Age. Dogs have been used for draft purposes a very long time and may have originally served as a source of food. There are now six distinct groups of dogs, namely, hound, greyhound, mastiff, spaniel, terrier, and wolflike breeds. Mason's "Dogs of All Nations" lists 183 breeds of dogs from 39 countries.

The variations in the different breeds of dogs as to size, form, color, coat, shape of jaw, length of body, temperament, vitality, and educability cover a wide range. All the breeds, from the huge Saint Bernard to the tiny Mexican Chihuahua, are the results of nature's variations plus man's selection. In *Life* magazine, January 31, 1949, there was a very interesting story entitled "Dogs of America," with many beautiful illustrations showing the probable origin of many of our present-day breeds.

THE HORSE

The name *horse* is equivalent to the Anglo-Saxon *hors,* which means swiftness, and it is logical to suppose that this genus was able to survive the vicissitudes of time and enemy attack chiefly because of its speed.

The only surviving genus of the family Equidae, to which the horse belongs, is that designated as *Equus.* There are at present four living types: horses, asses, half asses, and zebras. Besides the domesticated horses, *Equus caballus,* there is one wild type found in Mongolia, *E. przhevalski.* The domestic ass, *E. asinus,* has at least two wild varieties, *africanus* and *somaliensis.* In addition there are three Asiatic half asses, *E. kiang, E. onager,* and *E. hemionus.* Among zebras there are three types: *E. zebra, E. burchelli,* and *E. grevyi;* in addition there were several varieties of *E. quagga* (now extinct). All these species of horses, asses, and zebras are capable of producing hybrids among themselves, but the hybrids are generally sterile. The members of the genus *Equus* are generally taller and swifter animals than their near relatives the tapirs and rhinoceroses, their greater height being due to a lengthening of the cannon bones and middle toe. The head and the tail are also longer, and the body has a much denser hairy covering.

Origin and Domestication of the Horse. The horse was probably the last animal to be domesticated by man, but his immediate ancestry as well as the date of his domestication is still a matter of dispute. It seems probable that at least two, or perhaps three, wild types have made their contribution to our domestic horse. One of these was probably the *steppe horse,* now known as the fossil representative of Przhevalski's horse. This was a small, sturdy, short-legged horse with a moderately long, heavy head. Another was the so-called *desert horse,* standing, as did the steppe horse, about 13 hands high and corresponding closely to the now-extinct tarpan, or Mongolian horse. This strain was somewhat more slender than the steppe horse and had a shorter head. The third contribution was that of the *forest horse,* a type standing about 15 hands high with longer but stout limbs and having a long, narrow head and long body. It seems probable that all three types made a contribution to our modern breeds. The horse was apparently domesticated separately in Asia and Europe, probably earlier in Asia. A Turanian folk tamed the Przhevalski horse around 3000 B.C.

The earliest record of the horse dates back to Paleolithic times, about 25,000 years ago. Around an open camp at Solutré in France are found the remains of several thousand horses, indicating that horses may have served as a source of food. In later Paleolithic times, rock carvings of the horse were made, but they do not show him harnessed, ridden, or attached to any sort of vehicle; so we assume the horse was not yet fully domesticated.

The earliest trace of the horse hitched to a chariot goes back to about 2000 B.C. in Greece, whereas the first Egyptian records of the domestication of the horse date from about 1600 B.C. These were small horses, about 13 hands high, similar to Przhevalski's horse. The horse evidently grew in size and importance in Persia and Mesopotamia during the intervening years, and after about 750 B.C. it began to serve as a mount. Mounted horsemen were first given a place in the Olympian games in 648 B.C. The Arabs did not use horses until after the time of Christ.

The ancestry of the horse has been traced back about 55 million years by means of skeletons found in descending layers of the Earth's crust. The forerunner of our present horse was an animal about 10 to 20 inches tall with four toes on the front feet and three on the back. Twenty million years later he stood about 25 inches tall and had three toes fore and back. He has since been reduced to but one toe (with two splints), but his size has greatly increased, and his teeth have developed into more efficient tools for grinding feed. Series of skeletons showing these changes are on exhibit at Amherst College in Massachusetts and at the Museum of Natural History in New York.

CATTLE

The Pecora, or true ruminants, include the following families: the Cervidae (deer), Giraffidae (giraffes); and Bovidae, to which belong all types of cattle as well as sheep, goats, and true antelopes. Both sexes in these families are usually horned. The horns are hollow and arise from the frontal bones. The second and fifth digits are rudimentary or absent, the third and fourth fully developed. The Pecora are ungulates, or hoofed, and belong to the group known as *artiodactyls,* or even-toed. There are one or more enlargements, usually three, for food storage along the esophagus, and the members of these families ruminate, or chew their cuds.

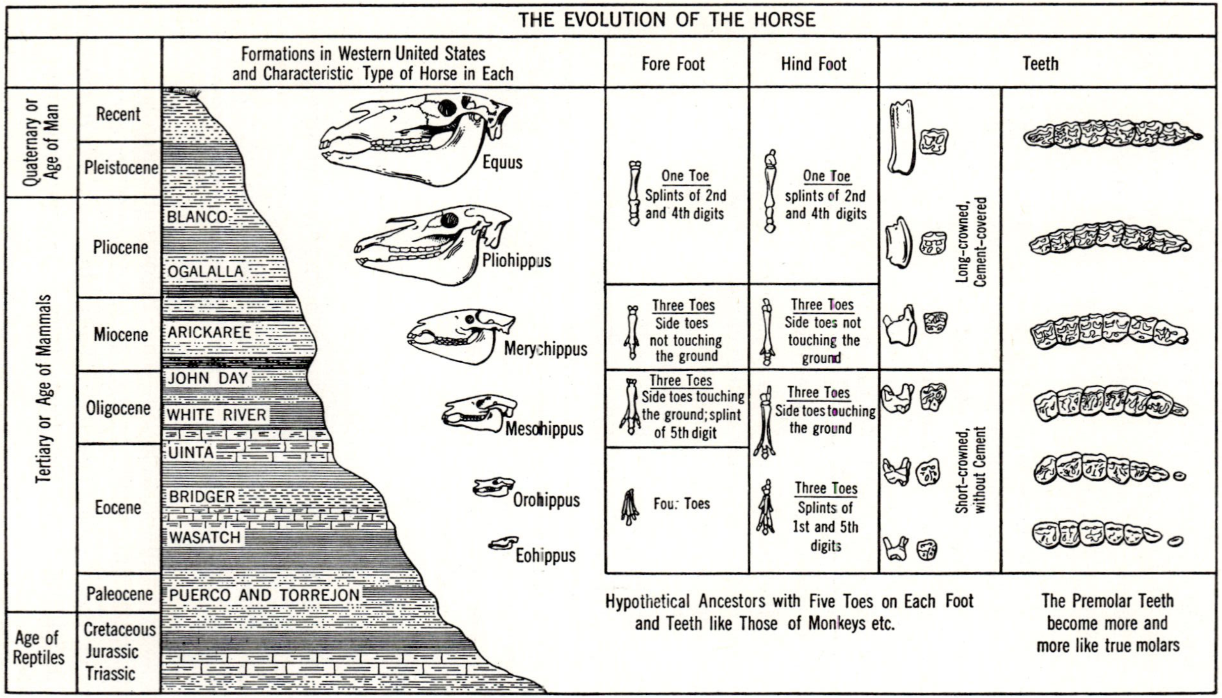

THE EVOLUTION OF THE HORSE

Fig. 1-1. Showing progressive changes in skull, feet, and teeth of the horse. (*From Matthew, Quarterly Review of Biology, Williams & Wilkins Company.*)

The genus *Bos* includes: (1) the taurine group, *B. taurus* (cattle) and *B. indicus* (humped cattle); (2) the bibovine group, *B. gaurus* (the gaur), *B. frontalis* (the gayal), and *B. sondaicus* (the banteng); (3) the bisotine group, *B. grunniens* (the yak), *B. bonasus* (the European bison), and *B. bison* (the American bison); (4) the bubaline group, *B. caffer* (the African buffalo), *B. bubalis* (the Indian buffalo), *B. mindorensis* (the Mindora buffalo), and *B. depressicornis* (the Celebes buffalo). In most of these species there are many varieties, both living and extinct.

Origin and Domestication of Cattle. It seems probable that cattle were domesticated during the New Stone Age in both Europe and Asia. There are two types of domestic cattle now living: *B. indicus,* the humped cattle of tropical countries, and *B. taurus,* of the more temperate zones. Humped cattle were domesticated as early as 2100 B.C. Cattle played an important part in Greek mythology; they were sacred animals in many older civilizations, and their slaughter was therefore forbidden. The great ox, or aurochs, *B. primigenius,* which Caesar mentioned in his writings, is generally considered to be one of the progenitors of our modern-day breeds. This was a very large animal, described by Caesar as "approaching the elephant in size but presenting the figure of a bull." The wild park cattle of Britain are considered by some authorities to be the direct descendants of *B. primigenius.* Another progenitor of our modern breeds is *B. longifrons,* a smaller type, with somewhat dished face. This is the Celtic Shorthorn, which has been found only in a state of domestication. It was the only ox in the British Isles until 500 A.D. when the Anglo-Saxons came and brought the great ox, or aurochs, of Europe.

It is doubtful whether any of our present-day European or American breeds trace back solely to either one of these ancient types. It seems much more probable that our present breeds are the result of various degrees of crossing between them. The cattle of India and Africa, *B. indicus,* are characterized by a lump of fleshy tissue over the withers sometimes weighing as much as 40 or 50 lb. They also have a very large dewlap, and the voice is more of a grunt than a low. They are thought to be descended from the wild Malayan banteng.

It seems probable that primitive man first used numbers of the family Bovidae as a source of food. Domestication perhaps began when these animals were used as draft animals, probably in the first steps of the tillage of the soil. In their wild state there was little tendency to store excess fat on the body, as this would have been a hindrance rather than a help under the conditions then existing. Milking qualities also were just sufficient for the rearing of the young. As civilization developed, feed became more abundant, methods of caring for livestock improved, and the latent possibilities for rapid growth, fat storage, and milk production began to be realized under man's selection.

That the ox played an important part in man's aesthetic development is attested by its use in architectural and interior mural decoration as well as by its frequent use as a subject of poetic fancy. The ox assumed great religious importance in many ancient civilizations; the best members of the breed were sacrificed to propitiate the gods. They were crowned with wreaths and honored in other ways during the pageants and holidays. To some extent we perpetuate this custom in our fairs and expositions today. The Romans' term for money was *pecunia*, a word derived from *pecus*, meaning cattle, and in ancient times a man's wealth was figured in terms of his cattle possessions.

SHEEP AND GOATS

These two genera of the family Bovidae are very closely related, so closely, in fact, that a naturalist never speaks lightly of "separating the sheep from the goats." The genus *Ovis* includes the sheep and its wild relatives, whereas goats and their kind make up the genus *Capra*. Sheep are distinguished from goats by glands in both forefeet and hind feet, by the absence of a true beard, and by the absence of the strong goaty odors in males. There are also marked differences in the skulls; and the horns generally spiral in opposite directions, the right horn of the sheep to the right like a corkscrew, and the goat's to the left. The sheep gets its Latin name *Ovis* from the Sanskrit *avi*, signifying to keep or to guard.

Our domestic sheep have a great variety of wild relatives. Lydekker divides these into four groups: (1) the bighorn (*Ovis canadensis*) of North America and Kamchatka, (2) the argali (*O. poli*) of Central Asia, (3) the urial (*O. vignei*) of Asia and the mouflon (*O. musimon*) of Asia Minor and Europe, and (4) the bharal (*O. nahura*) of Little Tibet and the Barbary (*O. tragelaphus*) of North Africa.

Origin and Domestication of Sheep. Sheep probably originated in Europe and in the cooler regions of Asia in the Pleistocene or later Pliocene era. Remains of a sheep or goatlike animal have been found at the sites of the Swiss lake dwellings of Neolithic times. Sheep are thought to have been derived from the antelopelike animals allied to the gazelles because of certain similarities of the molar teeth. It seems certain that our modern breeds trace back to at least two remote ancestors, the mouflon of Europe (*O. musimon*) and the Asiatic urial (*O. vignei*).

The sheep was originally a hairy animal with an underfur of wool. No doubt people living in cold climates who used skins as clothing were the first to begin the selection of sheep for wool production. As in all our domesticated animals, there is wide variation among sheep. Some, like the African long-legged and Abyssinian maned sheep, bear hair instead of wool; some have spiral horns 2 feet or more in length, others no horns at all. The tail of the common domesticated sheep is long and slender; in some other strains it is a fat depot about 1 foot in width; whereas still others have merely a vestige of a tail. The last sort often carry huge patches of fat on the rear quarters, the stored fat in all cases serving to tide the animal over periods of food shortage. The hunia, a tall, long-legged sheep, is used in India as a fighting animal.

Goats are also versatile in characteristics, yielding the underfur for Cashmere shawls and providing mohair, milk, meat, and draft power; they also provide one means of clearing up brush land because of their fondness for all sorts of tender shoots.

SWINE

Swine are ungulates belonging to the suborder artiodactyls (even-toed). They belong to the family Suidae. The Dicotylidae, or peccaries, and the Hippopotamidae, or hippopotamuses, are closely related families, these three families comprising the Suina. These animals have tubercles on the molar teeth, and in them there is not found a complete fusion of the third and fourth metapodials to form a cannon bone. The nose is elongated into a more or less mobile snout.

The genus *Sus* includes, besides the domesticated pig, several wild species. Among them are *Sus scrofa,* the European wild boar; *S. cristatus,* the Indian wild boar; *S. andamanensis* of the Andaman Islands; *S. salvanius* from the foot of the Himalaya; *S. vittatus, S. verrucosus,* and *S. barbatus* of the Malayan region; *S. africanus* and *S. procus* of Africa. The foregoing species vary greatly in size, color, length of coat, length of tail, presence or absence of crest on the neck, size of tusks, number of mammae, etc. The Indian wild boar stands 30 to 40 inches high, whereas *S. salvanius* measures only 11 inches over the shoulder. The tusks of the wild boar are several inches long; those of the red river hog, *S. africanus,* only as long as those of the domestic boar; and those of *S. salvanius* are very short. The young of all the above-mentioned species are born with longitudinal dark stripes on the body which disappear as the animal matures.

Subclass I. Prototheria.
  Order I. Monotremata—Duckbill mole, spiny anteater, etc.
Subclass II. Metatheria.
  Order II. Marsupialia—Opossum, kangaroo, etc.
Subclass III. Eutheria.
  Order III. Edentata—Sloths.
  Order IV. Sirenia—Manatees.
  Order V. Cetacea—Whales, porpoises, etc.
  Order VI. Ungulata (hoofed mammals).
    Suborder 1. Artiodactyles (even-toed).
      A. Suina.
        Families: a. Hippopotamidæ.
                      Hippopotamuses.

      b. Dicotylidæ.
           Peccaries.

      c. Suidæ.
           True pigs.
           1. Genus Sus—Sus scrofa—wild boar.
                 S. vittatus—collared Malayan pig, etc.
                 S. domesticus—domestic swine.
           2. Genus Babirusa.
                 Long curved upper tusks.
           3. Genus Phacochaerus.
                 Wart hogs—long upper tusks.

      B. Tragulina (chevrotains).

      C. Tylopoda (camels, etc.).

      D. Pecora (true ruminants).
           Families: a. Cervidæ—Deer.
                        b. Giraffidæ—Giraffes.
                        c. Antilocapridæ—Prongbuck (also hollow-horned).
                        d. Bovidæ (hollow-horned).
                             Genus Bos—Cattle
             1. Taurine group { B. taurus—ordinary cattle.
                                 B. indicus—humped cattle.
             2. Bibovine group { B. gaurus—the gaur.
                                  B. frontalis—the gayal.
                                  B. sondaicus—the banteng.
                                  B. grunniens—the yak.
             3. Bisontine group { B. bonasus—the European bison.
                                   B. bison—the American bison.
             4. Bubaline group { B. caffer—the African buffalo.
                                  B. bubalis—the Indian buffalo, etc.
                                  Etc.

             Genus Ovibos—Musk Ox

             Genus Ovis—Sheep
             1. Bighorn group—O. canadensis, etc.
             2. Argalis group—O. poli, etc.
             3. Urial—O. vignei, etc., and Mouflon group—O. musinon, etc.
             4. Bharal—O. nahura, etc., and Barbary group—O. tragelaphus, etc.
             5. Domestic group—O. aries.

             Genus Capra—Goats
             1. Ture group—C. caucasica, etc.
             2. Pasang group—C. aegagrus, etc.
             3. Ibex group—C. ibex, etc.
             4. Markhor group—C. salconeci.
             5. Domestic group—C. hircus.
             Genus Hemitragus—the tahr.

Also the various species of antelopes.

Suborder 2. Perissodactyles (uneven-toed).
Families: a. Tapiridæ. b. Rhinocerotidæ. c. Equidæ.
Tapirs.   Rhinoceroses.   Horses, asses, and zebras.
                                 Genus Equus.
                                 1. E. caballus—the horse.
                                 2. E. asinus—the ass.
                                 3. E. kiang, etc.—half asses.
                                 4. E. zebra—the zebra.

Order VII.   Rodentia (rabbits, mice, etc.).
Order VIII.   Carnivora.
Suborder 1.   Fissipedes.
Families: a. Felidæ—Cats.
          b. Hyænidæ—Hyænas.
          c. Proteleidæ—Earth wolf.
          d. Viverridæ—Civets.
          e. Ursidæ—Bears.
          f. Mustelidæ—Weasels.
          g. Procyonidæ—Racoons.
          h. Canidæ.
             Genus Canis.
             1. C. familiaris—dogs.
             2. C. lupus, etc.—wolves.
             3. C. aureus, etc.—jackals.
             4. C. decussatus—foxes.
Suborder 2.   Pinnipedes.
Families: a. Otariidæ—Eared seals.
          b. Trichechidæ—Walrus.
          c. Phocidæ—Seals.

Order IX.   Insectivora (moles, hedgehogs, etc.).
Order X.   Chiroptera (bats).
Order XI.   Primates.
Suborder 1.   Lemuroidea, lemurs.
Suborder 2.   Anthropoidea, monkeys, etc., and man.

Fig. 1-2.   Taxonomic arrangement of the larger farm animals.

15

Origin and Domestication of Swine. It appears that our modern breeds, S. *domesticus,* have descended from at least two wild stocks: the Northern European breeds from the wild boar S. *scrofa,* and those of Southern Europe, Asia, and Africa from one of the Malayan pigs, possibly the collared pig S. *vittatus.* The former was a larger, coarser animal throughout than the latter and had a denser covering of hair.

Present-day breeds are no doubt the result of varying degrees of crossing between the parent stocks and their offspring. It seems probable that the pig was domesticated later than cattle and sheep and earlier than the horse. Selected for his ability to fatten rapidly and economically, the pig is foremost in converting feed into flesh and is second to none in the economy of American meat production. Most of the breeds of hogs found in America are of our own breeders' making; for example, the Duroc, the Poland China, and the Chester White breeds are strictly American creations.

ANCIENT ANIMAL HUSBANDRY

The record of man's activities goes back some 5,000 to 6,000 years. Paintings, carvings, and sculptures from the Babylonian, Assyrian, and Egyptian civilizations show that animals played an important part in man's activities in ancient times. These peoples had herds of oxen, sheep, and goats, and used donkeys for draft purposes and horses when waging war. Cattle were used for draft purposes as well as for the production of milk and meat. Agriculture and stockbreeding were among the chief sources of wealth in Greece. When Alexander led the Greeks to Asia Minor and Egypt, a new and rich agricultural economy fell under Greek domination, including large studs of as many as 30,000 mares and large herds of camels and oxen. Improved methods of plant and animal production were developed during the ascension of the Roman Empire. However, with the fall of Rome, about 500 A.D., there ensued a period of about a thousand years known as the Dark and Middle Ages, during which agriculture and animal husbandry survived but improvement in them was halting or lacking.

EIGHTEENTH-CENTURY ANIMAL HUSBANDRY

Animal husbandry received a new impetus in England beginning about the year 1700. At this time from one-third to one-half the land was still cultivated on the open-field system, making it quite impossible for an individual to improve his herd or flock, since all the cattle and sheep grazed together on the same overcrowded commons. With the passing of the feudal system and the manors with their open fields, and with the change to individual ownership and the enclosing of fields, better methods of farming became possible. The cultivation of clover and improved grasses and the development of root crops made better wintering of livestock possible.

How great was the change can be seen in the weights of animals at the famous Smithfield Market. In 1710, beeves averaged 370 lb, calves 50 lb, sheep 28 lb, and lambs 18 lb; in 1795, they averaged 800, 148, 80, and 50 lb, respectively. Individual initiative, encouraged by enclosures, together with the introduction and use of turnips and clover and the labors of agriculturists like Tull and Townshend, helped enormously to bring about these changes.

EARLY AMERICAN AGRICULTURE

When the first permanent settlers came to America in the early 1600s there were no horses, cattle, sheep, or swine here. The early settlers therefore brought animals with them from various parts of Europe. At that time there were no distinct breeds of animals as we know them today, but there were in Europe various local varieties. The early settlers probably had some ideas and ideals about improved livestock, but their opportunities for selection were limited, since both the colonists and the animals were involved in a primitive struggle for existence. In time well-adapted types of livestock were developed in various parts of America. Selection is a never-ending process, and most of the early types disappeared or were merged with improved breeds. This period of development extended from 1620 to 1800.

LAYING THE FOUNDATION FOR PUREBREDS

The next period in American animal husbandry covers the years from 1800 to 1870, or up to the time of the establishment of the first purebred-recording associations. The most noteworthy development during this time was the spreading influence of the work of Robert Bakewell, an Englishman born in 1725. In 1760 Bakewell began his animal-breeding work at Dishley, England, with horses, sheep, and cattle. His success in improving his stock is said to have been due to three things:

1. He had definite ideals—in beef cattle, for example, a low-set, blocky, quick-maturing animal almost unknown at that time.
2. He leased rather than sold males and returned them to his own farm if they transmitted desirable qualities.

3. He bred the best to the best regardless of relationships, which often meant rather close inbreeding. Inbreeding led to the development of relatively true breeding strains, although there were no herdbooks until many years later.

Bakewell's methods were widely copied, and thus the foundations of the pure breeds were laid. Large numbers of these relatively purebred animals of various strains were brought to America to be bred along pure lines and to be mated to the local strains which had emerged from the test of "survival of the fittest." Thus pedigree breeding was initiated in Britain and brought to America.

During this period many significant changes occurred in America and its agriculture. The country expanded to the west, improved transportation systems were developed, cities grew (thus providing better markets), and improved farm machinery provided the means for an ever-increasing production of both crops and animals.

DEVELOPMENT OF BREED ASSOCIATIONS

The formation of the various breed associations during the period from 1870 to 1900 was a great impetus to livestock improvement. Registry books were set up to keep the breeds pure, and a vigorous competitive spirit was developed by the various breed associations. As we look back we can easily see that a better job could have been done. More rigid selection and refusal to register inferior purebreds would have strengthened all breeds, and much of the breed promotional activity was not genetically sound. Nevertheless, the purebred contribution to improved animal husbandry in America is beyond calculation. This is not to say that every breed has shown a steady march toward perfection. Most of the breeds have had ups and downs, and some breeds have met the test of time more successfully than have others.

ARTIFICIAL INSEMINATION

The most far-reaching development in animal husbandry since the establishment of the Purebred Breed Associations is undoubtedly the development of artificial insemination. This practice was successfully demonstrated by Spallanzani in dogs in 1780. In 1799 Hunter produced a pregnancy in the human by this method. It was used sporadically by horse breeders in the United States in the early part of this century. Ivanov began to experiment with farm animals in Russia in 1899, and by 1930 several European nations were using this technique.

The first large-scale organization for the artificial insemination of dairy cattle was founded at Clinton, New Jersey, in 1938 by Prof. E. J. Perry of the New Jersey Dairy Extension Service. The idea spread like wildfire and many organizations both public and private were organized.

In 1964, 2,316 sires were used to inseminate artificially 7,747,953 cows for 3,053 cows per sire. Of the total, 7,282,994 were dairy cows (41.4 per cent of all United States dairy cows), and 1,117,395, or 15.3 per cent of these, represented services to beef bulls. During the year, 464,959 beef cows were artificially inseminated with semen from beef bulls.

Artificial insemination is practiced also with swine, goats, horses, and poultry, but not yet to the extent that it is with cattle.

What artificial insemination does is to maximize the only two tools available to the breeder, namely, selection and breeding systems. With this system sires can be proved before being put into extensive use and can be used long after they are dead. One sire now has over 200,000 AI progeny, and semen has been stored for eight to nine years and then used successfully in service. Both outcrossing and inbreeding systems are readily available through artificial insemination.

We are now using only our genetically best sires, and we can look forward to the time when our best females will also leave more abundant offspring through artificial ovulation. These two processes will maximize the use of the best genetic material.

Then perhaps the final step of producing through chemistry better genes and combinations thereof will provide the capstone of animal breeding. Mendel in 1866 boldly forsook the prevalent blending nature of inheritance for an atomistic one involving relatively unchanging Mendelian factors as the mechanism of heritable entities. In 1912 Morgan, his students, and others began the studies which led eventually to the mapping of chromosomes with the hereditary genes.

Now we are beginning to learn the arrangement and other qualities of the nucleic acid DNA units and the protein molecules which apparently make up the genes. The usual method of science is first to discover the character, make-up, and activity of the basic units and then proceed to control these entities in some preordained fashion. So, strange as it may seem to some, in the lifetime of those now living, science will very likely have discovered how to make and package new genes leading to ever more efficient and beautiful animal and plant types.

LIVESTOCK TRENDS IN THE UNITED STATES

Animal breeding is one of the largest industries in the United States. The total national income in 1950 was $239 billion. Farm income was $28.7 billion, and of this income $16 billion came from livestock and livestock products. Animals of some sort are to be found on about 90 per cent of the farms in the United States. Livestock breeding involves nearly 200 million farm mammals and 700 million poultry. For some, animal breeding is a hobby, but many thousands of people depend upon it for their livelihood. In terms of food, clothing, and for the maintenance of soil fertility, every person in the United States has a stake in its success.

Table 1-1. Number and Average Value per Head of the Various Classes of Farm Livestock in the United States, 1880–1965

	Horses on farms		Mules on farms		Dairy cows	
Year	Number (000 omitted)	Av. value	Number (000 omitted)	Av. value	Number (000 omitted)	Av. value
1880	10,903	$ 53.74	1,878	$ 61.74	11,754	$ 23.31
1890	15,732	69.27	2,322	77.61	15,000	22.30
1900	17,856	43.56	3,139	51.46	16,544	31.30
1910	19,972	107.70	4,239	119.98	19,450	35.40
1920	20,091	96.45	5,651	148.29	21,455	81.51
1930	13,742	69.98	5,382	83.93	23,032	82.70
1940	10,444	77.30	4,034	116.00	24,940	57.30
1950	5,548	46.00	2,233	99.10	23,853	177.00
1955		4,309*	$ 56.20		23,462	134.00
1960		3,089*	113.00		19,527	210.00
1965					17,592	187.00

	Cattle other than milk cows		Stock sheep		Hogs including pigs	
Year	Number (000 omitted)	Av. value	Number (000 omitted)	Av. value	Number (000 omitted)	Av. value
1880	31,593	$ 15.75	44,867	$ 2.18	44,327	$ 4.40
1890	45,014	15.16	42,693	2.29	48,130	4.80
1900	43,195	24.67	45,065	2.97	51,055	5.36
1910	39,543	19.20	46,939	4.06	48,072	9.05
1920	48,945	39.99	37,328	10.59	60,159	20.00
1930	39,971	40.38	45,577	9.00	55,705	13.45
1940	43,369	30.95	46,266	6.35	61,165	7.78
1950	54,110	100.01	26,182	17.80	58,937	27.10
1955	73,130	73.41	27,137	14.90	50,474	30.60
1960	76,709	117.94	28,849	16.50	59,026	18.50
1965	89,592	99.73	23,299	15.80	53,132	24.70

* Horses and mules. Not available for 1965.

Table 1-2. Total Numbers and Value of Farm Mammals and Human Population, United States, 1880–1965*

Year	No. of all farm livestock†·‡	Value of all farm livestock†·‡	Human population United States*
1880	145,322,000	1,766,093,000	50,156,000
1890	168,891,000	2,615,796,000	62,948,000
1900	176,854,000	2,930,582,000	75,995,000
1910	178,215,000	4,733,083,000	91,972,000
1920	193,629,000	8,080,307,000	105,711,000
1930	181,409,000	6,011,216,000	122,775,000
1940	190,219,000	4,814,284,000	131,669,000
1950	170,863,000	12,172,798,000	150,697,000
1960	187,200,000	15,066,747,000	178,464,000
1965	183,615,000	13,896,922,000	192,942,000

* To even thousands.
† Includes horses, mules, cattle, swine, and stock sheep.
‡ Horses and mules are not included in 1965.

In Table 1-1 will be found an enumeration of the various farm mammals over the past 80 years. It is obvious from this table that mechanical power has been rapidly supplanting horse and mule power since 1920. Whether machines will entirely supplant horses and mules as power units the future will tell. Only from one-fourth to one-half as many work horses and mules are being produced as would be needed to maintain their numbers. On the other hand, there seems to be a growing interest in the horse for pleasure purposes. Dairy-cattle numbers have shown a decline since 1940. Meat animals have increased in numbers, and they now mature more rapidly than formerly, the quicker turnover actually meaning more product from a given number of animals. Many factors, of course, influence the numbers of animals kept on farms from year to year. The magnitude of the total value of American livestock and the meagerness of the average value stand in marked contrast to each other. Both these points indicate opportunity for the livestock breeder.

Table 1-2 shows the trend in total numbers and total value of farm livestock in the United States by decades for the past 80 years and also the trend in human population. It shows that the breeding of livestock is one of the most important types of endeavor in this country.

PRESENT STATUS AND PROBLEMS

Many animals are inefficient and unprofitable producers because they are poorly fed and poorly cared for. An optimum environment involves also the control or elimination of the various livestock pests and diseases which take an annual toll of millions of dollars in terms of lowered individual and breeding efficiency. The problem facing animal producers consists in feeding and managing more efficiently what we have as well as raising the level of potential performance through selective breeding.

The genetic task in the field of animal breeding is twofold: (1) to find out through performance records what we have genetically (analysis) and (2) to raise the potential of productive efficiency by making better combinations of genetic materials through systems of breeding and selection (synthesis).

REFERENCES

Allen, R. L. 1847. "Domestic Animals," Orange Judd Publishing Co., Inc., New York.

Boissonnade, P. 1927. "Life and Work in Medieval Europe," Alfred A. Knopf, Inc., New York.

Breasted, J. H. 1935. "Ancient Times," Ginn and Company, Boston.

Darwin, C. 1859. "The Origin of Species," A. L. Burt Company, New York.

Davenport, E. 1910. "Domesticated Animals and Plants," Ginn and Company, Boston.

Ernle, Lord. 1927. "English Farming: Past and Present," Longmans, Green & Co., Inc., New York.

Furnas, C. C., and Furnas, S. M. 1943. "The Story of Man and His Food," the New Home Library, New York.

Gras, N. S. B., 1925. "History of Agriculture," Appleton-Century-Crofts, Inc., New York.

Irvine, William. 1955. "Apes, Angels and Victorians," McGraw-Hill Book Company, New York.

Loomis, F. B. 1926. "The Evolution of the Horse," Marshall Jones Company, New Boston, N.H.

Lydekker, R. 1912. "The Horse and Its Relatives," George Allen & Unwin, Ltd., London.

———. 1912. "The Sheep and Its Cousins," E. P. Dutton & Co., Inc., New York.

———. 1912. "The Ox and Its Kindred," Methuen & Co., Ltd., London.

Osborn, H. F., 1929. "From the Greeks to Darwin," Charles Scribner's Sons, New York.

Shaler, N. S. 1895. "Domesticated Animals," Charles Scribner's Sons, New York.

CHAPTER TWO

Basic Process of Inheritance

For a complete understanding of the material in Chaps. 5 to 15, which make up the principal portions of this text, a knowledge of basic hereditary and reproductive processes of animals is assumed. Most students will have had courses in basic genetics prior to the use of this book. Most will have separate courses in reproductive physiology. Therefore, no attempt is made to go into these subjects thoroughly here.

However, it seems desirable to provide a refresher on basic facts for those having had the related courses. The material in this chapter and the two which follow will provide this and will also permit a general understanding of basic hereditary and reproductive processes for those not having had the detailed related courses. The brief coverage of these subjects will permit only a general exposition of principles and will not include details and special cases. Neither will it give detailed background information upon which current understanding rests.

NATURE OF INHERITANCE

Bodies of all higher organisms are composed of many billions of microscopic cells. Each of these body or somatic cells contains a large number of hereditary units called *genes*. These occur in pairs. Each cell contains a number of pairs of bodies known as *chromosomes* in which the genes are located. The number of chromosome pairs is characteristic of each species. The genes of each pair are located in a specific position in a specific pair of chromosomes.

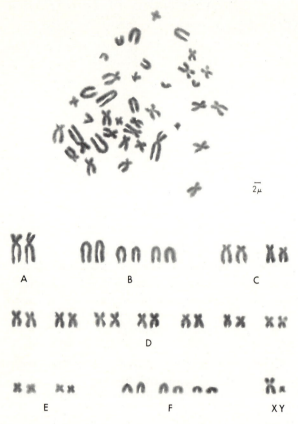

Fig. 2-1. Chromosomes of a male Yorkshire pig. *Above,* a photomicrograph of the chromosomes of a single cell. *Below,* chromosomes of a similar photo cut out and rearranged by pairs in a *karyotype.* It will be noted that all pairs are very similar in size and form except for the X and Y—the sex chromosome pair in which differences are readily apparent. (*Courtesy of Dr. Richard A. McFeely, University of Pennsylvania.*)

When reproductive cells are formed, the number of chromosomes is reduced to one of each pair. At fertilization one member of each pair is brought to the new individual by the male germ cell and one by the female germ cell. Thus, the double or diploid number is restored.

The fundamental processes of heredity were discovered over 100 years ago by an Austrian monk, Gregor Mendel, in experiments with garden peas. Because of this, the science of heredity is sometimes called *Mendel-*

ism, but the term *genetics* is more generally used. Mendel published his results in 1865, but their importance was not realized at the time. Others discovered the same principles independently in 1900. At this time Mendel's results were "rediscovered" and he was given credit— several years after his death. Knowledge of the hereditary process and of its physical and chemical bases has been greatly extended since Mendel's day. However, his essential principles have been proven correct and serve as the basis for classical genetics.

Fig. 2-1a. Diagram of maturation or mitosis of germ cells in animals. The process begins with the third row of cells. Maternal chromosomes white, paternal black. All chromosomes in the fertilized egg received from the mature egg are thereafter maternal, and those received from the spermatozoon are thereafter paternal, regardless of what they were in the mature germ cells. (*After Shull.*)

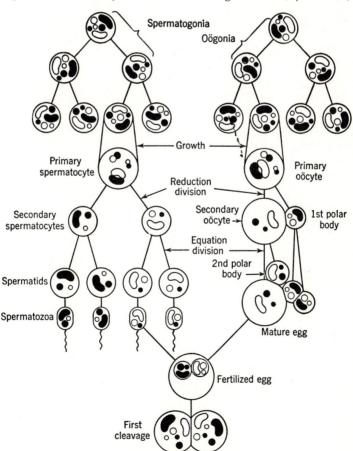

Table 2-1. Diploid Chrosome Numbers of Some Common Mammals

Common and scientific names	Chromosome number	Reference
Man (*Homo sapiens*)	46	Tjio and Puck, *Proc. Nat. Acad. Sci.*, **44**: 1229–37, 1958.
Horse (*Equus caballus*)	64	Benirschke et al., *Jour. Hered.*, **55**:31–38, 1964.
Ass (*Equus asinus*)	62	do
Cattle (*Bos taurus*)	60	Saski and Makino, *Jour. Hered.*, 53:157–162, 1962.
Swine (*Sus scrofa*)	38	McConnel et al., *Jour. Anim. Sci.*, **22**:374–379, 1963.
Sheep (*Ovis aries*)	54	Borland, *Jour. Hered.*, **55**:61–64, 1964.
Goat (*Capra hircus*)	60	Lopez-Saez and Gimenez-Martin, "Genetics Today," 1:137, 1963.
Mouse (*Mus musculus*)	40	Ford and Woollam, *Expt. Cell Res.*, **32**:320–326, 1963.
Rat (*Rattus norvegicus*)	42	Makino, "Chromosome Numbers of Animals," Iowa State College Press, 1951.

In recent years tremendous progress has been made in advancing knowledge of the chemical nature of hereditary material and the biochemical and physical pathways and mechanisms by which it exerts its effects. In addition to extending basic biological knowledge on the nature of living things, these discoveries may eventually permit the development of methods for controlling gene modification, transmission, and action. These possibilities are highly speculative at the present time. Since the newer knowledge of molecular genetics has as yet had little if any impact on practices available for genetic improvement of farm animals, we will not consider them further. However, references at the end of this chapter will provide an introduction to all phases of genetics for the inquiring student.

MONOFACTORIAL INHERITANCE

This is the simplest type of inheritance. It occurs when differences in some observable or measurable trait depend upon the presence of one or the other of a single pair of genes. In his original experiments Mendel found that tall peas and a form of dwarfism behaved in inheritance as if they differed by a single gene.

The hereditary relationships of an original cross of pure tall and pure dwarf peas and of intermating the offspring, or F_1 generation, can be shown schematically as follows:

Parents:

Phenotype	tall	$\times$	dwarf
Genotype	TT		tt
Gametes	all T		all t

Offspring, or F_1 generation:

Phenotype	tall	$\times$	tall
Genotype	Tt		Tt
Gametes	½ T ½ t		½ T ½ t

Offspring of F_1 mated to F_1, or the F_2 generation:

Phenotype	tall	tall	tall	dwarf
Genotype	TT	Tt	Tt	tt
Phenotypic ratio		3	:	1

Before going further with a discussion of a monofactorial cross such as that shown above, some definitions are needed:

Hybrid is the term used to describe the offspring of parents which are genetically pure for different characteristics. The first hybrid generation is known as the F_1, or first filial, generation. The word *phenotype* refers to the appearance or measurable characteristics of an individual. In this example only two phenotypes, tall and dwarf, are involved.

Genotype refers to the genetic constitution of an individual. In this example two different genotypes, TT and Tt, both have tall phenotypes. Therefore, we say the factor T is *dominant* over the factor t. Conversely, t is *recessive* to T.

Gamete is the general term used to refer to reproductive cells. These are also often called *germ cells*. In the above example we have not differentiated between male and female gametes since the outcome of this particular cross will be the same regardless of whether the tall parent is the male or female. However, for completeness we should indicate that male and female gametes have specific names. In plants they are the pollen cells and the ovules, respectively. In animals they are the sperm and the ovum. Sperm cells are also often referred to as *spermatozoa*.

The simple cross we have used here involves differences in only one pair of genes or hereditary factors. These alternative genetic forms, which are located at the same point on each one of a pair of chromosomes, are called *alleles* or an *allelic pair*. Another term sometimes used interchangeably with allele is *allelomorph*.

In our example all the F_1's are genetically Tt and phenotypically tall. Each parent was genetically pure or *homozygous* and produced only one type of gamete for this particular character.

When the F_1's are mated with each other to produce the F_2 generation, the situation is more complicated. Since the F_1's are genetically impure, or *heterozygous*, they each produce two kinds of gametes. These are produced in equal numbers and fertilization occurs at random with gametes of the other parent. That is, a T gamete of one parent has an equal probability of uniting with a T or a t gamete of the other parent. The four possible F_2 genotypic combinations as shown in the diagram occur with equal frequency. Thus the F_2 generation is composed of ¼ dwarf and ¾ tall individuals. Of the talls, ⅓ (¼ of all offspring) will be homozygous. They are thus genetically like the original tall parent. Two-thirds will be heterozygous like the F_1's. If these heterozygotes are used for breeding, their genetic performance will be exactly like that of the F_1 generation.

It should be emphasized that the F_2 ratios are what would be expected on the average if large numbers of offspring were produced. With only a few offspring there could be rather great deviations from the expected ratio. Male gametes are produced in great numbers in most species, and it is a matter of chance which ones eventually fertilize female gametes. This chance at fertilization is governed by the laws of probability, which are discussed in Chap. 6.

The first principle of heredity learned through studies of monofactorial inheritance and monohybrids is known as *segregation*. The genes remain as constant entities down through the generations and segregate at germ cell formation rather than combining or blending.

An organism gets one member of each pair of chromosomes from each of its parents, giving it the $2n$, or diploid, number of chromosomes. In turn it will pass on at random to each of its offspring one or the other member of each of its pairs of chromosomes, thus transmitting the n or haploid number. Heredity is a halving process.

There are a few well-known characters of farm animals which behave as unit characters, i.e., as though dependent on but one pair of genes with one allele being dominant. For example, black cattle crossed with red yield an F_1 generation which is black, and the F_2 has a ratio of 3 blacks to 1 red. In swine, erect ears show dominance over lop ears, white color over black, mule foot over normal foot; in sheep, white fleece is generally dominant over black, although a dominant gene for black is present in some breeds; in horses, gray color is dominant over chestnut.

Lack of Dominance. In other monofactorial cases the genes do not exhibit dominance. This is illustrated diagrammatically in the following case of a cross between pure red and pure white four-o'clocks:

Parents:

Phenotype		red	×	white
Genotype		RR		rr
Gametes		all R		all r

Offspring, or F_1 generation:

Phenotype		pink	×	pink
Genotype		Rr		Rr
Gametes		½ R ½ r		½ R ½ r

Offspring of F_1 mated to F_1,
or the F_2 generation:

Phenotype	red	pink	pink	white
Genotype	RR	Rr	Rr	rr
Phenotypic ratio	1 :	2	:	1

It can readily be seen that the genes behave exactly the same in this cross as in the cross of tall × dwarf peas. However, the heterozygote is intermediate in color.

The degree of dominance can vary greatly. In some cases it is complete or almost complete so that the heterozygote gives no phenotypic evidence of its genetic constitution. In other cases the heterozygote is a phenotypic intermediate.

Multiple Alleles. In the foregoing examples the different alleles were considered as pairs, or alternate forms of a gene which could occupy a certain spot in a given pair of chromosomes. Alleles also occur in series of three or more genes which can occupy a given chromosome locus. We call these *multiple alleles*. No individual can carry more than two members of a multiple allelic series, but in some cases very large numbers of alleles are present in different individuals in a population. Current genetic evidence indicates that genes are composed of desoxyribonucleic acid (DNA), which can occur in an almost infinite variety of chemical forms. Alleles and multiple alleles are presumably the result of rather small chemical changes.

One of the best-known series of multiple alleles occurs in rabbits. If we cross a colored rabbit with an albino, the F_1 are colored and we get three colored to 1 albino in the F_2. If we cross Himalayan with albino, the F_1 are Himalayan and we get 3 Himalayan to 1 albino in the F_2. If we cross a colored rabbit with the Himalayan type (which is white with dark nose, ears, tail, and feet), the F_1 are colored and we get 3 colored to 1 Himalayan in the F_2.

From this behavior in breeding it appears we are dealing with genes at a single locus having three alternative forms. There are, in other words, three genes in this multiple-allelic series. Since any rabbit can have any two of them and we know the dominance relations among

Fig. 2-2. Rabbits showing phenotypes resulting from a series of multiple alleles. *Top,* colored; *center,* Himalayan pattern; *bottom,* albino. (*From Castle, in Journal of Heredity.*)

the three genes, we can write the possible genotypes of the various colors of rabbits as follows, letting C stand for the color gene, c^h for the Himalayan gene, and c for the albino gene:

Colored rabbit	CC or Cc^h or Cc
Himalayan rabbit	$c^h c^h$ or $c^h c$
Albino rabbit	cc

There is a similar situation in man as regards blood type, although the dominance relations are different. We do not inherit blood from our parents (it is made by the embryo, just as are bone, muscle, etc.), but we do inherit genes which control our type of blood. There are four blood types in the human which are derived from the interaction of three allelic genes. The first gene is A, the second A^b, and the third a. Genes A and A^b are both dominant over gene a, but there is a lack of dominance between genes A and A^b, thus giving a new type as genes R (red) and r (white) in four-o'clocks give a new type, Rr (pink).

Thus, we can write the genetic make-up of men for this particular series of blood types as follows:

Type A	AA or Aa
Type B	$A^b A^b$ or $A^b a$
Type AB	AA^b
Type O	aa

The knowledge of the inheritance of these blood types (together with that of other types determined by genic series at other loci) is used in the courts in some states in the United States to establish possible paternity in disputed cases.

Sex-linked Heredity.   In the examples used thus far it has made no difference whether character A was in the male and character B in the female or vice versa. The crosses would work as well one way as the other. The reason for this lies in the fact that the qualities thus far studied have been determined by genes located in the *autosomes*. In the human, for example, there are 23 pairs of chromosomes. Of these, 22 pairs are autosomes and are alike in both sexes so far as their general form is concerned. The remaining pair, however is different in the two sexes. In the female there is a pair of straight chromosomes called the X chromosomes, making up this twenty-third pair. In the male there is one straight X chromosome, but its mate is a bent chromosome called the Y chromosome. These are the *sex chromosomes*. In the female there are two X chromosomes and in the male one X and one Y chromosome. The same type of XY heredity is present in all farm animals except poultry.

We are now ready to consider the matter of sex-linked heredity. *Sex-linked* simply means that the gene for the character being considered lies in one of the sex or X chromosomes. An example of this type of heredity is found in the fruit fly, *Drosophila melanogaster;* it has to do with the color of the eyes, the genes involved being R for red eyes and r for white eyes.

The genetic make-up of the flies for eye color is:

Male		Female	
RX Y	red	RX RX	red
rX Y	white	RX rX	red
		rX rX	white

In the above, X's and Y's refer to chromosomes and R and r refer to genes in these chromosomes. The sex chromosomes in *Drosophila* and also in farm mammals and man consist of a pair of X chromosomes in the female and an X and a Y chromosome in the male. A portion of the X and Y chromosomes in some species are homologous, with alleles in both types, so that crossing over between these portions of the X and Y is possible. Genes located in these segments behave in inheritance like autosomal genes. The remaining portions of the X and Y chromosomes are called the *differential* portions; in this portion of the Y chromosomes there are only a few active genes and those present do not have alleles in the X. Thus in diagraming sex-linked inheritance we will not indicate any genes in the Y chromosome. A cross of a red-eyed male and a white-eyed female is as follows:

Parents:
    Phenotype   red-eye (male) $\times$ white-eye (female)
    Genotype RX Y rX rX
    Gametes   ½RX ½Y   all rX
Offspring:
    Phenotype and sex   red-eye (female)   white-eye (male)
    Genotype RX rX rX Y

Sex-influenced Heredity. This type of inheritance has to do with genes which are in the autosomes and in which dominance is different in the two sexes. One simple sex-influenced character in man is a certain type of baldness. If we let B represent a gene for baldness and b its allele for normal hair, men who are BB or Bb become bald, whereas those who are bb do not. Women who are BB become bald, but those who are Bb or bb do not. Thus the gene B is dominant in men but recessive in women. This, undoubtedly, is due to the fact that different types of sex hormones are produced in the two sexes. It points up the importance of both heredity and environment, since an identical heredity expresses itself differently in the different internal environments of men and of women.

TWO-FACTOR INHERITANCE

A more complex type of inheritance than monofactorial inheritance occurs when two pairs of factors are considered concurrently. They may affect different phenotypic characters or some single character.

This can again be illustrated by characters with which Mendel worked in peas. He crossed round peas RR with wrinkled peas rr and got all round peas in the F_1 of this cross and 3 round to 1 wrinkled in the F_2 generation. He crossed plants having yellow cotyledons YY with those having green cotyledons yy and got all yellow cotyledons in the F_1 cross and 3 yellow to 1 green in the F_2. Mendel then crossed plants with round seed and yellow cotyledons with those which had wrinkled seed and green cotyledons. In the F_1 he got nothing but round seed and yellow cotyledons. But when he made the F_2 by crossing F_1 plants, he got the following results:

Approximately 9/16 round and yellow
Approximately 3/16 round and green
Approximately 3/16 wrinkled and yellow
Approximately 1/16 wrinkled and green

This cross can be represented diagrammatically as follows:

Parents:
  Phenotype   round, yellow $\times$ wrinkled, green
  Genotype $RRYY$ $rryy$
  Gametes   all RY all ry
F_1 generation:
  Phenotype   all round, yellow
  Genotype   all $RrYy$

F_1 gametes

♀ \ ♂	RY	Ry	rY	ry
RY	Round, yellow $RRYY$	Round, yellow $RRYy$	Round, yellow $RrYY$	Round, yellow $RrYy$
Ry	Round, yellow $RRYy$	Round, green $RRyy$	Round, yellow $RrYy$	Round, green $Rryy$
rY	Round, yellow $RrYY$	Round, yellow $RrYy$	Wrinkled, yellow $rrYY$	Wrinkled, yellow $rrYy$
ry	Round, yellow $RrYy$	Round, green $Rryy$	Wrinkled, yellow $rrYy$	Wrinkled, green $rryy$

F_2
Phenotypes
and
genotypes

The F_2 phenotypic ratio in a monohybrid showing dominance is 3:1, so in a dihybrid (if the respective sets of genes are in different chromosomes) we would expect the ratio in the F_2 to be $(3:1)^2$ or 9:3:3:1.

These results demonstrate the second law of inheritance, namely that of *independent assortment*. Independent assortment simply means that one member of a pair of genes going to one germ cell has no influence on which member of any other pair goes to that cell.

Depending upon dominance or lack of dominance in the gene pairs, the F_2 of a two-factor cross can deviate from the 9:3:3:1 ratio found when both pairs exhibit dominance. In cattle, polled is dominant to horns, but in Shorthorns, color exhibits no dominance. *RR* gives red, *rr* gives white, but *Rr* is an intermediate called roan. If we cross a polled, red Shorthorn with a horned, white one, we would have:

PPRR × *pprr*
giving *PpRr*, a polled roan

If we make the F_2 from this, we get the following ratio:

3 polled, red	:	1 horned, red
6 polled, roan	:	2 horned, roan
3 polled, white	:	1 horned, white

If we had a case in which both characters lacked dominance, our phenotypic ratio would become 1:2:2:4:1:2:1:2:1.

The above dihybrids involved two pairs of genes determining two different qualities. Cases are also known in which two pairs of genes act upon the same quality. Comb type in poultry is an example. If a bird has the double recessive *rrpp*, it is single-combed; if it is *RRpp* or *Rrpp*, it is rose-combed; if *rrPP* or *rrPp*, it is pea-combed; and if it has at least one *R* and one *P*, it will be walnut-combed.

Linkage. Independent assortment of two or more pairs of genes will occur only if the genes concerned are in different pairs of chromosomes. Since there are many hundreds of pairs of genes and only a relatively small number of pairs of chromosomes, it follows that each chromosome must carry many genes. Genes occurring in the same chromosome are said to be *linked*. They do not assort independently. The classical example of linkage is in the sweet pea.

In 1906 Bateson and Punnett in England were breeding sweet peas. Because crosses of purple (dominant) and red had given a 3 purple:1 red ratio in the F_2, and long pollen grains (dominant) and round had also given a 3 long:1 round ratio in F_2, it was expected that a cross

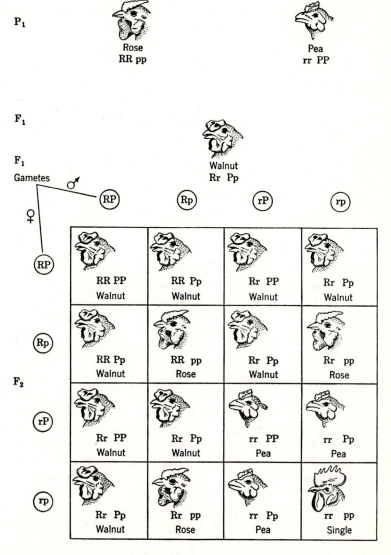

Fig. 2-3. Diagram showing interaction of factors for comb form in fowls. The cross of a pure rose-comb bird with a pure pea-comb one gives all walnut-combed offspring. The 16 possible combinations of the F_1 gametes, with their genotypes and the phenotypes resulting from factor interaction, are shown in the F_2 checkerboard. (*From Sinnott and Dunn.*)

of purple-long with red-round would give a 9:3:3:1 ratio in F_2. Instead of this, Bateson and Punnett obtained the following results:

Characteristics	Actually obtained		Expected	
	Number	Ratio	Number	Ratio
Purple, long	4,831	12	3,910.5	9
Purple, round	390	1	1,303.5	3
Red, long	393	1	1,303.5	3
Red, round	1,338	3	434.5	1
Total	6,952		6,952.0	

It is evident that these two characters of sweet peas did not follow the second Mendelian law since they did not assort into all the possible combinations in an independent fashion. The two things that went in together from the grandparents, purple and long from one grandparent and red and round from the other, tended to be held together, so that there were more of these combinations than expected in the F_2 and fewer than expected of the new combinations (purple-round and red-long). Since the characters tend to stay together, this feature of heredity is called *linkage*.

The genes for color of flowers (purple or red) and for shape of pollen grain (long or round) are located in the same chromosome pair and therefore tend to be passed along together. Bateson and Punnett did get some new combinations in the cross although not as many as they had expected according to the law of independent assortment. They expected 2,607 and got only 783. The new combinations did occur, but less frequently than expected. This makes it appear that the chromosomes do not always remain intact, but that part of one chromosome has joined up with part of its mate, and vice versa, in some cases. This is known as *crossing over*.

Linkage is a conservative force in heredity. It does not prevent the formation of new genetic combinations, but it reduces their frequency.

The frequency of crossovers is a function of distance apart on the chromosomes. Information for linkage groups and crossover frequency is used to determine the locus of specific gene pairs and thus to map chromosomes. These maps are quite extensive for several species of plants and laboratory animals. In farm animals almost nothing is known about linkage groups and the specific location of the individual genes. This is because the high costs of rearing large animals prevent the rear-

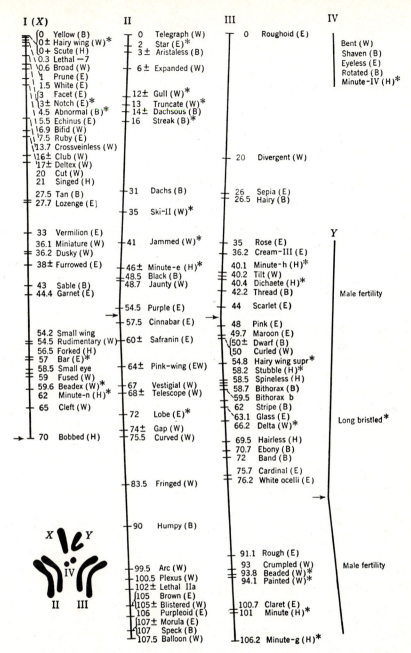

Fig. 2-4. Chromosomes of *Drosophilia melanogaster* (lower left) with linkage map showing relative positions of many of the known genes in the chromosomes as determined genetically. The letters in parentheses indicate the portion of the fly in which the characters appear: *B*, body; *E*, eye; *H*, hairs; *W*, wings. The arrows indicate positions of spindle-attachment regions. All genes in chromosome IV are very closely linked. (*From Sharp, adapted from Morgan, Sturtevant, and Bridges, 1925, and Stern, 1929.*)

ing of large enough numbers in planned experiments to establish linkage groups and crossover frequencies.

INHERITANCE INVOLVING THREE OR MORE GENE PAIRS

Phenotypic and genotypic ratios have been determined for several cases of inheritance involving three or more pairs of genes influencing different qualitative characters. Such studies are difficult because the number of possible gametes, genotypes, and phenotypes increases rapidly with increases in number of gene pairs. The number of individuals required in controlled matings for firm establishment of ratios becomes astronomical if many gene pairs are involved. These facts are illustrated in Table 6-2 in Chap. 6.

Most traits of a quantitative nature are under the influence of several pairs of hereditary factors. Individually their effects are small, and most such traits are also influenced by environmental factors. These facts, together with the number of gene pairs, make it impossible in most cases to identify individual genes and determine specific genetic ratios. However, the principles of segregation and recombination apply in such cases just as in the qualitative traits we have been using as illustrations thus far. Quantitative inheritance will be discussed in more detail in a subsequent section on the Multiple-gene Heredity.

EPISTASIS

Interactions between nonallelic genes are called *epistasis*. Several types of epistasis are known for specific cases of two-factor inheritance. It is probably a widespread phenomenon. Indeed, in a broad sense, the expression of any gene in inheritance is dependent upon interactions and interrelationships with others.

If one gene of a pair masks the presence and prevents the manifestation of its allele, we say it is *dominant*. Likewise, a gene or genes of one allelic pair may mask the presence and manifestations of those of another pair. Several kinds of epistatis gene action are known, and the epistatic genus themselves may be either dominant or recessive.

If we mate a black rat $AABB$ (gene A for color, gene a for diluted color, gene B for expression of any color, and genes bb masking all color, that is, epistatic to A) to an albino rat $aabb$, all the F_1 are black, $AaBb$; but the F_2 will appear as 9 blacks, 3 creams, and 4 albinos. This is due to the fact that the presence of at least one A and one

B produces black, two *a*'s and at least one *B* produce cream, but either *AA, Aa,* or *aa,* together with *b*'s, result in albino. This is because the *b*'s mask the expression of *A* or *a*; that is, *b* is epistatic to *A* or *a*, so that the last two portions of the usual 9:3:3:1 ratio are thrown together phenotypically.

Another manifestation of epistasis is one that throws the two middle portions of the 9:3:3:1 ratio together phenotypically. An example has been found in the fruit shape of a certain type of squash in which genes *A* and *B* together give disk shape, genes *AA* or *Aa* with *bb* or genes *aa* with *BB* or *Bb* give spherical shape, and the double recessive *aabb* results in an elongated shape.

It has been found that a certain type of white dogs, *AABB* (with color in their eyes), if mated to brown dogs, *aabb*, produce all white offspring, and that when these F_1 dogs are mated, the ratio of 12:3:1 appears in F_2. This apparently is due to the fact that the white dog carries color genes *A* that are inhibited from forming color in the hair by another gene *B*. The brown dogs, *aabb*, lack the gene for black color *A*, being, therefore, brown in color, for they also lack the inhibitor *B*. The F_1 cross is *AaBb* and is white because of the dominant epistatic effect of gene *B* over genes *A* or *a*, whereas in the F_2 all those getting at least one *B* are white, those getting *b*'s and at least one *A* are black, and the rest, *aabb*, are brown. Genes that behave in this fashion are often called *inhibitors.*

The foregoing examples will serve to illustrate the varying nature of epistasis or factor interaction. The accompanying figure gives in diagrammatic form various modifications of the classical 9:3:3:1 F_2 ratio which can occur. Names have been given to the different kinds of two-

Fig. 2-5. Summary of phenotypic classes in two-gene F_2, resulting from dominance and different types of inter-allelic interactions or epistasis.

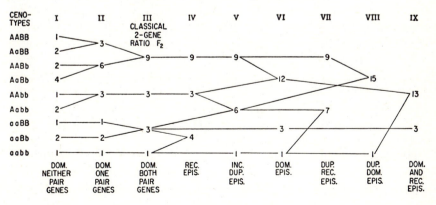

factor epistasis. These are rather infrequently used now, but are included in the figure for their historical interest.

At present we have very few clear-cut examples of epistasis in farm mammals, but it is not difficult to imagine that there may be almost limitless numbers of possible epistatic reactions actually present in them. Dominance is a hindrance in breeding because it makes it impossible to separate the homozygotes and heterozygotes by inspection. Likewise, epistasis is a hindrance in breeding since, if any desirable qualities of an animal are due to epistatic combinations, they may not be passed on intact to offspring because of the halving nature of heredity.

MULTIPLE-GENE HEREDITY

Early in this century it was often believed that Mendelian-type inheritance was limited to qualitative traits such as color, horns, etc. Hereditary differences in quantitative traits such as size, milk yield, and others were believed to depend upon some other mechanism. It has been found, however, that characters such as these are dependent upon the action of many pairs of genes, each of which behaves in Mendelian fashion. Usually we are not able to identify individual genes, but relatively simple cases have been established. These establish the basic principle.

The multiple-gene hypothesis grew out of the work of Nilsson-Ehle in 1908 with wheat color and East's 1910 work with ear length of corn. The former crossed several strains of red and white wheats. In general the red color was only partially dominant over white, for the F_1 was not as dark red as the red parent. In the F_2 of one cross there were 3 reds (1 dark red, 2 lighter red) to one white. This indicated a one-gene pair situation. Another cross of red and white wheats gave a similar F_1 but an F_2 of 15 reds (of varying shades) to one white, thus indicating a two-gene pair situation. Still a third cross gave the usual F_1 but an F_2 of 63 reds (again of varying shades) to one white. This indicated a three-gene pair situation. Breeding tests revealed that there were one, two, and three pairs of genes segregating in these cases. In the two-gene-pair cross it was shown that a wheat with four genes for red was redder than one with three, this latter one was redder than one with two, and this, in turn, redder than a wheat with only one gene for red. In these cases there is not complete dominance, but the genes act in a cumulative or *additive* fashion.

The mode of inheritance involving several gene pairs with cumulative but individually indistinguishable effects is now thought to be operative (along with dominance, overdominance, epistasis, and other types) in the inheritance of most of the commercially important animal characters.

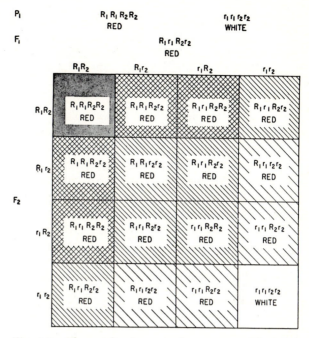

Fig. 2-6. The result of a cross between a red-kerneled and a white-kerneled wheat, in which the red color is a result of the operation of either or both of two genes, R_1 and R_2. The intensity of the red color is indicated by the density of the crosshatching. (*From Sinnott, Dunn, and Dobzhansky, Principles of Genetics, 4th ed., p. 123.*)

In other words, instead of color of wheat we might substitute size, weight, yield, or many other quantitative traits. This will explain the difficulties encountered in attempting to unravel the exact mode of inheritance of many characters in farm animals.

Contemporaneously with Nilsson-Ehle, East was working with corn. He crossed one variety having an ear length of 5 to 8 cm with a variety having an ear length of 13 to 21 cm (on the average 6.63 cm and 17.31 cm respectively). The F_1 ranged from 9 to 15 with an average ear length of 12.11 cm. When the F_2 was produced, the range extended from 7 to 21 cm with an average of 12.57 cm. The variability in the parent types and the F_1 was presumably due to environmental differences. The increased variability in the F_2 was probably due to the segregation of several gene pairs. East's work with corn parallels that of Nilsson-Ehle with wheat, and both find their simplest explanation on the basis of multiple-gene effects.

TYPES OF GENE ACTION

Where many genes influence the expression of a quantitative character, it is extremely difficult to determine the mode of gene action. It may be additive with the genes showing no dominance and each "plus" gene adding a given increment to the character. This is termed an *additive effect* and is typified by the wheat-color results of Nilsson-Ehle. Alternatively, the many gene pairs may exhibit varying degrees of dominance.

A broad range of epistatic interactions may occur. Among these is the possibility of multiplicative gene action. Essentially this theory suggests that each plus gene exerts an influence leading to a certain percentage change in a character rather than a certain number of units, as is postulated by the additive theory.

We can seldom identify single gene effects for quantitative traits in farm animals. However, statistical methods are available for estimating average types of gene effects. These methods are covered in Chap. 7. Essentially, they estimate the average effects of large numbers of gene pairs.

SUMMARY

Hereditary processes in animals and plants depend upon the action of genes which occur in pairs in each individual. There may be multiple allelic series of genes which can pair, but only two normally occur in any one individual. Genes of a given pair or allelic series are located at specific loci in paired chromosomes within each body cell. In the formation of reproductive cells, chromosome and gene numbers are reduced with only one member of each pair or allelic series going to each gamete. Random chance at segregation determines which member of a pair goes to each gamete. Chance also determines which gamete fertilizes a particular gamete from the other sex. Two or more pairs of genes (or allelic series) assort independently if they are in different pairs of chromosomes, but they are linked and exhibit a varying amount of recombination if located in the same chromosome pair. Quantitative characters are usually under the control of many pairs of genes which may act additively, multiplicatively, exhibit dominance, or interact epistatically with each other.

REFERENCES

Auerbach, Charlotte. 1961. "The Science of Genetics," Harper & Row, Publishers, Incorporated, New York.

Gardner, Eldon J. 1964. "Principles of Genetics," 2d ed., John Wiley & Sons, Inc., New York.

Hartman, Philip E., and Suskin, Sigmund R. 1965. "Gene Action," Prentice-Hall, Inc., Englewood Cliffs, N.J.

Hutt, Frederick B. 1964. "Animal Genetics," The Ronald Press Company, New York.

Mittwoch, Ursula. 1967. "Sex Chromosomes," Academic Press, New York.

Sager, Ruth, and Ryan, Francis J. 1961. "Cell Heredity," John Wiley & Sons, Inc., New York.

Scheinfeld, Amram. 1965. "Your Heredity and Environment," J. B. Lippincott Company, Philadelphia.

Singleton, W. Ralph. 1962. "Elementary Genetics," D. Van Nostrand Company, Inc., Princeton, N.J.

Sonneborn, T. M. 1965. "The Control of Human Heredity and Evolution," The Macmillan Company, New York.

Srb, Adrian M., Owen, Ray D., and Edgar, Robert S. 1965. "General Genetics," 2d ed., W. H. Freeman and Company, San Francisco.

Wagner, Robert P., and Mitchell, Herschel K. 1964. "Genetics and Metabolism," John Wiley & Sons, Inc., New York.

CHAPTER THREE

The Anatomy of the Reproductive System

The reproductive process is the means by which hereditary material is passed from generation to generation. Reproductive rate is the key to the success or failure of specific breeding systems. It is also of fundamental importance to profitable commercial livestock production. In this chapter and the one which follows we will present basic information on the reproductive processes of male and female farm animals. An understanding of the anatomy of the reproductive organs of the male and female is essential to an understanding of their function.

Normal development and function of the reproductive organs depends upon the action and interaction of many factors. Many hereditary defects affect the anatomy or function of the reproductive systems. The success of the reproductive process depends upon proper nutrition, management, and disease control. Detailed consideration of these latter factors is outside the scope of this particular text; however, they are of basic importance to the livestock breeder.

THE MALE GENITALIA

In the male the principal reproductive organs are the paired testes, the accessory glands, and a system of ducts through which the spermatozoa and seminal fluids are discharged at the time of ejaculation.

Anatomy of the Testis. The testes are the primary reproductive organs of the male. In all mammals except the whale, elephant, rhinoceros, and seal, the testes are suspended outside the body cavity in the scrotum. In birds and in the four mammalian species mentioned, the testes are maintained in the abdominal cavity adjacent to the kidneys.

The testis is covered by a dense connective-tissue capsule, the tunica albuginea, from which connective-tissue extensions, the septa, radiate to divide the testis into lobules. Within each lobule are long, coiled ducts called the *seminiferous tubules*. The seminiferous tubules converge at the apexes of the lobules, where they straighten to form the tubuli recti. The tubuli recti enter the rete testis, a series of irregular anastomosing ducts. Connecting the rete testis and the head of the epididymis there is a series of ducts called the *vasa efferentia*. Between the seminiferous tubules there are connective tissue, blood vessels, and nerves, and the interstitial, or Leydig, cells, which secrete the male hormone testosterone.

The Scrotum. The testes are formed in the abdominal cavity, and in some species they remain there throughout life. In the course of evolutionary development they gradually came to lie at the posterior end of the body, pressing against the body wall. A continuation of this process finally led to an outpouching at this region, forming a sac, the scrotum, outside the body proper.

In addition to serving as a covering for the testicles, the scrotum functions as a thermoregulatory mechanism. This function of the scrotum is shown by actual differences in temperature, the temperature in the scrotum being from 1° to 8°C lower than that of the abdominal cavity. It has been shown by several workers that insulation or the application of heat to the testicles results in a degeneration of the spermatogenetic tissue, the production of abnormal sperm, and temporary sterility. These results are also known to follow periods of intense fever.

The Epididymis. The seminiferous tubules converge at the rete testis, from which there is a series of ducts, the vasa efferentia, which convey the sperm to the epididymis. The epididymis is a continuous, greatly coiled duct varying from a few feet to several hundred feet in length in some species. It is lined with ciliated, tall columnar cells and has some smooth-muscle cells in the tubule walls. The epididymis has three regions: the head, adjacent to the vasa efferentia; the mid-portion; and the tail, which is continuous with the vas deferens.

The Vasa Deferentia. The passage of sperm from the tail of the epididymis to the urethra is provided by the vasa deferentia. Each vas deferens consists of a firm tubular structure lined with ciliated columnar epithelial cells and a relatively thick wall composed of smooth-muscle and connective-tissue cells. In the bull, ram, and stallion the wall of the vas deferens is greatly thickened near its junction with the urethra. This area is called the *ampulla*. It is not present in the boar.

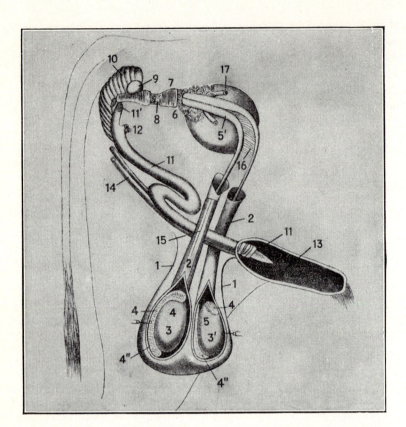

Fig. 3-1. Generative and urinary organs of bull. (1) scrotum; (2) spermatic cord; (3) testicle; (4) epididymis (globus major and minor); (5) vas deferens; (6) vesicula seminalis; (7) membranous portion of urethral canal covered by Wilson's muscle; (8) part of prostate gland covered by Wilson's muscle; (9) Cowper's gland; (10) accelerator urinae muscle; (11) penis; (12) cut suspensory ligaments of penis; (13) sheath, laid open; (17) ureters. (*From U.S. Department of Agriculture, Diseases. of Cattle.*)

The vas deferens, in combination with the arteries, veins, and nerve supply of each testis, plus smooth-muscle and connective-tissue cells, constitutes the spermatic cord. The vas deferens is a passageway for sperm; the spermatic cord provides the vascular and nervous regulation of the testis.

The Seminal Vesicles. The seminal vesicles are the largest of the accessory glands of reproduction in the male. They are located in the pelvic cavity at the ends of the vasa deferentia, being an outgrowth of the latter at their urethral end, and each connects with the urethra by means of a duct.

The seminal vesicles vary in size; they are several inches in length in the stallion and have a bladderlike appearance. In the other farm animals they are elongated, firm, and lobulated, being about 4 or 5 inches long in the bull and the boar and smaller in the ram.

The Prostate Gland.   The prostate is a compound tubuloalveolar gland made up of a large number of lobules. It is located at the neck of the bladder and tends to surround the urethra. It is not prominent, as are the seminal vesicles, and is easily overlooked.

The Cowper's Glands.   The Cowper's glands, often called the *bulbo-urethral glands,* are small, firm, oval glands located on either side of the urethra posterior to the prostate and covered by the urethral muscle in the stallion, bull, and ram. In the boar they are elongated, cigar-shaped structures 4 or 5 inches in length.

The Penis.   The remaining portion of the male genitalia consists of the penis. This organ, besides conveying urine to the exterior, has the additional function of conveying spermatozoa into the female genital tract. The penis is made up of muscular and erectile tissue that becomes engorged with blood during the process of erection.

Farm animals exhibit a wide variation in the structure of this organ. In the horse the end of the penis is rounded, and the organ nearly fills the vagina at copulation. In the bull the diameter is much smaller, with the external urethral orifice situated in the urethral papilla at the end and on the left side. The urethral papilla of the bull represents a vestigial filiform appendage. This is well developed in the ram in the form of a slender appendage which protrudes beyond the glans penis in a twisted manner. The penis of the boar is not provided with an appendage or lateral papilla, and the opening to the exterior is situated at the end of the organ and in the center.

THE FEMALE GENITALIA

In the female the principal reproductive organs are the paired ovaries and a system of ducts which are concerned with the transport of sperm, the transport of the ovum, the development of the fertilized ovum, and the delivery of the fetus at parturition.

Anatomy of the Ovary.   The ovaries are the primary reproductive organs of the female and are homologous to the testes in the male. The

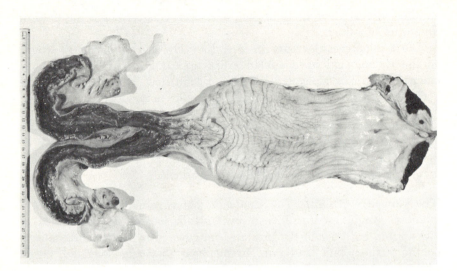

Fig. 3-2. Dissected genital organs of the cow. A fresh corpus luteum appears in the left ovary, and a corpus luteum from the previous estrual cycle is in the right ovary. (*Courtesy of Dairy Breeding Research Center, the Pennsylvania State University.*)

ovaries are suspended in the broad ligament of the uterus in the dorsal part of the abdominal or pelvic cavity. In the mare the ovaries are bean-shaped and 2 to 4 inches in the longest diameter, in the cow 1 to 1½ inches, and in the ewe ½ to 1 inch. The ovaries of the sow are usually larger than those of the cow and are irregular in shape because of the large number of follicles or corpora lutea which occur in all species that produce litters.

Microscopically the ovary consists of two regions, the outer shell, or cortex, and a central portion, the medulla. The ovary is almost completely covered with a single layer of cuboidal epithelial cells, the germinal epithelium. This is a distinct contrast to the testis in which the germinal epithelium is located deep within the seminiferous tubules. All functional ova arise from the germinal epithelium through a series of mitotic and meiotic cell divisions. The ova develop in structural units called *follicles.* Usually one ovum is produced in each follicle, but thousands of cells are contained in each follicle and contribute to the development of the ovum. The larger follicles are filled with a watery or serous secretion called the *follicular fluid,* and as these follicles increase in size and protrude above the surface of the ovary they are called *Graafian follicles.* At certain stages of ovarian activity the ovary contains one or more large solid oval structures called *corpora lutea.*

Immediately beneath the germinal epithelium is a layer of dense fibrous connective tissue called the *tunica albuginea.* Throughout the outer layer of the ovary, the cortex, are follicles in all stages of development and, at certain times, corpora lutea. The medullary (internal) portion of the ovary contains the larger blood vessels as well as the nerve supply and a mixture of connective tissue.

The Fallopian Tubes.　The Fallopian tubes, or oviducts, are modified at the ovarian end into a thin, funnel-shaped fimbriated membrane. This portion of each tube is in close association with, and partly surrounds, the ovary but is not attached to it.. Because of its funnel shape it is called the *infundibulum* or the fimbriated end of the Fallopian tube. This structure is especially designed to catch the ovum at the time of ovulation. Following ovulation the ovum rapidly descends into the Fallopian tube proper, where fertilization usually takes place. This portion of the tube is of small total diameter, being about 0.1 inch in the cow. Its wall contains smooth-muscle cells, and the lining epithelium is simple or ciliated columnar epithelium. Both the muscle cells and the cilia participate in the transport of the ovum and the sperm.

Uterus and Horns.　In cattle, sheep, and swine, the horns of the uterus are well developed, since fetal development occurs here. In swine the horns are folded, or convoluted, and may be 4 or 5 feet in length. This is an obvious adaptation for multiple fetation. In cattle and sheep the horns are curved somewhat like the horns of a ram; hence the descriptive name.

In the mare and in woman the uterine horns are poorly developed, and the body of the uterus is correspondingly larger, since fetal development takes place there. The lining of the entire uterus is called the *mucosa.* It consists of a row of columnar epithelial cells beneath which are large numbers of uterine glands. These glands produce the secretion which is evident at estrus and participate in the nutrition of the blastocyst prior to implantation. In cattle and sheep the mucosa contains from 70 to 130 specialized areas for the attachment of the fetal placenta. These structures are called the *cotyledons.* In the sow and the mare the placenta is of the diffuse type, is in more or less continuous contact, and there are no cotyledonary areas.

The muscular structures of the uterus are well developed. There is an inner layer of circular and an outer layer of longitudinal smooth muscle. These muscle cells assist, by rhythmic contractions, in the transport of sperm to the Fallopian tubes, in the support of the fetus during pregnancy, and in contracting vigorously at parturition to expel the fetus.

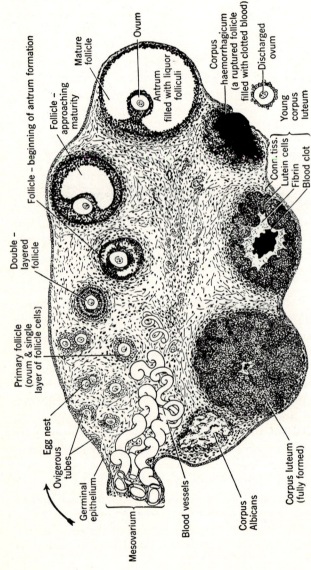

Fig. 3-3. Schematic diagram of mammalian ovary showing the sequence of events in the origin, growth, and rupture of the ovarian (Graafian) follicle and the formation and retrogression of the corpus luteum. Follow clockwise around the ovary, starting at the arrow. (*From Patten, Embryology of the Pig, 2d ed., McGraw-Hill Book Company.*)

Ovum

Mature follicle

Follicle – approaching maturity

Follicle – beginning of antrum formation

Antrum filled with liquor folliculi

Corpus haemorrhagicum (a ruptured follicle filled with clotted blood)

Discharged ovum

Young corpus luteum

Conn. tiss.
Lutein cells
Fibrin
Blood clot

Double – layered follicle

Primary follicle (ovum & single layer of follicle cells)

Egg nest

Ovigerous tubes

Germinal epithelium

Mesovarium

Blood vessels

Corpus Albicans

Corpus luteum (fully formed)

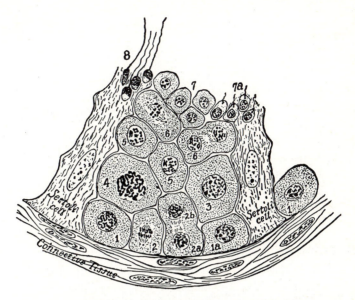

Fig. 3-4. Semischematic figure showing small segment of the wall of an active seminiferous tubule. The sequence of events in the production of spermia is indicated by the numbers. A spermatogonium (1) goes into mitosis (2) producing two daughter cells (2a and 2b). One daughter cell (2a) may remain peripherally located as a new spermatogonium eventually coming to occupy such a position as 1a. The other daughter cell (2b) may grow into a primary spermatocyte (3) being crowded meanwhile nearer the lumen of the tubule. When fully grown, the primary spermatocyte will go into mitosis again (4) and produce two secondary spermatocytes (5,5). Each secondary spermatocyte at once divides again (6,6) producing spermatids (7). The spermatids become embedded in the tip of a Sertoli cell (7a), there undergoing their metamorphosis and becoming spermia (8), which when mature are detached into the lumen of the seminiferous tubule. (*From Patten, Embryology of the Pig, 2d ed., McGraw-Hill Book Company.*)

The Cervix. The terminal portion of the uterus is called the *cervix* or *neck* of the womb. This part of the reproductive tract has thick walls, is generally rather difficult to penetrate, and is usually filled with a thick, sticky secretion during pregnancy.

The Vagina. From the cervix posteriorly to the urogenital sinus, or vestibule, the female genital tract is termed the *vagina*. In farm animals it lies horizontally below the rectum and above the bladder. This portion of the tract is relatively simple, both in structure and in function. It consists anatomically of the usual three layers of tissue, viz., mucosa, muscularis, and serosa.

The Vulva and Clitoris.   The female genital tract terminates posteriorly in the vulva, or external orifice. With the onset of heat, or estrus, these structures become congested and enlarged, particularly in the sow, somewhat less so in the cow, the mare, and the ewe. The clitoris, the small erectile organ homologous to the penis in the male, is situated just inside the portion of the vulva farthest removed from the anus. Stimulation of the clitoris results in a sexual orgasm in the female.

SUMMARY

The basic anatomy of the reproductive systems of male and female farm animals has been described. In the male, spermatozoa are produced in the testis. The remainder of the system has the purposes of providing for maturation and storage of spermatozoa and for their deposition in the female. The basic female organ is the ovary, in which the female reproductive cells, the ova, are produced. Other ducts and organs exist for the purpose of transporting the ova to the site of fertilization and receiving the sperm cells and transporting them to the site of fertilization. In addition, the uterus provides for the growth and development of the embryo and fetus until it reaches a stage of development compatible with independent life.

REFERENCE

Sisson, S., and Grossman, J. D. 1953. "The Anatomy of the Domestic Animals," 4th ed., W. B. Saunders Company, Philadelphia.

CHAPTER FOUR

The Physiology of the Reproductive System

In those species characterized by sexual reproduction, the most important functional reproductive processes are the production of the spermatozoa and ova; fertilization, pregnancy, and parturition; the estrual cycle; an understanding of fertility and sterility; and the basic breeding management practices which insure a high level of reproductive efficiency.

TESTICULAR FUNCTION

The testicle has a dual function. Its primary function is the production of male germ cells. The secondary function is the production of the male sex hormone, testosterone, which in turn stimulates development and maintenance of other male reproductive organs. It is responsible for the sex drive or libido leading to the mating act necessary for reproduction. Secondarily, the male hormone affects the growth process and the anatomy of body parts and features not directly related to reproduction.

Spermatogenic Function of the Testis. The spermatozoa or sperm are produced in the seminiferous tubules by a series of cell divisions. These include a reduction of chromosome numbers and cellular modifications which result in the production of the characteristically motile sperm cells.

During the period of sexual maturity, after attainment of puberty, the spermatogonia divide by ordinary mitosis to form primary spermatocytes in some cases. In other cases division results in additional spermatogonia in order that this cell form can maintain itself as a continuous source of primary spermatocytes. In order to maintain a constant chromosome number in each species, a reduction in chromosome number takes place in the formation of both sperm and ova. The type of cell division which accomplishes this is called *meiosis*. In it one member of each pair of chromosomes goes to each resulting cell. This is the mechanism by which gametes each containing only one member of each gene pair are produced, as discussed in Chap. 2.

Each primary spermatocyte divides meiotically to form two secondary spermatocytes, each of which contains only one of each pair of chromosomes. Each secondary spermatocyte then divides mitotically to form two spermatids. Each spermatid has the reduced, or haploid, number of chromosomes. No further cell divisions occur, but each spermatid undergoes a marked change in shape as it develops into a spermatozoon. The nuclear material becomes condensed, the chromosomes becoming indistinguishable, and the cytoplasm appears to shrink and is no longer recognizable. The nucleus forms the sperm head, and a flagellum, composed of a thickened middle piece and a tail, is developed. Each primary spermatocyte thus gives rise to four spermatozoa.

The seminiferous tubules contain one additional type of cell, the Sertoli, or nurse, cell. The spermatids become attached to the Sertoli cells and presumably are nourished by them during the final stages of spermatogenesis. When the spermatozoa are completely formed, they are detached from the Sertoli cells and pass through the lumen in each seminiferous tubule into the rete testis and eventually to the epididymis where further maturation occurs.

Anatomy of Spermatozoa. Following Leeuwenhoek's development of the microscope, spermatozoa were first observed by one of Leeuwenhoek's students, Ham, in 1677. Their significance in reproduction was not realized at the time. About the middle of the eighteenth century, Spallanzani showed that filtered semen was incapable of initiating pregnancy, and Barry, in 1843, observed the actual fertilization of a rabbit ovum by a sperm. There are many species differences in sperm morphology. The general pattern in man and in the farm mammals is very similar. Normal sperm consist of a head, a neck, a body, and a tail portion. The head in profile is flattened but, as ordinarily seen under the microscope, is rounded or oval. It consists chiefly of nuclear material and is covered by a thin layer of cytoplasm.

Semen Volume and Sperm Numbers. There are considerable species and individual differences in semen volume and sperm numbers. They are influenced by age, season of the year, general health, and sexual use. The ram is capable of repeated service without noticeable decline in volume and sperm concentration. In boars and stallions, service rates

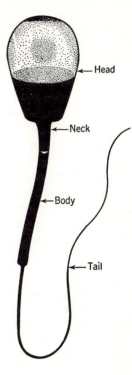

Fig. 4-1. Generalized anatomy of the spermatozoon.

higher than twice daily usually affect semen quality. The best general rule is that semen characteristics should be expected to be equal to the species averages shown in Table 4-1.

Endocrine Function of the Testis. Scattered between the seminiferous tubules are the highly specialized interstitial cells, or cells of Leydig. These cells secrete the male sex hormone. There are several substances

Table 4-1. Species Differences in Semen Quality

Animal	Volume per ejaculate (ml)	Sperm per mm (1,000's)	Total sperm per ejaculate (billions)
Boar	150–250	100	15–25
Bull	3–5	1,000	3–5
Cock	0.6–0.8	3,000	1.8–2.4
Dog	4–6	3,000	12–18
Man	3–5	100	0.3–0.5
Ram	0.8–1.0	1,000	0.8–1.0
Stallion	70–100	100	7–10

in nature which have male-hormone activity, and in the laboratory several specific chemical compounds have been synthesized which have varying degrees of male-hormone activity. Substances which produce the effects of the male hormone are called *androgens* or *androgenic hormones*. The specific androgenic compound produced by the interstitial cells of the testis is called *testosterone*.

The androgens have many functions and produce somewhat diverse effects in the normal male. The testicular hormone assists in the development and maintenance of the entire male reproductive system. At puberty increasing amounts of testosterone bring about growth of the penis, scrotum, the seminal vesicles, Cowper's and prostate glands, and assist in the final stages of seminiferous-tubule development. Sexual drive and secondary sexual characteristics, such as the growth of facial hair and the change of voice in man, the tusks of the boar, and the comb and spurs of the rooster, are the result of increased levels of male-hormone production.

Androgens are of considerable importance in general body growth. It is well known that these substances increase nitrogen retention, facilitate muscular and skeletal growth, and usually reduce fat deposition. When administered to growing and fattening lambs and cattle they tend to increase growth rate and improve feed efficiency. In swine, at certain levels, androgens appear to increase muscular development and to reduce fat deposition.

Hormonal Regulation of the Testis.   The primary regulating agents of the testis are the gonadotropic hormones produced by the anterior pituitary gland. If these hormones fail to be produced in sufficient amounts in the young animal, sexual maturity does not occur. Failure of gonadotropic-hormone production in the mature animal is followed by both spermatogenic and endocrine dysfunction of the testis. There are two gonadotropic hormones, the follicle-stimulating hormone (FSH) and the luteinizing hormone (LH). These gonadotropins also regulate ovarian function; their names are actually descriptive of the effects in the female. Some workers refer to the luteinizing hormone as the interstitial-cell-stimulating hormone (ICSH). Both hormones are proteins and have been isolated from the anterior pituitary in relatively pure form. Both FSH and LH produce enlargement of the testis when injected, but each acts in a different manner.

FSH is concerned primarily with the development of the seminiferous tubules and the production of sperm. ICSH (LH) is primarily responsible for the development of the interstitial cells, which in turn secrete the male sex hormone testosterone. However, relatively pure preparations of FSH are not as effective in tubule development as FSH plus traces of ICSH or androgens. In experiments with laboratory animals from which the anterior pituitary gland has been removed, ICSH or androgens are capable of maintaining or repairing the seminiferous tubules in a reasonably functional state. The best evidence is, therefore, that complete seminiferous-tubule development requires FSH, ICSH, and testosterone.

OVARIAN FUNCTION

Like the testicle, the ovary has the dual function of gamete production and the production of hormones. The hormones are responsible for inducing sexual receptivity and for maintaining the remainder of the reproductive tract in the states required at the proper times for fertilization, pregnancy, parturition, and lactation.

Oögenic Functions of the Ovary. All ova originate from the germinal epithelium covering the ovary. During reproductive life, individual germinal epithelial cells migrate from the surface of the ovary, become surrounded by the thousands of specialized cells which compose the ovarian follicle, and, if the process goes to completion, are liberated from Graafian follicles by the process of ovulation. Oögenesis is a more or less continuous process in the female, just as is spermatogenesis in the male. There is a vast difference, however, in the number of functional gametes which are produced by the two sexes. In the male, one primary spermatocyte is capable of producing four functional sperm. In the female, a primary oöcyte can at best produce one functional ovum and three polar bodies.

In ova production the primary oöcyte divides by meiosis to form a secondary oöcyte with the reduced or haploid chromosome number and a polar body also having a haploid chromosome complement. In this division most of the cytoplasmic material goes to the secondary oöcyte. The polar body may divide again by mitosis to form two polar bodies, but it has little cytoplasm and is nonfunctional, not being capable of fertilization. The secondary oöcyte divides mitotically to produce the mature ovum or egg and the second polar body. Again, the polar body gets little cytoplasm and is nonfunctional. This process of ova formation in the female results in a germ cell many times the size of the sperm. It contains food material in the cytoplasm which nourishes the embryo during the first few cell divisions after fertilization.

The reproductive process in the female is characterized by a series of distinct cycles called the *estrual cycles*. Some wild mammals have only one cycle each year. In cows the cycles are normally repeated at intervals of 19 to 22 days. Since most cows produce only one calf per gestation, only one ovum is produced during each estrual cycle. It is obvious, therefore, that a nonpregnant cow produces only 18 to 20 ova annually.

Anatomy of Ova. The ovum is more typical of body cells than is the sperm. Ova are normally spherical, have a relatively large amount of cytoplasm, have a typical nucleus with a recognizable nucleolus, and are immobile. The ova of mammals are small, but when they have been identified under the microscope, they can be seen without magnification. The size of the ovum is 0.07 mm in the mouse and from 0.14 to 0.16 mm in sheep, cattle, and the human. In egg-laying species, the ovary produces large amounts of yolk and the oviduct contributes albumen and the shell.

Endocrine Function of the Ovary. The primary hormones secreted by the ovary are the female sex hormone, estrogen, and the hormone of the corpus luteum, progesterone. Estrogens are widely distributed in nature, and their chemical structures differ greatly. The specific estrogen secreted by the ovary, largely by the Graafian follicle, is called estradiol.

A number of compounds which do not occur in nature but which have very high estrogenic activity have been synthesized. These should be called *hormonelike substances* rather than hormones, but they are members of the estrogen group. The more common of these are diethylstilbestrol, dienestrol, and hexestrol. The estrogens produce both psychologic and physiologic effects. These hormones were named estrogens because of their role in inducing estrus, sexual receptivity. One of their primary functions, therefore, is the regulation of sexual behavior. The estrogens are also concerned with the development and function of the tubular portion of the female genitalia. The most easily recognized effect of the estrogens is their stimulatory action on the vagina and vulva. The development of the uterus is dependent upon both estrogen and progesterone. Development is initiated by estrogen and completed by progesterone. The estrogens influence both growth rate and type of growth. In chickens the administration of estrogens by injection or feeding brings about increased fat deposition and improvement in carcass quality of broilers and fryers. In cattle and sheep, estrogen administration consistently increases growth rate and the efficiency of feed utilization.

The second ovarian hormone is progesterone. Progesterone is a steroid hormone closely related chemically to estrogens and androgens. It is secreted primarily by the functional corpus luteum but is present in the ovary before ovulation and in some species is secreted in large amounts by the placenta. Estrogens and progesterone often act in combination or in sequence. The development of the uterine mucosa in preparation for pregnancy is initiated by estrogens and continued by progesterone. Mammary development is likewise initiated by estrogens and further stimulated by progesterone. Although estrus can be induced by estrogen alone, in some species the presence of minute amounts of progesterone increases the effectiveness of estrogen.

Progesterone is of great importance in the regulation of ovarian activity. The continuous secretion of progesterone at a high level by a corpus luteum will prevent follicular development and the occurrence of estrus. This situation often occurs in nature. The development of the corpus luteum of pregnancy is important in the maintenance of the pregnancy and is usually accompanied by the cessation of the heat periods. In all except a few species, viz., guinea pig, man, monkey, the horse, and probably the sheep after the first third of pregnancy, failure of the corpus luteum to secrete progesterone during pregnancy is followed by abortion. It is believed that the animals mentioned produce sufficient progesterone in the placenta to maintain pregnancy even after ovariectomy.

Hormonal Regulation of the Ovary. The distinctly cyclic nature of ovarian function, requiring follicular growth, estrus, ovulation, corpus luteum formation, and progesterone secretion, creates a problem in both timing and hormone levels. The follicle-stimulating hormone (FSH) produces growth of the small follicles and the development of Graafian follicles, but it alone does not bring about estrogen production in the follicles. Small amounts of the luteinizing hormone (LH) act in combination with FSH and cause the follicle both to grow and to produce estrogen. When sufficient LH is produced, ovulation occurs and the corpus luteum is formed by luteinization of the cells in the wall of the Graafian follicle. Heat, or estrus, is brought about by increasing estrogen levels supplemented by trace amounts of progesterone which are secreted before ovulation. A third pituitary gonadotropin, the luteotropic hormone, brings about the secretion of progesterone by the corpus luteum following luteinization of the ovaries.

Ovulation. Ovulation is the discharge of the ovum from the Graafian follicle. In most animals ovulation occurs near the end of estrus. Most animals, including all the farm animals, ovulate spontaneously; that is, ovulation does not depend upon copulation. The rabbit differs in that some type of stimulation, such as copulation, mechanical or electrical stimulation of the cervix, or another afferent stimulus, is necessary if ovulation is to occur. In the species in which the mechanisms of ovulation have been studied, whether or not they ovulate spontaneously or are similar to the rabbit, it appears that the primary regulation is the hormonal mechanism just described.

THE ESTRUAL CYCLE

The outstanding characteristic of the reproductive process in the female is its rhythmicity. All of the vertebrates which have been studied have definite reproductive cycles. There are wide species differences in the cycles, but each species has some basic pattern.

In most mammals the cycle is called the *estrous* or *estrual cycle*. Some animals, such as the fox, have only one heat period each year and are called *monestrous*. The farm animals all have more than one cycle per year and are called *polyestrous*. Females which have a limited reproductive period, such as the ewe, are called seasonal breeders in contrast to the cow, which is a continuous breeder. The estrous cycle can be divided into several distinct phases. The most easily recognized phase is the period of sexual receptivity, or heat, and is technically known as *estrus*. The stage following estrus is called *metestrus*. It is during this period that the corpus luteum forms and secretes progesterone, which brings about the final development of the uterus for pregnancy. If pregnancy does not occur, the corpus luteum regresses and there is a short period of genital rest called diestrus. Diestrus is followed by proestrus, which is characterized by follicular growth and generally heightened reproductive activity. Follicular growth is accompanied by the secretion of estrogen and traces of progesterone which gradually induce estrus. During heat, the follicle usually reaches its maximum size, and ovulation occurs shortly before or just following the end of estrus. In seasonal breeders, the genitalia have a long period of quiescence which is called *anestrus*. In cattle and swine, the first part of lactation is usually accompanied by a period of sexual inactivity called *lactational diestrus*.

Gonadotrophic hormone production and/or release by the pituitary gland is regulated to a considerable degree by ovarian hormones. When FSH has stimulated the growth of follicles to near maturity, they produce estrogen in increasing amounts. It in turn acts back on the pituitary to slow the release of FSH and to stimulate release of LH. Progesterone produced by the corpus luteum acts on the pituitary to inhibit release of FSH. It is only when the corpus luteum regresses near the end of the estrual cycle that FSH is released in quantities necessary for rapid follicular growth.

The normal estrual cycle is thus the result of a delicate interplay and coordination of hormones produced by both the pituitary gland and the ovary.

FERTILIZATION AND PREGNANCY

Fertilization may be defined as the union of the male and female gametes. There are several distinct steps in the process of fertilization:

1. Insemination, the deposition of sperm in the female genital organs by either natural or artificial means
2. The transport or travel of sperm from the insemination site to the ovarian end of the Fallopian tubes
3. Ovulation
4. Penetration of the ovum by a sperm
5. The union of the male and female pronuclei, forming the single nucleus of the zygote and restoring the species number of chromosomes

It is clearly established that only one sperm participates in the last stage of fertilization, the union of the male and female pronuclei. Many philosophical explanations have been given, and many physiological studies made, to explain the tremendous difference in numbers of gametes contributed by the sexes. Perhaps the best reason for the ejaculation of millions or billions of sperm is that most of the sperm never reach the immediate vicinity of the egg. The ovum, upon recovery from the ovarian end of the Fallopian tube, is usually surrounded by a few hundred or, at most, one or two thousand sperm.

One of the more important stages of embryonic development is the formation of the primary germ layers: the ectoderm, mesoderm, and entoderm. The germ layers differentiate into special tissues, such as epithelial cells, nerve cells, and connective tissue, and these tissues multiply and associate in the formation of the various organs of the body. All embryos, whether bird, mammal, or some lower form, follow the same general pattern of organization. The nervous tissues, derived from the ectoderm, give rise to the brain, spinal cord, and the various nerves. The entoderm is the origin of the gut and its associated digestive glands and the respiratory tract. The heart, the circulatory system, the kidneys, the genital organs, the supporting structures such as cartilage, bone, and connective tissue, and the various types of muscle are derived from the mesoderm. The embryonic period is completed when the various organ systems have been developed and general body form is recognizable. This requires about 35 days in sheep, 45 days in cattle, and about one month in swine.

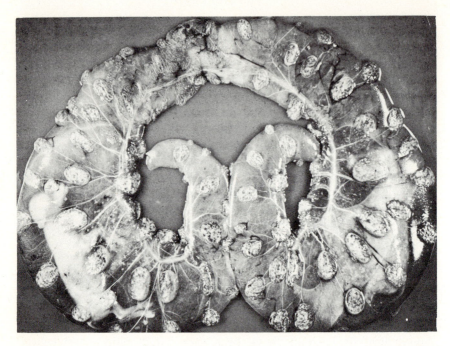

Fig. 4-2. Twin calf fetuses at the 106th day of gestation. (*Courtesy of Dairy Breeding Research Center, the Pennsylvania State University.*)

The developing embryo, and the fetus, as it is known after differentiation has taken place, is nourished by the fetal membranes and the placenta. The fetal membranes are made up of the yolk sac, amnion, allantois, chorion, and umbilical cord. In farm mammals there is no recognizable yolk, but in the early development of the pig, for example, the yolk sac is very large. As embryonic development progresses, the yolk sac regresses.

The amnion is the inner, fluid-filled membrane which surrounds the embryo and presumably protects the embryo from shock and from adhesions as growth occurs. The allantois is a saclike outgrowth of the hind gut. It is associated with the chorion in forming the fetal placenta. The chorion is the outer sac or membrane surrounding the early embryo.

The umbilical cord joins the body of the embryo to the fetal membranes. It is made up of the stalk of the yolk sac, the stalk of the allantois, and the umbilical vein and artery.

The placenta is composed of structures associated with both the embryo and the uterus. These are sometimes called the *fetal placenta* and the *maternal placenta*. The fetal portion is composed of the chorion and the allantois, which fuses with it; the lining of the uterus is modified during early pregnancy, and this modified portion which is in contact with the chorion is called the maternal placenta. There are several distinct types of placenta. In swine and the horse, small villi are scattered over the entire chorion. These villi contact the uterine mucosa, forming what is called the *diffuse placenta*. In cattle and sheep, the chorion contacts the uterine mucosa only at definite areas. These areas are called *cotyledons,* number from 70 to 130 in sheep and cattle, and are made up of groups of villi.

PARTURITION

Why parturition occurs at the moment it does remains one of nature's yet unexplained mysteries. Many theories have been developed, but it is still impossible to induce parturition at will in farm mammals. It is obvious that the survival of the newborn does not depend upon birth at a particular moment. There are many standard gestation tables, similar to Table 4-2, which assist in predicting the date of birth. Individual animals may deviate widely from the average gestation period for the species, and there are some variations between breeds or even within lines of a breed.

The relative distention of the uterus, pressure of the fetus on the cervix, stimulation of the cervical ganglia, and changes in endocrine balance may all contribute to the termination of pregnancy. In recent years there has been increasing evidence that the hormonal relationships are the determining factor, but endocrinologists agree that much is yet to be learned about this process. Estrogen, progesterone, relaxin, and the posterior pituitary oxytocic factor appear to be the chief hormones involved.

Parturition is divided into several phases. The first phase is the dilation of the cervix and the loosening of the pelvic region. The second phase is the gradually accelerating contractions of the uterine and abdominal muscles, causing the actual expulsion of the fetus. The third stage is a continuation of the second; the muscular contractions continue at a lesser rate and result in the expulsion of the afterbirth. A convalescent phase covers several days, during which the uterus, cervix, and pelvis gradually return to the normal nonpregnant state.

LOWERED FERTILITY AND STERILITY

Sterility is complete and permanent reproductive failure. In farm animals the completion of the reproductive process depends on fertilization, normal embryonic and fetal growth, and the birth of young capable of

Table 4-2. Gestation Table

Date bred	Mare, 336 days	Cow, 281 days	Sow, 114 days	Ewe, 147 days
Jan. 1	Dec. 3	Oct. 9	Apr. 25	May 28
Jan. 6	Dec. 8	Oct. 14	Apr. 30	June 2
Jan. 11	Dec. 13	Oct. 19	May 5	June 7
Jan. 16	Dec. 18	Oct. 24	May 10	June 12
Jan. 21	Dec. 23	Oct. 29	May 15	June 17
Jan. 26	Dec. 28	Nov. 3	May 20	June 22
Jan. 31	Jan. 2	Nov. 8	May 25	June 27
Feb. 5	Jan. 7	Nov. 13	May 30	July 2
Feb. 10	Jan. 12	Nov. 18	June 4	July 7
Feb. 15	Jan. 17	Nov. 23	June 9	July 12
Feb. 20	Jan. 22	Nov. 28	June 14	July 17
Feb. 25	Jan. 27	Dec. 3	June 19	July 23
Mar. 2	Feb. 1	Dec. 8	June 24	July 27
Mar. 7	Feb. 6	Dec. 13	June 29	Aug. 1
Mar. 12	Feb. 11	Dec. 18	July 4	Aug. 6
Mar. 17	Feb. 16	Dec. 23	July 9	Aug. 11
Mar. 22	Feb. 21	Dec. 28	July 14	Aug. 16
Mar. 27	Feb. 26	Jan. 2	July 19	Aug. 21
Apr. 1	Mar. 3	Jan. 7	July 24	Aug. 26
Apr. 6	Mar. 8	Jan. 12	July 29	Aug. 31
Apr. 11	Mar. 13	Jan. 17	Aug. 3	Sept. 5
Apr. 16	Mar. 18	Jan. 22	Aug. 8	Sept. 10
Apr. 21	Mar. 23	Jan. 27	Aug. 13	Sept. 15
Apr. 26	Mar. 28	Feb. 1	Aug. 17	Sept. 20
May 1	Apr. 2	Feb. 6	Aug. 23	Sept. 25
May 6	Apr. 7	Feb. 11	Aug. 28	Sept. 30
May 11	Apr. 12	Feb. 16	Sept. 2	Oct. 5
May 16	Apr. 17	Feb. 21	Sept. 7	Oct. 10
May 21	Apr. 22	Feb. 26	Sept. 12	Oct. 15
May 26	Apr. 27	Mar. 3	Sept. 17	Oct. 20
May 31	May 2	Mar. 8	Sept. 22	Oct. 25
June 5	May 7	Mar. 13	Sept. 27	Oct. 30
June 10	May 12	Mar. 18	Oct. 2	Nov. 4
June 15	May 17	Mar. 23	Oct. 7	Nov. 9
June 20	May 22	Mar. 28	Oct. 12	Nov. 14
June 25	May 27	Apr. 2	Oct. 17	Nov. 19
June 30	June 1	Apr. 7	Oct. 22	Nov. 24
July 5	June 6	Apr. 12	Oct. 27	Nov. 29
July 10	June 11	Apr. 17	Nov. 1	Dec. 4
July 15	June 16	Apr. 22	Nov. 6	Dec. 9
July 20	June 21	Apr. 27	Nov. 11	Dec. 14
July 25	June 26	May 2	Nov. 16	Dec. 19
July 30	July 1	May 7	Nov. 21	Dec. 24
Aug. 4	July 6	May 12	Nov. 26	Dec. 29
Aug. 9	July 11	May 17	Nov. 31	Jan. 3
Aug. 14	July 16	May 22	Dec. 6	Jan. 8

Table 4-2. Gestation Table (*Continued*)

Date bred	Mare, 336 days	Cow, 281 days	Sow, 114 days	Ewe, 147 days
Aug. 19	July 21	May 27	Dec. 11	Jan. 13
Aug. 24	July 26	June 1	Dec. 16	Jan. 18
Aug. 29	July 31	June 6	Dec. 21	Jan. 23
Sept. 3	Aug. 5	June 11	Dec. 26	Jan. 28
Sept. 8	Aug. 10	June 16	Dec. 31	Feb. 2
Sept. 13	Aug. 15	June 21	Jan. 5	Feb. 7
Sept. 18	Aug. 20	June 26	Jan. 10	Feb. 12
Sept. 23	Aug. 25	July 1	Jan. 15	Feb. 17
Sept. 28	Aug. 30	July 6	Jan. 20	Feb. 22
Oct. 3	Sept. 4	July 11	Jan. 25	Feb. 27
Oct. 8	Sept. 9	July 16	Jan. 30	Mar. 4
Oct. 13	Sept. 14	July 21	Feb. 4	Mar. 9
Oct. 18	Sept. 19	July 26	Feb. 9	Mar. 14
Oct. 23	Sept. 24	July 31	Feb. 14	Mar. 19
Oct. 28	Sept. 29	Aug. 5	Feb. 19	Mar. 24
Nov. 2	Oct. 4	Aug. 10	Feb. 24	Mar. 29
Nov. 7	Oct. 9	Aug. 15	Mar. 1	Apr. 3
Nov. 12	Oct. 14	Aug. 20	Mar. 6	Apr. 8
Nov. 17	Oct. 19	Aug. 25	Mar. 11	Apr. 13
Nov. 22	Oct. 24	Aug. 30	Mar. 16	Apr. 18
Nov. 27	Oct. 29	Sept. 4	Mar. 21	Apr. 23
Dec. 2	Nov. 3	Sept. 9	Mar. 26	Apr. 28
Dec. 7	Nov. 8	Sept. 14	Mar. 31	May 3
Dec. 12	Nov. 13	Sept. 19	Apr. 5	May 8
Dec. 17	Nov. 18	Sept. 24	Apr. 10	May 13
Dec. 22	Nov. 23	Sept. 29	Apr. 15	May 18
Dec. 27	Nov. 28	Oct. 4	Apr. 20	May 23

survival in a favorable environment. If the reproductive process is interfered with during early embryonic differentiation or during the act of parturition and the young are not born alive, *that mating* must be regarded as a sterile one. Improper development of the young which prevents survival during the birth process as well as defects in the genitalia of the dam which will not permit birth, should be included in a definition of sterility. Our definition requires that reproductive failure be of a permanent nature.

It is obvious that not all matings between normal males and females result in the birth of normal young. In well-managed cattle herds, for example, only about 70 per cent of the normal cows conceive on the first service. The fact that 30 per cent of the individual matings are sterile matings should not be interpreted to mean that the participating individuals are sterile.

Fertility is the degree of reproductive ability. In animals characterized by sexual reproduction, fertility may be limited by either sex. The mating of a highly fertile male to a sterile female, e.g., a freemartin, has the same result as the mating of two sterile animals. Since fertility is a measure of the reproductive ability of both sexes, it must express reproductive ability in terms of numbers of offspring for a specific period of time.

Prolificacy is the production of large numbers of young. There are wide species differences in prolificacy, the sow being more prolific than the mare, the ewe more prolific than the cow, and the hen being the most prolific of all farm animals.

Fecundity is the ability of the male or female to produce large numbers of normal sperm or ova. This term is not in common use by livestock breeders, and even among research workers there are variations in usage.

Potency is the ability of the male to perform copulation. Highly potent males have a strong sex drive and mate readily. Males which will not copulate at all are impotent.

Causes of Reproductive Failure. There is a common erroneous belief that sterility is a single problem, a specific disease which can be corrected by a specific treatment. In some instances a specific defect or a particular disease can be identified as the cause of reproductive failure. In many cases there is no apparent reason for sterility or lowered fertility. In some individuals a variety of minor disturbances may, in combination, limit reproduction. However, the causes of reproductive failure can most commonly be classified as

1. Anatomical defects
2. Mechanical injury of the genitalia
3. Genital diseases
4. Nutritional deficiencies
5. Endocrine disturbances
6. Genetic factors
7. Miscellaneous or unknown causes

Anatomic Defects of the Genitalia. Literally hundreds of types of anatomic defects of the genital organs have been reported. Some of these are of sufficient severity to cause sterility, and others affect the degree of fertility.

Probably the most common male genital defect is cryptorchidism, the failure of the testes to descend fully into the scrotum. If one testis is in a scrotal position, the male is usually fertile, but if both are retained in the abdominal cavity, sterility usually results. Since this defect is heritable, it should be guarded against in breeding animals. Another genital defect common in males is the occurrence of scrotal hernia.

Probably the best-known anatomic abnormality in the female is the freemartin condition. In cattle, when twins of opposite sex are produced, a high proportion of the genetic female calves have abnormal genital organs. These abnormal females are sterile and are called *freemartins*. About 10 per cent of the females born twin with a male are normal, will reproduce, and should not be called freemartins. In the freemartin the external genitalia are generally of the female type, the vagina is present but greatly shortened, the gonads tend to resemble nonfunctional testes, and the Fallopian tubes and the uterus may be present in a rudimentary state.

Some of the other congenital defects of the female genitalia are unilateral or bilateral absence of the ovaries, Fallopian tubes, or uterine horns, closure of the cervix or tubes, persistent hymen, and miscellaneous developmental abnormalities.

Genital Diseases. Of the farm animals, cattle are most likely to be affected by specific transmissible genital diseases. The economic losses incurred as the result of such diseases in cattle are probably many times as great as those for all other farm animals combined.

Brucellosis. Brucellosis is a contagious disease of cattle, swine, goats, and man. The *Brucella* bacteria have been classified into three types: *Br. abortus,* which chiefly affect cattle; *Br. suis,* which are adapted to swine; and *Br. melitensis,* which chiefly affect goats. Man is affected by all types of *Brucella* organisms.

In cattle the *Br. abortus* organisms tend to localize in the uterus of the pregnant cow and produce abortion. Abortion may occur at any stage of pregnancy but is most common between the 5th and 8th months. Most cattle develop a degree of resistance to the disease after exposure and usually do not abort more than once. This may develop a false sense of confidence among cattlemen who regard abortion as the only symptom of the disease. Cattle which are infected, which may or may not abort, may be dangerous spreaders of the disease. The placenta and genital discharge of such cows and their milk may contain large numbers of the organisms. In bulls the organisms may localize in the testes, seminal vesicles, or vas deferens, and males may spread the disease by copulation, although libido is generally reduced if testicular involvement occurs.

Brucellosis in swine is most often caused by *Br. suis.* Swine are sometimes infected by *Br. abortus,* and cattle may be affected by both the *abortus* and *suis* types. Swine brucellosis is often a serious economic problem to the swine producer and may occasionally prevent the profitable operation of a swine enterprise. The incidence of abortions may vary greatly between swine herds, ranging from an occasional abortion to the majority of pregnant females. In the boar the organisms may localize in the testes and produce a severe orchitis. *Brucella* may be eliminated in large numbers in the semen, and it is well known that the boar is frequently a spreader of the disease.

Vibriosis. Cattle and sheep may become infected with the bacterium *Vibrio foetus*. The first evidence of infection with this organism is abortion. In badly infected cattle herds, abortion by about 20 per cent of the cows may occur. Abortion occurs most commonly in the 5th and 6th months of gestation. In sheep, abortion usually occurs during the last 6 to 8 weeks of gestation, and the lambs may be delivered dead at term. As many as 60 per cent of the ewes may abort. *V. foetus* is spread by infected bulls or rams, but whether or not this is the only route of infection is not known for either cattle or sheep.

Trichomoniasis. Trichomoniasis is caused by the flagellated motile parasite *Trichomonas foetus,* and the chief symptom is abortion or embryonic death. The usual spreader of the disease is the bull. The parasites live, more or less permanently when established, in the sheath. They cause no damage to the bull and do not affect semen quality or conception rate. Since the bull is the primary source of infection, it is extremely important that all sires used in artificial insemination be free of the disease.

Leptospirosis. The *Leptospira* are saprophytic microorganisms which are widely distributed in fresh and salt water, especially in water contaminated with fecal material. These organisms are of several types and have been recovered from the human, rodents, dogs, cattle, swine, and the horse. Serological tests indicate *L. pomona* commonly infect cattle and swine, although some research workers have referred to the cattle strains as *L. bovis. Leptospira* may cause abortion in both cattle and swine. Abortion usually occurs late in the gestation period, but it may occur at any stage of pregnancy. Often no symptoms other than abortion are observed. Diagnosis requires extremely careful laboratory procedures.

Nutritional Deficiencies. Because lowered fertility often occurs in the absence of recognizable anatomical defects or genital disease, there is a tendency to explain it in terms of nutritional deficiencies. It is entirely possible that reproduction may be interfered with by inadequate amounts of nutrients or by a lack of specific nutrients. It is likewise true that under average farm conditions nutritional deficiencies which affect only reproduction and produce no other symptoms are unlikely to occur. Vitamin-A deficiency occurs under farm conditions more widely and in more species than any other vitamin deficiency. In the male, it causes inhibition or cessation of spermatogenesis, a reduction in sex drive, and, in advanced deficiencies, general debility. In the female, estrus and ovulation occur in a relatively normal pattern and fertilization takes place. Embryonic death, abortion, and the birth of dead or weak young usually occur. In cattle, the occurrence of cysts in the anterior pituitary glands of calves from cows deficient in vitamin A is common. It is reported that these cysts do not disappear following vitamin-A therapy, but the significance of this is not known. Vitamin-A deficiency in pregnant swine may result in fetal resorption or in abnormal embryonic development. The birth of weak, malformed, blind, or even eyeless pigs has been reported in experimentally induced vitamin-A deficiency.

Vitamin E is extremely important in certain metabolic functions. Fortunately it is a fairly common constituent of natural feedstuffs and is not often a limiting factor in livestock rations. The selection of the descriptive name of vitamin E as the antisterility factor was unfortunate, for in the farm animals there is little evidence that the supplementation of practical livestock rations with vitamin E will improve fertility or correct infertility. Vitamin E has a role in normal muscle development. In both lambs and calves, vitamin-E deficiency may result in muscular dystrophy. This condition is called *stiff-lamb disease* or *white-muscle disease* in calves.

Phosphorus deficiency is common in cattle, especially in some western and southern regions of the United States. Unfortunately, in certain range areas phosphorus deficiency is accompanied by protein deficiency. This dual lack may result in poor growth and general unthriftiness in both cattle and sheep, and under such conditions estrus may not occur. Calcium deficiency is not common in cattle, sheep, or swine, but it may limit reproduction in poultry.

Although cobalt does not have a direct effect on reproduction, its role in rumen function gives it a place as an essential element in cattle and sheep. In the Great Lakes basin and in some other areas, supplementary cobalt may be required.

Iodine is essential for normal thyroid function in all animals, but its lack is most likely to be shown in pregnant sheep and swine. Iodine deficiency is manifested by the birth of weak, dead, hairless pigs or woolless lambs with greatly enlarged thyroid glands (goiter). Prior to the widespread use of stabilized salt in breeding females, iodine deficiency limited or even prevented sheep and swine production in the

Great Lakes area, in part of the Great Plains, and the Rocky Mountain and Pacific Coast states.

Copper and iron, in a broad sense, are important for the completion of the reproductive process in swine. Unless supplementary sources of iron and copper are available, newborn pigs may become anemic and there may be excessive postnatal death losses. This can be prevented by giving the pigs access to soil containing trace amounts of iron and copper or by the use of proprietary antianemic preparations of copper-iron salts.

Endocrine Disturbances. Failure of the genitalia to develop is called *sexual infantilism*. The entire reproductive tract remains small, and in males the testes are visibly of reduced size. Females do not have regular heat periods, and males lack sex drive. Sometimes, especially in cattle, the animals become excessively fat and sluggish and resemble spayed heifers or steers. This condition presumably is the result of decreased gonadotropic-hormone secretion by the anterior pituitary gland.

Abnormal corpus luteum function can affect fertility in at least several ways. The corpus luteum may persist in the ovary and continue to secrete progesterone. If progesterone is being produced in large amounts, new Graafian follicles are not formed and estrus and ovulation do not take place. In dairy cows the estrual cycle should be reestablished within the first 60 days following calving. In some animals in which no evidences of estrus have been observed, rectal palpation of the ovaries reveals a prominent and presumably functional corpus luteum. Manual removal of the corpus luteum is usually followed by heat within one to four days. It is well known that the removal of the corpus luteum of cattle during pregnancy may produce abortion. Because of the role of the corpus luteum in the preparation and maintenance of the endometrium, there is presumptive evidence that corpus luteum failure may sometimes cause early embryonic death.

Ovarian cysts are common in cows, mares, and sows. Ovarian cysts are ordinarily considerably larger than normal Graafian follicles and with time may develop thickened walls. In cows and mares, persistent cystic follicles often produce nymphomania or constant estrus. However, some animals with ovarian cysts may show no signs of heat. In cattle, when the condition is allowed to persist, there may be pronounced anatomic and psychologic changes. The pelvic ligaments relax and the tail head is elevated. The animal may acquire such male characteristics as thickened forequarters and may bellow and behave like a bull.

Genetic Factors in Fertility and Sterility. The development of breeds or of lines of animals within breeds which differ in prolificacy is good evidence that fertility may have a genetic basis. It is common knowledge that some once-popular families have become extinct because of a high incidence of sterility. Perhaps the best-known example is the Duchess family of Shorthorns. Strangely enough, although the high rate of sterility was known, breeders considered it an advantage. The Duchess Shorthorns were considered an excellent type in their day, and since infertility limited their propagation, they sold at high prices and were in considerable demand. It is self-evident today that, regardless of other traits, animals with subnormal fertility cannot be considered efficient.

It has long been recognized that intense inbreeding of cattle, sheep, swine, or poultry is usually followed by decreased litter size, or decreased fertility. As will be emphasized in subsequent chapters, inbreeding may be a valuable procedure in the establishment of new lines or breeds, but it cannot be expected to be entirely free of undesirable effects. A number of well-known laboratory strains of rats and mice have been inbred for many generations. It should be recognized that the better strains have been subjected to prolonged critical selection and that factors for infertility have been largely eliminated.

BREEDING MANAGEMENT PRACTICES

As previously stressed, there is a very great difference between sterility and fertility, and there are many distinct causes of sterility or suboptimum fertility. It is not within the scope of this book to discuss the question in depth, but it is appropriate to make brief mention of some of the practices which are known to be associated with optimum reproductive performance.

Cattle-breeding Practices. Although the use of artificial insemination has increased rapidly during the past decade, natural service is still the norm through most of the world. A well-developed yearling bull in good health can be expected to service at least 25 cows between his first and second years and after two years of age should be able to mate with 50 to 75 cows annually. When bulls are used under range conditions during a limited breeding season, a mature male should be able to service 25 to 35 cows under average pasture conditions. Bulls kept in specialized artificial breeding studs are usually ejaculated at 5- to 7-day intervals, and under experimental conditions vigorous individuals have been ejaculated several times per week for long periods with no effect on fertility level.

Since both sperm and ova have a relatively short survival time once they have entered the tubular structures of the uterus, one of the factors in a high conception rate is the proper timing of natural service or artificial insemination with ovulation. As shown in Table 4-3, estrus usually lasts for 16 to 20 hours in the cow, and ovulation occurs about 10 to 14 hours after heat ends. In natural service it is desirable that mating be timed from about the middle until the end of the heat period. Artificial insemination can be carried out with reasonable assurance of pregnancy from the middle of estrus until about 6 hours after heat ends. Cows are functionally capable of conception at 12 to 14 months of age and may be bred to calve at two years of age if the owner is prepared to cope with any calving problems which may occur. When early calving is practiced, the cattle must be adequately fed during pregnancy, must be of normal size for age, and the herdsman must know how to assist at parturition.

Cows should not be rebred too soon after calving, since the uterus must undergo a characteristic involution and preparation for pregnancy. Rebreeding is usually delayed until 60 to 70 days after parturition.

Swine-breeding Practices. The reproductive characteristics of boars and sows are greatly different from bulls and cows. Unless the rate of service of boars is controlled, both conception rate and litter size may be affected. Although some boars may be willing to copulate at 4 months of age, this is not an indication that they are ready for practical service. The average boar should be well grown out and 8 or 9 months of age before he is turned in with sows for use in a herd-mating system. He should not be expected to serve more than 7 to 8 females per week, and a vigorous, mature boar should usually be limited to 12 sows per week. Because the boar has much less mating capacity than the bull, there is considerable interest in the use of performance-tested boars by artificial insemination. Although artificial insemination has been practiced in swine for many years, its use has been limited by the fact that boar semen can, as yet, neither be extended by dilution nor stored with anywhere near the success that is possible with bull semen.

As shown in Table 4-3, the average sow remains in heat 2 or 3 days. Ovulation may occur over a wide range, but most ova will be liberated from the ovaries between the 18th and 48th hours of estrus. When copulation is limited to one natural service, it is best to mate late on the first or early during the second day of heat. In commercial swine herds, 1 boar is often kept with a group of 10 sows for 24 hours and then rotated to a second group of females when a second boar is placed with the first group. Such systems of rotation tend to reduce the effects which a boar of low fertility might have on total conception rate.

Sheep-breeding Practices. The most important factor to consider in practical management of the breeding flock is the seasonal occurrence

Table 4-3. The Reproductive Cycle in Farm Animals

Species	Length of estrual cycle, days		Length of estrus		Usual time of ovulation	Length of gestation, days		Age at puberty, months
	Av.	Range	Av.	Range		Av.	Range	
Mare	21	10–37	5–6 days	1–14 days	24–48 hr before end of estrus	336	310–350	10–12
Sow	21	18–24	2–3 days	1–5 days	18–60 hr after estrus begins	112	111–115	5–7
Ewe	16	14–20	30 hr	20–42 hr	1 hr before end of estrus	150	140–160	4–8
Goat	20	12–25	36–48 hr	20–80 hr	Near end of estrus	151	140–160	4–8
Cow	19–21	16–24	16–20 hr	8–30 hr	10–14 hr after end of estrus	281	274–291	4–8

of estrus. Although there are seasonal fluctuations in semen quality of the ram, the limiting factor is the seasonal periodicity of the female. Although breeds such as the Dorset and the Merino have a longer breeding season than the other common breeds, it is not possible to produce lambs throughout the year in the temperate zones of either hemisphere. In the United States, most mutton breeds exhibit estrus from August through January. In the Southwest and Pacific Coast states, lambing is most common from January through March. In colder areas, most range ewes lamb in May.

As shown in Table 4-3, the ewe has an average heat period of about 30 hours and ovulates about one hour before the end of heat. When hand mating is practiced, excellent fertility is usually obtained by breeding at approximately the middle of the heat period. In practice, a great share of all breeding ewes are flock-mated. A normal fertile ram one year or older should be able to service 40 to 50 ewes during a breeding season of one or two months in a flock-mating system. A vigorous ram, when allowed to run with the ewes, will usually serve each ewe more than once during each heat period.

Horse-breeding Practices.   Although horse breeding is now carried on almost exclusively by specialists rather than livestock farmers, it is appropriate to describe one or two unique problems. There is more variation in the estrual cycle of the mare than in any other farm animal, and the problem of correlating mating with ovulation is a very difficult one. As shown in Table 4-3, the average mare remains in heat 5 or 6 days, but a significant number have shorter periods of 2 or 3 days, and others typically remain in heat for 7 or 8 days. The stallion produces a rather large volume of semen, and the number per unit volume is much less than in the case of the bull or the ram. Like the boar, stallion semen is difficult to preserve under artificial storage conditions, and the horse breeder must usually rely on natural mating at selected stages of estrus. If the exact time of ovulation could be predicted in advance, service about 24 hours prior to ovulation would probably be the system of choice. Since this is not possible, conception rate can be improved by mating more than once during estrus. One common system is service every other day, beginning on the 3d day of heat, and some successful breeders mate daily on the 3d, 4th, and 5th days if the mare remains in heat.

ARTIFICIAL INSEMINATION

Few advances in the genetic improvement of livestock have been as widely accepted as the artificial insemination of dairy cattle, and the practice is now world-wide. It is emphasized that artificial insemination is not recommended because it is a novel way of initiating pregnancy, but because it is the only practical method of extending the germ plasm of outstanding proved bulls. In recent years there has been a steadily increasing use of performance- or progeny-tested meat-type bulls by artificial insemination. In some species, and in the case of certain breeds within species, the use of artificial insemination has been discouraged by breed association legislation. The recognition that significant livestock improvement depends on the identification and use of superior genetic material is now gaining acceptance by most livestock breeders, and we may expect a steady increase in the practice of artificial insemination.

The genetic implications of world-wide use of frozen semen are staggering to the imagination. It is possible to extend the germ cells of a single bull to a half million cows and perhaps more, and it is already commonplace to use stored frozen semen for several years after a sire's death. The selection of sires which may improve or damage the genetic make-up of domestic animals for future generations is a responsibility which should not be taken lightly. It is obvious that the semen must be free of organisms such as *Brucella abortus*, *Vibrio foetus*, and *Trichomonas foetus*, and other pathogens which might be transmitted in semen. It is necessary to test for the presence of certain harmful recessive genes, especially those associated with dwarfism, cryptorchidism, scrotal and umbilical hernias, and other undesirable traits.

Most characteristics of economic importance such as milk and egg production, growth rate, efficiency of feed utilization, and meat qualities are influenced by a great many pairs of genes. It is difficult to select for a large number of traits simultaneously, and it is even more difficult to carry out a selection program when the importance of some of the economic traits undergo radical change. The animal breeding specialist must provide some means of preserving a great deal of genetic material that may have no particular use at the moment but which may be extremely important in the future. We need only consider some of the traits which were emphasized during the first part of the twentieth century but which are out of fashion at present—high butterfat percentage in cattle, the lard-type hog, compact or comprest beef cattle, and a full covering of wool on the face of sheep. It is indeed fortunate that we preserved, by accident, genes which are now making it possible to select for high protein and high solids-not-fat in milk, a type of swine with a maximum of muscle and a minimum of fat, and a variety of genes in cattle which are associated with rapid skeletal growth and greater muscle mass.

IMPORTANCE OF REPRODUCTIVE RATE TO SELECTION PROGRAMS

In the improvement of population of farm animals through breeding, it is important that the superiority of those selected as parents of the next generation be as great as possible relative to the population average. The possible magnitude of the selection differential is related to the proportion which must be selected as breeding stock. The principles of selection will be discussed in Chap. 10. For present purposes it is sufficient to point out that, as an example, the selection differential can be approximately 1¾ times as great if only 10 per cent of a population has to be retained as compared to retaining 40 per cent.

The genetic usefulness of artificial insemination is based upon the greater selection differentials possible. When each sire can fertilize hundreds or thousands of females, fewer sires are needed and only the very best can be chosen.

In general, the higher the reproductive rate, the lower will be the percentage of animals which must be retained for breeding purposes. The percentages which must be saved for breeding purposes in populations being held constant in numbers will vary greatly from herd to herd within species. It will depend upon age at which offspring are first produced and maximum age to which breeding animals are retained. Both these things are partially under control of the herd manager. It will also depend upon the amount of culling done among breeding animals on the basis of their own production or the performance of their progeny.

SUMMARY

The testis and the ovary both have dual functions. They produce spermatozoa and ova, respectively. They also produce sex hormones which aid in developing and maintaining the remainder of the reproductive tract and in inducing sex drive in the male and receptivity in the female. The hormones have other effects on growth and development. The testis and ovary are under the control of the pituitary hormones. In the female the regulation of the estrual cycle requires a close synchronization of pituitary and ovarian hormones.

Lowered fertility and sterility are important economic problems and reduce potential genetic improvement for other traits, since fewer animals are available for selection. Infertility can be due to many factors including genetically influenced anatomical or physiological defects, disease, inadequate nutrition, or unskilled management. Each class of stock requires different managerial approaches for maximum fertility. Artificial insemination offers a means of greatly extending the usefulness of outstanding sires.

REFERENCES

Asdell, S. A. 1964. "Patterns of Mammalian Reproduction," 2d ed., Cornell University Press, Ithaca, N.Y.

Cole, H. H., and Cupps, P. T. 1959. "Reproduction in Domestic Animals," vols. 1 and 2, Academic Press Inc., New York.

Nalbandov, A. V. 1964. "Reproductive Physiology," 2d ed., W. H. Freeman and Company, San Francisco.

Salisbury, G. W., and VanDemark, N. L. 1961. "Physiology of Reproduction and Artificial Insemination of Cattle," W. H. Freeman and Company, San Francisco.

Young, W. C. 1961. "Sex and Internal Secretions," 3d ed., vols. 1 and 2, The Williams & Wilkins Company, Baltimore.

CHAPTER FIVE

A Potpourri—Lethals, Antigens, and Others

Many hundreds of qualitative characters have been identified as having a genetic basis in farm animals. As a somewhat oversimplified generalization, we define a qualitative character as one in which a given animal falls in one discrete class or another as contrasted to quantitative characters in which all grades of measurable variation exist.

The mode of inheritance of many qualitative characters is well established. In other cases a hereditary basis is known to exist, but the exact mode of inheritance is not certainly known.

The processes of development are very incompletely understood, but it is clear that all externally visible characters are affected by many genes probably acting in a sequential fashion. The failure of any gene in the chain to exert its effect, probably by failing to produce a needed enzyme, can have marked effects on the development of an organism. We might liken these steps to links in a chain—if all are of good steel, we have a dependable chain, but if one is of putty, it is worthless. Similarly, the genetic effect can be disastrous as is true of those genes which result in death (lethals) or the production of an animal so abnormal that its usefulness is impaired.

In other cases genetic differences merely modify an animal's appearance or other physical characteristics but do not affect vigor or survival. Some such modifications have economic value, as for example, horned vs. polled in sheep or cattle, wool color, etc.

Still other genetically controlled differences among animals have no visible outward effects and no known relationship to performance or productivity.

We often speak of a gene or genes for a given character such as horns, color, or a physical abnormality. Strictly speaking, this is not correct, since the gene we are talking of is only one of many necessary for development of the character. It is, however, one of the gene pair, or an allelic series in which variation has occurred and enabled us to identify an effect. By analogy, it may be correct for us to say that a burned-out tube is responsible for a television set's failure to operate, but we can hardly say that the replacement tube is responsible for the successful operation of the set. Rather, it operates as the result of the successful, coordinated function of this tube and many other components.

In this chapter it is not our purpose to attempt a cataloging of genetically controlled qualitative traits. We hope, rather, to illustrate some of the many forms which can occur and to discuss some of the problems a breeder may face in connection with them. Some useful aspects of genetic variants will also be considered.

LETHALS AND GENETIC ABNORMALITIES

Although wide anatomical and physiological differences occur within species, there is an eventual limit of deviation from the norm beyond which the organism cannot survive. Death of the organism may occur at any stage of development—immediately following fertilization, during embryonic differentiation, at parturition, or postnatally. Death may be due to a variety of causes, such as injury, disease, malnutrition, and harmful irradiations, such as X rays and gamma rays. We speak of any cause of death as a *lethal effect.* Among the many causes of death are gene changes which are incompatible with development or survival. These genes are known as *lethal* genes. There are some genes which are deleterious to the organism but not lethal, provided environmental factors are especially favorable. These genes are called *semilethals.* There are undoubtedly many undetected lethal genes, but there are likewise many unexplained deaths due to environmental factors.

The development of a new individual from a tiny fertilized egg is one of the most interesting and complicated processes in all of nature. What any individual will eventually become is obviously dependent on both heredity and environment. Heredity provides the basic specifications; the environment, both internal and external, provides the wherewithal for fulfilling the specifications. When an organism succumbs before the usual time for its species, it is often exceedingly difficult to determine whether the cause of its early demise is hereditary or environmental, or perhaps often some combination of these.

Is Abnormality Heritable?   Typically, a lethal or other abnormality will first come to the attention of a breeder when one or more defective

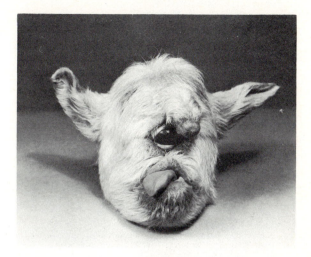

Fig. 5-1. Head of a cyclopian-type malformed lamb with a centrally located single eye, a proboscislike protuberance originating just above the eye and absence of the premaxilla bone. This condition and variations of it have been produced experimentally by feeding the weed *Veratrum californicum* early in gestation. It is thought the sensitive period of the embryo is the 14th day of gestation. (*Courtesy of Dr. Wayne Binns, Animal Disease and Parasite Research Division, U.S. Department of Agriculture, Logan, Utah.*)

individuals appear in his herd or flock. There are no absolute rules for determining whether the abnormality is hereditary or environmental in origin, whether it is due to some combination of hereditary and environmental influences, or whether it is merely an "accident of development."

The following would indicate a hereditary basis:

1. If previous studies on a scale large enough to be conclusive had shown a hereditary basis for a phenotypically similar condition in the same species or breed. This would not necessarily be a completely reliable guide because some characters, almost or completely phenotypically indistinguishable, may be caused by either heredity or environment. Hydrocephalus in cattle is a probable example.
2. If the condition appeared only in some breeding groups. An example would be a condition appearing in some sire progenies but not in others.
3. If it occurred in herds where there had been inbreeding. Inbreeding does not create genetic abnormalities, but most are recessive, and it tends to bring them to light as a result of the increased homozygosity resulting from inbreeding.

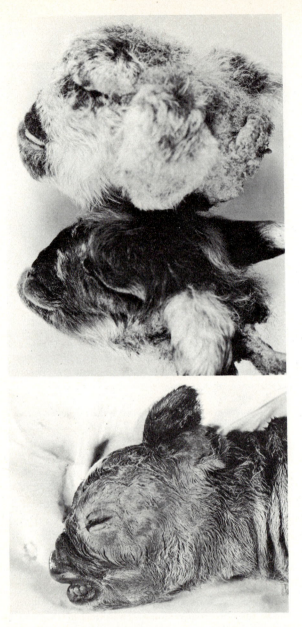

Fig. 5-2. *Above,* two "monkey faced" lambs thought to be less severe expressions of the same environmentally induced condition shown in Fig. 5-1. *Below,* a calf homozygous for the gene which conditions the trait adenohypophyseal aplasia. It was born dead after a gestation period 100 days longer than normal. Note the striking resemblance in profile of the lambs and the calf. (*Courtesy of Dr. Clyde Stormont, University of California.*)

4. If it occurred in more than one season when rations and environment differed.

The following would indicate an environmental basis:

1. If it had previously been reliably reported as due to ration or environment.
2. If it occurred when the ration of the dam was known to have been deficient or she had been under other stress.
3. If it did not recur after rations or environments were changed or improved.

If a condition met none of the above criteria and occurred only once or very rarely, it should be looked on as an accident of development. No action to prevent reoccurrence should be undertaken until facts are available to provide a sound basis for action.

The foregoing represents an oversimplification in that we have arbitrarily classified abnormalities into those due to (1) genetic causes, (2) environmental effects, and (3) accidents of development, i.e., those for which we have no explanation. In a strict sense every abnormality is the product of heredity *and* environment. If the gene or genes condition-

Fig. 5-3. A "crooked calf." Conditions similar to this have been observed in many Western states. Deformities of the front legs are most consistent, but the neck and back and the hind legs can be affected. A similar condition accompanied by hydrocephalus is also known. The condition has been produced experimentally by feeding lupine (*L. sericeus*) and high levels of lead during gestation. It is believed that other nutritional imbalances or toxic materials may also cause it. (*Courtesy of Dr. Wayne Binns, Animal Disease and Parasite Research Division, U.S. Department of Agriculture, Logan, Utah.*)

Fig. 5-4. A "crooked calf" typical of those produced experimentally by feeding diets low in manganese to pregnant cows. (*Courtesy of Dr. I. A. Dyer, Washington State University.*)

ing an abnormality are uniformly expressed over a range of environments which includes the "normal" range, we see only the genetic effects and think of it as genetic.

Genetic Abnormalities Affected by Environment. The expression of several abnormalities in laboratory animals and plants varies within the range of normal environments. The bar-eye condition in the fruit fly *Drosophila melanogaster* is one of the best known of these. The normal compound eye of this insect has many subunits or facets—usually 800 or more. In bar-eye individuals the number is greatly reduced, but the reduction is much greater at high rearing temperatures than at low. Expression of abnormalities of this type are said to exhibit "genotype-environmental interactions." In swine, scrotal hernia has a heritable basis, but its incidence is also influenced by a maternal effect.[1] The nature of the effect is not known, but it must have differential effects on different genotypes.

Several defects in farm animals are conditioned partially by hereditary variations of a quantitative nature in resistance and partially by environmental factors. The best known of these is cancer eye of cattle.[2] It

[1] Magee, W. T., *Jour. Anim. Sci.*, 10:516–522, 1951.
[2] Vogt, D. W., and Anderson, David E., *Jour. Hered.*, 55:133–135, 1962.

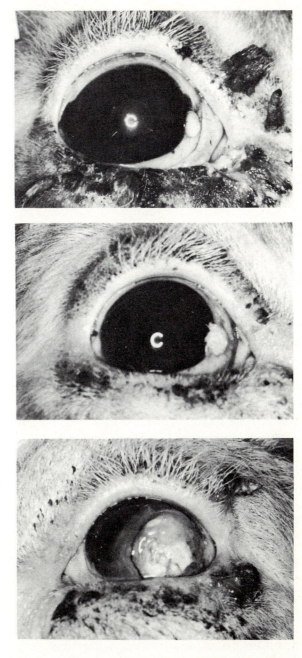

Fig. 5-5. Three stages in the development of "cancer eye" in cattle. *Above,* a plaque. *Middle,* a slightly more advanced but still benign stage or papilloma. *Below,* a carcinoma. (*Courtesy of Dr. David E. Anderson, University of Texas Medical School, Houston.*)

occurs more frequently in Herefords than in other breeds. Although it occurs in most locations, its frequency within this breed is greatest in geographic locations with high-average annual hours of sunshine. Latitude and altitude are also related to incidence, probably as a result of differences in the ultraviolet component of sunlight. It is usually an affliction of older cattle. Its incidence increases and it occurs at younger ages in cattle maintained on high levels of nutrition. Hereditary variation affects age of cancer development as well as occurrence or nonoccurrence. Hereford cattle with pigmented eyelids and corneoscleral areas are less susceptible than white-eyed types.[1] Preference in the United States has generally tended to be for white-eyed types. This is a probable case of selection for a "fancy point" having harmful effects.

Vaginal and uterine prolapse in cattle appears to be of a somewhat similar nature in that its occurrence is influenced by both heredity and environmental factors.[2]

Selection against defects of these types is difficult because heritability is quantitative in nature and because they are usually expressed at rather advanced ages. A sire may have left many offspring before it is discovered that his daughters have an unusually high susceptibility to a condition such as cancer eye or uterine prolapse. Selection for pigmented eyes should be an effective indirect method of selecting against cancer-eye susceptibility, but unfortunately, pigment in the corneoscleral areas does not develop fully until the age of five years or more. Thus, selection for it is not fully effective at young ages.

Losses from Genetic Abnormalities. No accurate data are available on the total losses from hereditary abnormalities. Usually, any one abnormality, although it may be serious in a given herd, will not be a serious industry-wide problem. Total losses from all such conditions may be more serious than generally realized. In most farm animals, prenatal death losses range from 20 to possibly over 50 per cent of all fertilized eggs. Causes are not well understood, but increases with inbreeding suggest that many have a genetic basis.

Recessive Genetic Defects with Some Expression in Heterozygote. Most lethals and abnormalities are recessive in inheritance, whether due to one or several genes. The death or culling of affected individuals usually keeps the frequency of the gene or genes at low levels and in equilibrium with mutation rate as discussed in the chapter on Population Genetics. A few cases are known in which recessiveness is not complete, and the heterozygotes or "carriers" have characteristics which

[1] Anderson, David E., *Ann. N.Y. Acad. Sci.*, **100**:436–446, 1963.
[2] Woodward, R. R., and Quesonberry, J. R., *Jour. Anim. Sci.*, **15**:119–124, 1956.

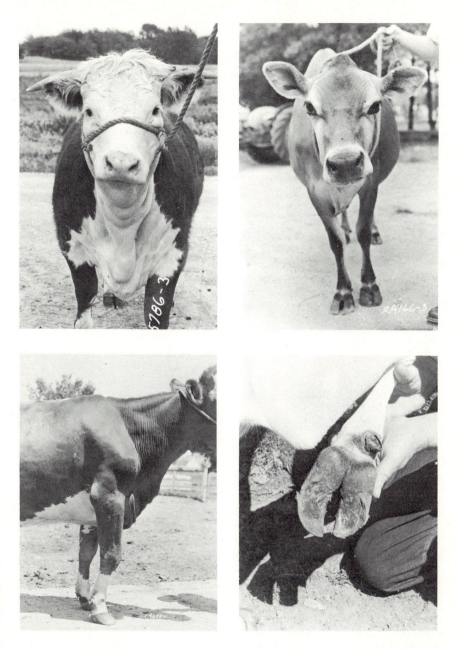

Fig. 5-6. *Top,* Hereford and Jersey females with twisted nose, or "wry face." *Below left,* a Guernsey with winged shoulders. *Below right,* a Jersey with a missing dew claw. These defects may or may not have hereditary bases, but none are simple one-gene pair conditions. (*Courtesy of Dr. K. A. Huston, Kansas State University.*)

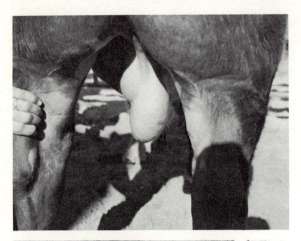

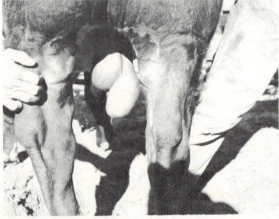

Fig. 5-7. Two yearling Hereford bulls showing incomplete descent and imperfect development of left testicles. This defect appears to have a heritable basis since it has occurred in only a single inbred line and not in other Herefords maintained in the same environment. It is not, however, simply inherited and it also varies in expression from cases like those shown to cryptorchidism. In the latter condition, the testicle is retained in the body cavity. (*Courtesy of Dr. R. S. Temple, Animal Husbandry Research Division, U.S. Department of Agriculture.*)

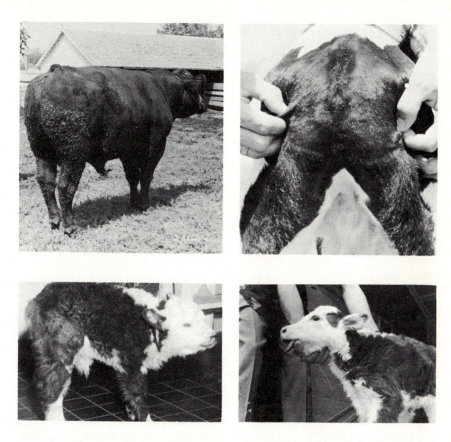

Fig. 5-8. Top left, a tailless Angus bull. *Top right,* a Holstein calf born tailless and with no anal opening (atresia ani). *Below left,* a Hereford calf born tailless, eyeless (microopthalmia), and with multiple genital abnormalities. *Below right,* a calf with a short lower jaw (brachygnathia inferior). The latter condition has a hereditary basis in at least some breeds. The others are of unknown etiology. These and other photos of abnormalities are printed to illustrate the wide variety of things which can go wrong in developmental processes. (*Courtesy of Dr. K. A. Huston, Kansas State University.*)

make them more desirable[1] to breeders than homozygous normals. The classic example of this type of thing is the Dexter breed of cattle. Cattle of this type are always heterozygous for a semidominant gene which when homozygous produces a lethal achondroplasia (bulldog calves). Dexters themselves show the effects of this gene by shortness of leg. When intermated, Dexters produce about ¼ long-legged individuals known as Kerrys, ½ short-legged Dexters, and ¼ bulldog calves.

[1] We here use the term *desirable* to mean things favored by breeders. These may or may not be economically desirable things.

Apparently the hereditary situation is as follows:

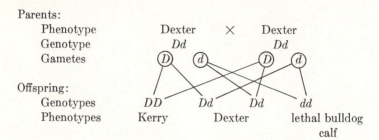

Parents:
Phenotype   Dexter   ×   Dexter
Genotype   *Dd*   *Dd*
Gametes   Ⓓ ⓓ   Ⓓ ⓓ

Offspring:
Genotypes   *DD*   *Dd*   *Dd*   *dd*
Phenotypes   Kerry   Dexter   lethal bulldog
                                                                  calf

Preference of British breeders for the short-legged Dexters has resulted in the development of a breed carrying a lethal gene at a frequency of 0.5. Since the heterozygotes are easily identified, it would be an easy matter for breeders to cull them if they so desired. The Dexter type can be propagated without the production of lethals by avoiding the intermating of Dexters. Kerry × Dexter matings give 50 per cent of the latter type.

A more insidious situation occurs in Swedish dairy cattle. A type of infertility characterized by gonadal hypoplasia was found to be highly hereditary, although not a clear-cut case of a single pair of genes. Although it was not demonstrated beyond all reasonable doubt, evidence indicated that unilaterally affected cows on the average produced milk with a higher-than-average fat test and were favored in selection. The gene or genes responsible for the condition attained such frequency in the breed that bilateral gonadal hypoplasia (with resultant sterility) reached a level constituting a serious problem to the breed. Intentional selection against the condition has since reduced the frequency greatly.[1]

In the midyears of the twentieth century, a type of dwarfism characterized by small size, high mortality, bulging foreheads, undershot jaws, difficult breathing, a tendency to bloat, and poor coordination (Fig. 5-9) reached a frequency in at least two breeds of beef cattle in the United States high enough to constitute an economic problem. Initially it appeared to be inherited as a simple recessive. Although definite proof is lacking, the apparent increase in frequency of this defect strongly suggests that the gene is not completely recessive but has some effect in the heterozygous condition which led breeders in many cases to select heterozygous (or "carrier") animals for breeding purposes in more than a random proportion of cases.[2]

[1] Ericksson, K., "Hereditary Forms of Sterility in Cattle," Hakan Ohlossons Boktyckeri, Lund, Sweden, 1943.
[2] See Marlowe, T. J., *Jour. Anim. Sci.*, **23**:454–460, 1964; and Bovard, K. P., *Anim. Breed. Abstracts*, **28**:223–237, 1960.

Fig. 5-9. *Above*, a "snorter" or brachycephalic dwarf Hereford calf nine months of age with a normal Jersey heifer of the same age. *Below*, a crossbred Hereford × Angus dwarf of the same type with her Angus mother. Since this type of dwarfism is inherited as a one-gene recessive, the occurrence of a crossbred dwarf indicates both of the parental breeds include animals carrying the same recessive gene. (*Courtesy of Dr. K. A. Huston, Kansas State University, and Dr. K. P. Bovard, Virginia Agricultural Experiment Station, Front Royal.*)

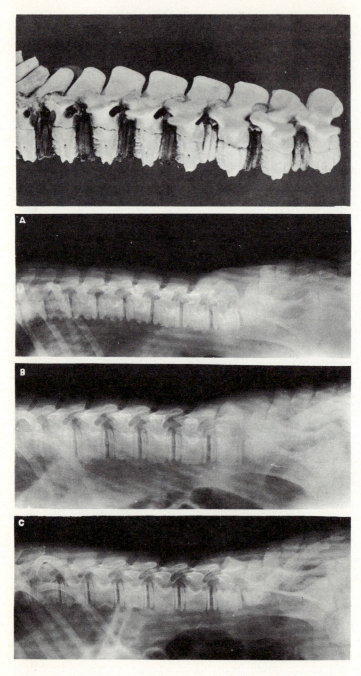

Fig. 5-10. One of the characteristics of young snorter-type dwarfs is a vertebral malformation in the lumbar region resulting in jagged

Situations such as the foregoing are difficult to cope with. The heterozygous animals apparently have an *average* superiority, but they overlap so much with homozygous normals that they cannot be accurately identified phenotypically. Research may develop methods for identifying them. Until that time breeders will have to use pedigree information and progeny-test methods for reducing the frequency of such conditions.

Eliminating Genetic Defects. Action to be taken if a lethal or genetic abnormality is discovered in a herd will depend upon the type of herd and the seriousness of the abnormality.

In a commercial herd, the only action usually necessary is to cull the sire or sires which sired the defective offspring and replace them with unrelated males. For most traits, frequency is so low that probability of obtaining new sires which also carry the undesired gene or genes is very low. Only normal offspring will likely be produced from new sires even though some females in the herd still carry the defective germ plasm. Probability of acquiring replacement sires which do not carry the deleterious gene or genes can be increased by a knowledge of the bloodlines to avoid in a breed. A few breed associations are now making information of this kind available on a restricted basis for a few defects.

Corrective measures should be more drastic in seedstock herds since the owner has an obligation to produce the best possible stocks for his future customers. For these herds the following measures will be helpful in elimination of the defect or in reducing its frequency.

1. Cull all sires which have produced defective offspring.
2. Remove all females which have produced defective offspring from the seedstock herd itself. They may either be culled or placed in an auxiliary herd and used to progeny test future herd sires to determine whether or not they are heterozygous for the gene responsible for the defect.
3. Cull other close relatives of affected individuals including normal offspring of sires and dams which have produced defective individuals.
4. If the affected individuals are viable and fertile, retain them for progeny testing prospective breeding animals.

ventral surfaces. This characteristic has some expression in the heterozygote. Vertebral type can be determined by X rays of young calves. At top is a photograph of a two-day-old dwarf's vertebral column. Immediately below are three X rays showing: A—a dwarf; B—a normal calf with slight abnormalities of the ventral surfaces of the vertebrae; and C—a normal calf with normal vertebrae. Although the slight abnormalities such as shown in B are indicators of heterozygosity and currently the best physical indicator available, there are many errors in estimates of genotype from X rays. (*Courtesy of Dr. K. A. Huston, Kansas State University, and Dr. L. N. Hazel, Iowa State University.*)

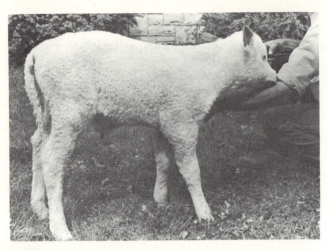

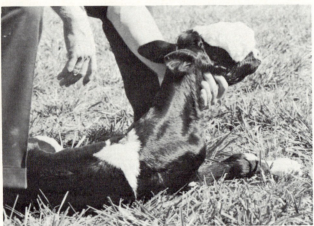

Fig. 5-11. Above, an Ayrshire calf homozygous for a recessive gene resulting in an extremely curly hair coat known as "Karakul Curl." *Below,* a Holstein calf with brain hernia—thought to be caused by a recessive lethal. (*Courtesy of Dr. K. A. Huston, Kansas State University.*)

How far to go on the foregoing program will depend on the economic importance of the defect and the frequency in the breed. If the defect is lethal or seriously reduces production, and if it is known to have a wide distribution in a breed, all steps should be taken if at all possible. If the defect is more esthetic than economic (an example might be hair swirls in swine), culling of sires and dams of affected individuals is probably the only step which could be justified. Even this could not be justified if the proved carriers were especially outstanding in other characters.

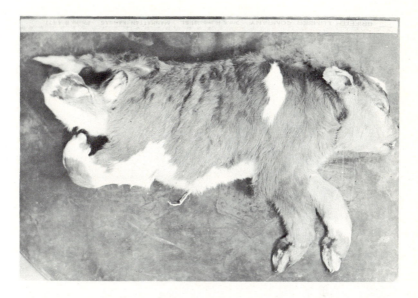

Fig. 5-12. An achondroplastic or "bulldog" calf. Several hereditary forms of this general type of abnormality are known. They vary in severity when homozygous. They occur in several breeds of cattle. Some types are recessive while others have the effect of shortening the long bones of normal but heterozygous animals. (*Courtesy of Dr. K. A. Huston, Kansas State University.*)

Progeny Testing for Recessive Genetic Defects. The principles of progeny testing for deleterious recessives should be clearly understood. If a condition is known to depend upon a single recessive gene, several possible progeny-testing procedures are available for estimating genotypes of potential herd sires: (assume that r = a recessive deleterious gene and R its dominant normal allele in all cases)

1. Mating a prospective phenotypically normal but genotypically unknown sire to known female carriers, matings will be:

Normal male   Normal, carrier female

$R?$ $\times$ Rr

If the genotype of the male is homozygous normal (RR), all his germ cells will carry R and all offspring will be normal, i.e., either RR or Rr.

If, however, the male is a carrier, half his germ cells will contain r and on the average ¼ of the offspring from the test matings will be homozygous rr and will exhibit the abnormality. Even though ¼ of the offspring are *on the average* expected to be affected, the occurrence will be on a random basis and the probability of any given

Fig. 5-13. A recessive form of achondroplasia. This calf lived for three days. The white ball of hair covers a cartilaginous type of growth from the calf's forehead.

individual being abnormal is only 1 in 4, or 0.25. Conversely, the probability of any given individual being normal is 3 in 4, or 0.75. The probability of all of n offspring being normal is $(0.75)^n$. For any given number of offspring, the foregoing expression represents the probability of failing to detect the sire as a carrier, since all the test offspring are normal even though the sire is really a carrier.

For some given numbers of offspring, the probabilities are as follows:

Number of offspring	Probability of failure to detect a carrier sire
1	0.75
2	0.56
3	0.42
4	0.32
5	0.24
8	0.10
10	0.05
12	0.03
15	0.01
20	0.003

From the foregoing, it is clear that a sire can never be proven to be a noncarrier. We can only say that the odds of his being a carrier are only 1 in 5, 1 in 10, 1 in 100, etc.

2. Mating a sire to his own daughters:

If a sire is a carrier but has been mated only to noncarrier females, half his daughters on the average will be carriers. Since there is no way of knowing which daughters are carriers (assuming the sire is), it is necessary to mate all or a random sample to the sire for a progeny test. Thus, approximately twice the number of sire-daughter matings is necessary to reach the same probability for a particular deleterious gene as when known carrier females are used. The principal disadvantage of this test is that the sire will necessarily be well advanced in age before offspring of his daughters can be observed. In a few cases (the mule-foot condition in cattle is an example) defects are expressed early in embryology, and it has been possible to speed up progeny tests by removing embryos surgically and classifying them.

The progeny test of a sire by mating to his own daughters has the advantage, however, of testing simultaneously for all recessives and is an especially valuable tool for use with sires being considered for wide use in artificial insemination.

3. Mating a sire to affected females, matings will be:

$$\begin{array}{ccc} \text{Normal male} & & \text{Affected females} \\ R? & \times & rr \end{array}$$

If the male is RR, all offspring will be Rr and phenotypically normal. If he is Rr, half the offspring *on the average* will be rr. Again this will be on a random basis, but in only $(0.5)^n$ of the cases will all offspring be normal even if the sire is Rr.

For some given numbers of offspring, the probabilities are:

Number of offspring	Probability of failure to detect a carrier sire
1	0.50
2	0.25
3	0.12
4	0.06
5	0.03
8	0.004
10	0.001

It can readily be seen that this is a much more powerful test than mating to either known carriers or daughters. It is to be recommended if affected females are viable and fertile.

An adaptation of this test has sometimes been used with color in cattle. In some breeds of cattle (the Angus and Holstein in the United

Fig. 5-14.   Upper left, a yearling Angus bull affected with a muscular hypertrophy condition often referred to as "double-muscled," or "doppelender." *Upper right*, the carcass of an affected new-born calf showing extreme muscle development in the rear quarters. *Lower left*, an affected and a normal calf each six weeks of age. *Lower right*, the carcasses of the calves shown at left. This condition is found in several

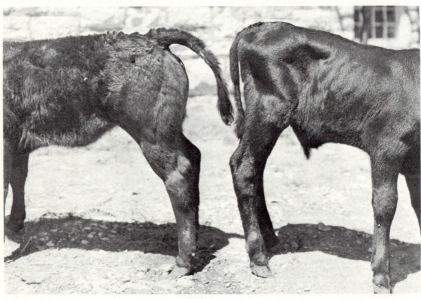

cattle breeds. In most cases it behaves as a recessive with some expression often apparent in the heterozygote. Other cases, possibly involving modifying genes, do not satisfactorily fit a single-gene hypothesis. (*Courtesy of Prof. Walter H. Smith, Kansas State University.*)

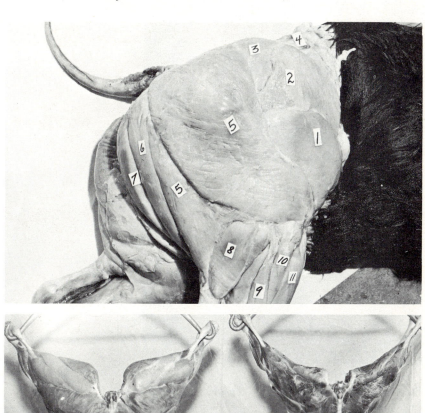

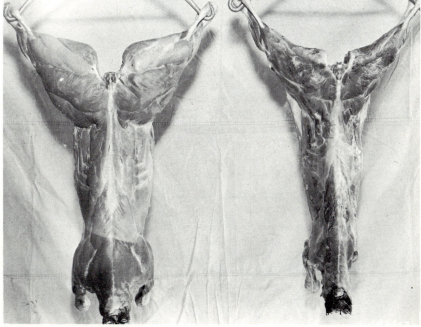

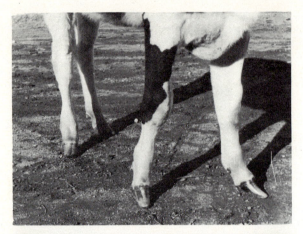

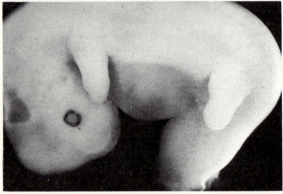

Fig. 5-15. Above, a Holstein with the single-toe condition, syndactylism, often popularly called "mule foot." It is inherited as a simple recessive. *Below,* a 37-day embryo removed by Caesarean section which already shows the syndactylous condition. This technique permits a rapid progeny test. (*Courtesy of Dr. K. A. Huston, Kansas State University.*)

States) pigmented skin is black. Black color is dominant and is a breed standard. Red is recessive and is considered undesirable even though of no known economic significance. Bulls can be tested for presence or absence of the recessive gene for red by breeding to a number of females of a breed in which pigmented skin areas are red. To shorten the time required, feed-lot heifers of a breed in which pigmented areas are red are sometimes used for the test and slaughtered at about five months of pregnancy. Color can be determined from the fetus at this stage of gestation.

Fig. 5-16. Pigs homozygous for a recessive gene causing "brain hernia." This gene is usually lethal when homozygous, but by special care a few affected animals have been successfully reared and used in breeding tests. (*Courtesy of Dr. Clyde Stormont, University of California.*)

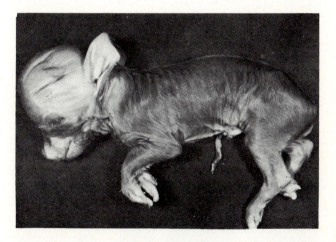

Fig. 5-17. A Duroc pig homozygous for a recessive lethal gene causing hydrocephalus and taillessness. Some affected pigs are born alive but die early in life.

Fig. 5-18. An Ayrshire calf homozygous for a recessive gene resulting in congenital dropsy. Most cases are lethal but a few survive. (*Courtesy of Dr. K. A. Huston, Kansas State University.*)

The above outline of progeny-testing procedures has emphasized the probability of *failing to detect* a heterozygous sire. Since the birth of a single defective offspring identifies a sire as a carrier, the probability of *detection* of a carrier equals one minus the probabilities given. Thus, mating of a sire to even a few known carrier females or a few of his daughters is of considerable help in finding most carriers. Culling them will be a real aid in reducing the frequency of the undesired gene.

Fig. 5-19. A bovine monster. Abnormalities of this type occur rather rarely and are not known to have hereditary bases.

The same principles outlined for males will apply to progeny testing females. Cows and ewes produce so few offspring that they cannot be adequately progeny tested in a normal lifetime. The practice may sometimes be useful with litter-bearing animals.

GENETIC DIFFERENCES IN DISEASE AND PARASITE RESISTANCE

The success of plant breeders in developing crops which are highly resistant to specific diseases has long been an important factor in maintaining yields at high levels.

Genetic differences in resistance to specific diseases in laboratory animals and in poultry have been repeatedly demonstrated. Further, some of the apparent physiological characters associated with disease resistance have been identified. Examples are the relation of number of leucocytes to resistance of mice to *Salmonella typhimurium* and the genetic control of the thermoregulatory mechanism to resistance of chickens to S. *pullorum*.

In chickens, genetic differences in resistance to lymphomatosis have been exploited in commercial stocks. However, as yet this approach has not provided a complete answer to the problem.

In larger farm animals genetic differences in resistance to specific diseases undoubtedly exist. They have been demonstrated conclusively for mastitis[1] and much suggestive evidence is available on other diseases. Genetic differences have been demonstrated in resistance of sheep to the common stomach worm *Haemonchus contortus*.[2]

In plants, resistance to disease has often been found to depend upon a single gene. Introduction of these single desired genes into commercial varieties can be accomplished rather rapidly when needed. To date, genetic differences found in resistance to disease or parasites in animals have not been of this type. They are apparently due to many gene pairs. Results give the impression of additive-type genetic variation with incomplete heritability. In other words, effective use of genetically controlled disease resistance appears to present many of the same problems as do breeding for improvement of quantitative traits such as milk production or growth rates. This situation, together with the fact that low reproductive rates (as compared to plants) would make it difficult to incorporate genes for disease resistance into entire populations quickly, has not encouraged research on the problem.

[1] Legates, J. E., and Grinnels, C. D., *Jour. Dairy Sci.*, **35**:829–833, 1952.
[2] Warwick, Bruce L., et al., *Jour. Anim. Sci.*, 8:609–610, 1949 (abstract).

Fig. 5-20. A screw-tail condition in a Holstein cow. None of her offspring exhibited the condition. It is not known whether it is hereditary, environmental, or an "accident of development."

Philosophy to date has generally been that diseases in large farm animals can be controlled more effectively and economically by vaccines, sanitary procedures, test and slaughter, etc., than by genetic approaches. Thus, relatively little study has been put into potential effectiveness of genetic procedures.

Breeding for disease resistance would likely be very expensive. Unless physiological indicators of resistance could be used in selection, a system of challenging animals of every generation through use of a live, virulent organism would be necessary. High death or debility rates would be almost certain until resistant animals were located. An indirect expense would be the fact that selection for disease resistance would reduce or eliminate possibilities of selecting for other desirable traits. This would almost certainly be true if concurrent selection was made for resistance to two or more diseases. It is thus possible that a highly disease-resistant, but otherwise useless, stock might be developed.

In spite of the apparent difficulties, the field of genetic resistance to disease and parasites is one which would probably justify more research with farm animals than has been put on it to date. It might be especially fruitful for diseases and parasites for which other control methods are unsatisfactory. However, until proved methods of selecting for disease resistance have been developed, breeders should depend upon other methods of control and concentrate on selection for traits on which progress is probable.

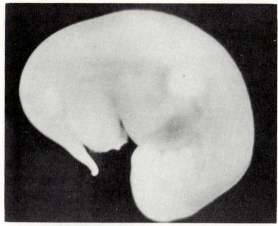

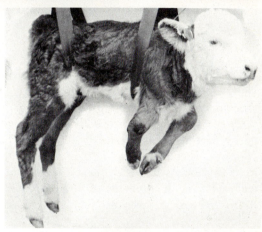

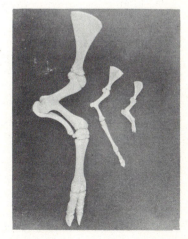

Fig. 5-21. Under experimental conditions irradiation at early embryonic stages can result in anatomical abnormalities. *Above,* a normal 32-day calf embryo—note the developing front limb bud; *Center,* a calf with malformed, shortened forelegs born to a cow given 400 r air-dose of gamma radiation from Co⁶⁰ on the 32nd day of gestation; *Below,* front leg bone (left) of the calf shown in center photo and leg bones of a lamb and a pig whose dams were irradiated on the 23rd and 21st days of gestation, respectively. Note ankylosis of radial-humeral joint in all three sets of leg bones. (*Courtesy of Dr. R. L. Murphree University of Tennessee.*)

COLOR AND HORNS

Most livestock breeds have standard colors or color patterns which serve as breed trademarks. Most of these are unrelated to productivity, but a few are of economic worth. As an example, white udders in beef cows may lead to "snow burn" if cows calve in the spring before snow is gone. As discussed earlier, pigment in and around the eyes of white-faced cattle reduces the incidence of eye and lid cancers. In hot areas with intense sunlight, light coat colors in cattle reflect more heat and thus are an aid in maintaining normal body temperatures. Black wool will not take dyes of other colors and is of very low value.

Most basic color variations are inherited in a fairly simple fashion, and in many cases maintenance of the basic breed trademark constitutes no special problem. However, modifications of basic patterns sometimes depend upon additive-type gene action with considerable nonhereditary variation also sometimes occurring. Examples of this type of thing are the amount and distribution (i.e., color pattern) in breeds such as Hereford cattle and Hampshire and Poland China swine.

The Hereford color pattern is distinctive—red body with white face, white underline, and other white markings. The white face is dominant and crosses of the Hereford with other breeds exhibit it to varying degrees—the degree apparently depending upon the kind of modifying genes present in the breed with which the cross is made. Within the Hereford breed, the exact color pattern desired has proved difficult to fix with animals in which less or more than the desired amount of white still occurs with appreciable frequency despite generations of selection. A careful study[1] indicates that the preferred amount of white "is the result of an intermediate genetic situation and that it is not likely that selection of breeding stock for this trait will fix the color pattern for this desired intermediate."

Shade of color (i.e., light or dark) has been a problem in several breeds of both cattle and swine—especially red breeds. In spite of the fact that there is no known relationship between shade of color and productivity, fads for certain shades have sometimes developed.

Selection for unfixable color patterns, specific shades of color, or other details of appearance reduces the intensity of selection for important traits and is to be avoided if at all possible.

In Shorthorn cattle, a single pair of alleles with one homozygote giving red and the other white explains most color inheritance. The heterozygote is intermediate, being a mixture of red and white known as *roan*. It is fortunate that the breed has never adopted roan as a standard

[1] Stanley, Marion E., et al., *Okla. Agr. Expt. Sta. Misc. Publ.* MP-51, 1958, pp. 50–54.

since it would obviously not be fixable. In some breeds of European sheep, a genetic factor which when homozygous gives black wool has been kept at high frequencies, because when heterozygous it results in a desired dark facial coloration.[1] Selection intensity for the colored faces is intense enough to result in maintaining the gene at a high enough frequency that a good many black lambs are born.

Some standard colors in breeds are dominant. The black color of the Angus is an example already mentioned. Complete elimination of the recessive gene for red in the breed is very difficult (actually almost impossible) without a general system of progeny testing. This would probably not be economically feasible. In spite of many years of eliminating reds as they occur, and probably some culling of bulls and cows producing reds, various estimates are that 1 in 200 to 1 in 400 births are of red calves.

In most breeds of sheep and cattle, presence or absence of horns depends upon fairly simple genetic patterns. Size and shape of horns are apparently modified by many pairs of genes each with minor effects.

Breed standards often specify horn size and shape as well as presence or absence. Horns are of economic importance in cattle, being a detriment under farm conditions because of the trouble of dehorning. They sometimes have advantages under range conditions.

In European breeds of cattle[2] presence or absence of horns usually (there are some unexplained exceptions) behaves as if under the control of a single pair of allelic factors with the dominant allele resulting in the absence of horns or *polled.* Some of the traditionally polled breeds such as the Angus are apparently pure, or nearly so, for the factor for polled and can be depended upon to produce 100 per cent polled calves in crosses with other European breeds. Polled strains of other breeds developed in the United States during this century from polled mutants have been forced to draw heavily on horned stocks for foundation material. They are still far from pure. Making them homozygous for polled will be a difficult task—having the same problems as eliminating genes for red from a black breed.

In most fine-wool sheep, presence or absence of horns depends upon a single pair of alleles with heterozygotes being horned in males. Females of all genotypes are polled or have a slight amount of horny growth known as knobs. The polled gene is related in some way in breeds of this type to cryptorchidism, a defect in which testicles are retained in the abdominal cavity rather than descending into the scrotum. Whether the cryptorchidism is due to a pleiotropic effect of

[1] Dry, F. W., *Jour. Genet.*, **45**:265–268, 1943.
[2] There are additional genetic factors controlling horn development in Zebu-type cattle.

the same gene or to a closely linked gene is not known with certainty. A few normal rams have been progeny tested at high levels of probability and are apparently homozygous polled. This would indicate either (1) close linkage which has been broken or (2) presence of modifying genes which prevented the expression of cryptorchidism even though this is a normal pleiotropic effect of the gene.

Interestingly enough, a gene for polled in some breeds of goats is associated with a form of intersexuality.

BLOOD ANTIGENS, TRANSFERRINS, AND OTHER CHARACTERS

Many genetically controlled substances have been studied in blood of farm animals. In cattle a large number of antigenic red blood cell substances under the control of genes at 11 loci have been discovered. Multiple allelic series are known to occur at many of the loci. It has been estimated that more than 156 billion possible blood types could occur in cattle.

The antigenic substances can be detected only by detailed laboratory procedures involving agglutination or lysis of cells of the animal under test when mixed with a series of specific antibodies developed in the blood of other cattle after transfusion of blood containing an antigen not present in the blood of the recipient.

Similar systems have been discovered in sheep, swine, and poultry.

These substances have been of great usefulness in cases of disputed or uncertain parentage. As is the case with human beings, these tests cannot prove that a certain animal *is* a parent, but merely that it *could be*. Other animals can be excluded as possible parents. This follows from the fact that an animal can transmit only genes for antigens or groups of antigens (phenogroups) it possesses. For instance, assume a calf whose sire is unknown has antigen A and its dam does not. This means that the sire had the antigen. Thus, the test excludes as possible sires all males not carrying this antigen.

Blood antigens have also been very useful as an aid in determining monozygosity of bovine twins. An outgrowth of blood-antigen studies with twins was the discovery[1] that interchange of blood-forming elements takes place in about 90 per cent of dizygotic twins early in embryonic development. The transferred elements persist, and as a result these twins upon gross examination have the same blood antigens. However, they have two types of red cells—one descending from embryonic tissue of the animal itself and the other from tissue of its twin. This

[1] Owen, R. D., *Science*, **102**:400–401, 1945.

admixture, when found by appropriate techniques, proves dizygosity. Twins with the same antigens and without admixture can be monozygotic but are not necessarily so since they could be dizygotic and merely happened to have gotten the same genes from their parents.

In bovine twins of unlike sex, it has long been observed that there is sterility in the female (associated with major abnormalities in development of the reproductive tract) in about 90 per cent of the cases. The sterile females are termed *freemartins*. Lillie hypothesized many years ago that the freemartin condition was due to a vascular interchange between twins and that hormones of the male exerted a deleterious influence on the early development of the female reproductive system. Owen's work proved that vascular interchanges occurred, and subsequent studies have shown that the freemartin condition occurs only when blood admixture has also occurred. Thus, the observations strengthen Lillie's theory but still do not provide absolute proof.

Blood antigens have been studied intensively for relationships with productive characters. Some real relationships apparently exist, but it is uncertain if any characters or combination of characters studied thus far are closely enough related to milk production, fat percentage, or growth to be effective aids in selection. Associations could be due to (1) pleiotropic effects of the antigen genes themselves or (2) linkage of antigen-determining genes with other genes influencing performance. If the latter occurred, the relationship would be transitory and would be retained only until linkage equilibrium was reached. In other words, in time, crossover and recombination would be expected to break up desirable combinations. In view of the behavior of most productive traits, we have good reason to believe they are influenced by many pairs of factors. It thus appears unlikely that any one of the few antigen genes would have major effects on productivity.

In chickens it has been found that heterozygosity at the *B locus* is associated with greater viability than is found in any homozygote of the rather extensive series of alleles at that locus. Thus, there appears to be an economic incentive to develop lines or strains homozygous for different alleles and cross them for commercial production. It is not known why the improved viability occurs. It may be due to action of the blood antigen genes per se. It is also possible that each gene is linked with a block of genes in which crossover is suppressed and that a heterotic response occurs as a result of many genes being heterozygous.

Many proteins in body fluids have been studied by means of electrophoresis, and many genetically controlled variations have been discovered. Two forms of hemoglobin, A and B, controlled by a single allelic pair of factors, have been identified in cattle. The genes do not exhibit

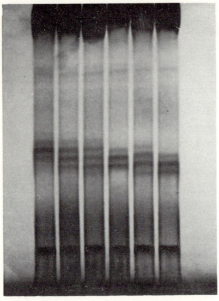

AA AD AE DD DE EE

Fig. 5-22. Center bands show separation of genetically different β globulins (or transferrins) of cattle blood serum by polyacrylamide gel electrophoresis. The β globulins migrate at different rates when an electrical potential is applied to a gel upon which a sample of serum has been placed. After a suitable time interval, staining is used to bring out the bands. The phenotypes are determined on the basis of number, positioning, and intensity of bands. Phenotypes shown from left to right correspond to genotypes Tf^{AA}, Tf^{AD}, Tf^{AE}, Tf^{DD}, Tf^{DE}, and Tf^{EE}. (*Courtesy of Dr. C. A. Kiddy and Mr. W. W. Thatcher, Animal Husbandry Research Division, U.S. Department of Agriculture.*)

dominance. Heterozygotes have both types and homozygotes only one. A third member of the allelic series has been tentatively identified.

Another type of blood substance investigated has to do with heritable variation in the serum β-globulin. These substances are often termed *transferrins* because the function of the β-globulin is to transport iron in the body. Several variants controlled by single gene differences have been found in cattle, swine, and sheep. In cattle, five alleles at a single locus are known and 13 of the theoretically possible 15 phenotypes have been identified.

There is some evidence, although incomplete and somewhat contradictory, of relationships between fertility and transferrin type of animals mated, and between transferrin type and milk production in dairy cattle.

Variation in type of milk protein is known to occur, with at least four loci having been identified.

Blood antigens and other heritable variation in components of body fluids pose many interesting questions which are as yet unanswered. Are they related to physiological function in ways yet unknown? Why have the genetic variants, or polymorphisms as they are often called, persisted? If they had no function or if there were no selective advantage of one over another, it seems probable that genetic drift would long ago have resulted in populations becoming homozygous for one type. These and related questions pose intriguing problems for further study.

SEX RATIOS AND POSSIBILITY
OF SEX CONTROL

The development of functionally normal males and females in farm animals is a complicated process involving the coordinated interaction of genetic, hormonal, and environmental factors. Mechanisms of sex determination are not completely understood. It is known that they vary widely between species in both the plant and animal kingdoms.

In farm animals, sexuality is associated with chromosomal differences. In mammals, there is one pair of chromosomes called the *sex chromosomes*. For this pair, the female has two like chromosomes (usually called X chromosomes). The male has one X chromosome and one different in size and shape, called the Y chromosome. The female produces only one type of gamete or germ cell and it contains one X chromosome plus one member of each pair of autosomes. The male produces two types of germ cells, half containing one X and the other half one Y chromosome in addition to one member of each pair of autosomes in each case. These unite at random with female germ cells at fertilization to produce approximately half females (XX) and half males (XY) in the next generation. In poultry, sex is also related to chromosome complement, but females have the unlike and males the like pair.

In all species of farm animals, the sex ratio at birth is on the average very close to 1:1. Further, in spite of apparent deviations from this ratio in the progeny of some sires and dams and even in entire herds over a period of time, it appears that deviations can be accounted for by the laws of probability as discussed in the chapter on Variation. In practice, it sometimes seems there are more deviations than could be accounted for by chance deviations, but it should be remembered that we usually talk about only the exceptional cases noted out of the millions of herds in existence. Thus, the cases we hear about represent very incomplete samples of the total population.

In spite of the fact that breeders often feel some sires tend to beget an undue proportion of one sex or the other, there is no evidence to substantiate this. Thus, breeders are to be discouraged from eliminating otherwise highly desirable sires because they have in the past produced a preponderance of one sex or from making other changes in breeding programs based on past sex ratios.

The possibility of controlling sex has intrigued man from antiquity, and literally hundreds of methods have been proposed to accomplish it. Since the two types of sperm produced by mammals contain different members of the sex chromosome pair, there is a possibility of discovering a physical or chemical difference between them which would permit their separation or the inactivation of one type. If this could be accomplished, only X- or only Y-bearing sperm cells could be used for insemination and only the desired sex produced. This could be a marked advantage under many conditions.

To date, in addition to the many proposals for sex control based on folklore or superstition, studies aimed at separating sperm types by centrifugation, electrolysis, and many chemical approaches have all been unsuccessful or at least not proven to the satisfaction of qualified scientists.

Thus, no method of sex control is now available to breeders. The possibility remains an intriguing one, however, and it is hoped that future studies may be successful in developing reliable methods.

In concluding this brief discussion of sex ratios and possibilities of sex control, it should be emphasized that much is still unknown about sex determination. At the earliest stages of embryology at which sex of embryos can be determined, there are more males than females. During fetal life, more males than females die, so that at birth in most species the ratio approaches 1:1.

If, as an overwhelming amount of evidence indicates, the chromosome theory of sex determination is basically correct, then either (1) Y-bearing sperm must fertilize more than half the ova, so that the primary sex ratio is well above 50 per cent or (2) there is a differential death rate of female embryos during the period from fertilization until the sex of embryos can be determined. Experimental evidence is not available to indicate which of these two factors may be operative. There is a heavy embryonic mortality during the period immediately following fertilization, but we have no way of knowing whether or not one sex is dying in greater numbers than the other.

The chromosome theory of sex determination in mammals is strengthened by studies in humans and mice in which a variety of aberrations of the sex chromosomes have been found to be related to abnormal development of sexual characteristics. For example, human individuals with Turner's syndrome have a single X chromosome but lack a Y. They are female but are unusually short, have tiny ovaries, and lack many sexual characters that normally develop at puberty. Individuals with two X's and a Y are male but are characterized by underdevelopment of the male sex glands, enlargement of the breasts, and often mental retardation—a condition known as Klinefelter's syndrome. Apparently the Y chromosome is a strong factor in developing maleness. This is in contrast to *Drosophila*, the organism on which much of the pioneer work on sex determination was done. In this insect, the X chromosomes exert a strong influence toward femaleness, but the Y seems not to have an influence toward maleness—this influence coming from the autosomes.

Sexuality in farm animals in relation to sex chromosomes has been studied very little. On the assumption that mechanisms are similar to those in humans, it is probable that sex is determined at conception. Early in embryology the sex glands have the potentiality of both sexes. At some critical stage the presence of a Y chromosome is all-important. If present, the sex glands develop into testes. If there is no Y chromosome, ovaries develop. Once testes are present, they produce hormones which result in male development. In the absence of either testes or ovaries, development tends to be in the female direction.

Thus, it would seem that the function of the sex chromosomes is to switch the embryo into one or the other channel of sex development. From this stage the hormones take over and are responsible for further differentiation.

Many types of faulty sex development are known in farm animals. Intensive studies of these in relation to sex-chromosome distribution would appear to be a promising field both for extending knowledge of basic mechanisms of sex determination and for developing methods for reducing their occurrence.

MISCELLANEOUS TERMS AND NOTIONS ABOUT BREEDING

For the most part, the terms, expressions, and beliefs which follow are outmoded or not believed to be based on fact. However, many will be found in some popular writing and older literature. We believe a student should be aware of them.

Acquired Characters. This term refers to the possibility of a character induced in the body (soma) of an individual by environmental forces becoming part of its germ plasm and being transmitted to future generations. This topic has been one of the historic battlegrounds of biology. Speaking generally, things which happen to the soma, either favorable or unfavorable, do not in any way affect the genes, in which reside the potentialities for future generations. Environment does permit (or prohibit) the development of existing potentialities but in general does not alter the genes.

Atavism or Reversion. The terms *atavism* and *reversion*, meaning the reappearance of some ancestral trait or character after a skip of one or several generations, are often encountered in the older literature on animal breeding. Such reappearances were more or less mysterious before the physical basis of heredity was understood. The birth of a red Angus when the past few generations have been black is known to be because of each parent's supplying the gene for red. Recessives may be carried along, hidden by dominants, for any number of generations. Whenever two recessives come together, or, in other words, whenever the dominant gene is lacking, the "atavistic" character will be evident. Atavism or reversion is the sudden reappearance of some ancestral trait, but there is nothing mysterious about it, for it is one of the normal manifestations of the hereditary mechanism.

Nicking. The terms *nick* or *nicking* are sometimes used to describe an individual mating or the mating of an individual or members of a family with members of another family, in which the progeny turn out to be either better or worse than would have been expected from the merit of the parents. Usually they are used to describe better-than-expected results.

Genetically, a favorable nick *could* occur in an individual mating purely as a result of chance. The offspring could get a greater-than-average number of favorable genes from each parent as the result of chance at segregation and fertilization. If more productive offspring than expected were produced from matings between different families, chance could still be a factor if only small numbers were raised. However, if larger numbers were produced, the probability of this would be reduced. Likely explanations with large numbers are (1) the two families carried genes of different allelic pairs which complemented each other and produced a favorable epistatic combination in most offspring or (2) if overdominance is important, favorable results could occur if the families were different in the alleles carried for many gene pairs so that the offspring were above average in heterozygosity. One or the other or both of these explanations could also account for favorable results of a single mating.

Nicking has not been demonstrated to be very important in several studies on outbred animals within pure breeds. It is almost certain that many breeders put more emphasis than can be justified in looking for favorable nicks. Families in such populations are usually only as much related to each other as half sibs, or perhaps a little more. Therefore, the proportion of their genes which are similar because of relationship could not be great enough to establish a very predictable breeding behavior when mated to animals of another such family. Sometimes ideas gain general acceptance among breeders that such and such a family "always" or "never" crosses well with another. Usually, there is little or no basis for such ideas.

With highly inbred lines, possibilities of nicking being important are greatly increased, but it is then usually referred to as *specific combining ability.*

Prepotency. The term *prepotency* means the ability of an animal, either male or female, to stamp a given set of characteristics on its offspring to the exclusion of effects of genes from the other parent. The term is usually applied to males and is usually thought of in relation to a sire stamping his own characteristics on the offspring. It is also applied to cases in which there is a greater than expected resemblance or uniformity among the offspring even if the resemblance is not to the sire himself. This is the only way prepotency can be estimated for a trait not expressed in the male, such as milk production.

Genetically, prepotency depends primarily upon an animal being homozygous for dominant genes. Epistasis can also be a factor if the animal carries genes which give uniform progeny when combined with genes of the particular kind of mates he receives.

The term prepotency is often misused and exaggerated ideas are often held about the prepotency of particular sires. Sometimes a rugged, masculine appearance is thought to be associated with prepotency. There is no evidence for this. Since prepotency depends mostly on dominance and homozygosity, it is unlikely that an animal would be prepotent for all characters. For example, in a Hereford × Angus cross, the Hereford would appear to be prepotent for white face while the Angus would appear to be prepotent for black body and polled since offspring will have these characters.

It is sometimes implied that prepotency is transmitted and that prepotent sires have prepotent sons. This may well be true, but the only way it can occur is for the prepotent sire to have been mated to females carrying the same dominant genes he carried so that his sons will also be homozygous for the genes. Obviously, offspring of a cross to a homozygous recessive could not be prepotent for the same character.

Maternal Impressions. Belief in maternal impressions assumes that what a pregnant mother sees, hears, or experiences may influence her offspring. In general, this old belief can be repudiated, because, as Marshall points out, if it were true, all calves born in the spring up north would tend to be white because the mothers have viewed a white landscape all winter, and, similarly, calves born in the fall would be green. There is no direct nerve connection between parent and offspring to afford means for transporting the effects of experiences. Animal experiments have all given negative results in this field. It is indeed a fortunate provision of nature that in the higher animals the embryo is so well protected from all external influences. If maternal impressions were actually registered on the offspring, all types and races would have long ago become a hideous conglomerate mess.

Telegony. Telegony is the more or less common belief that, after a female has borne young by a certain male, her subsequent offspring will show characteristics derived from the previous sire. The classic example is that of a mare which bore offspring by a quagga and later produced horse colts said to show some striping. Numerous attempts have been made to confirm this by crossing mares and zebras, but in all cases they have failed. The basic elements that determine the characters of any individual are the ovum and the spermatozoon that unite to produce it. Spermatozoa from one service could not possibly live through a gestation period in the female organs of any higher species to fertilize some future ovum, for a spermatozoon lives a very few days at the most. If this first offspring had any influence on other undeveloped ova in the ovaries, it would come under the heading of the inheritance of acquired characters, for which there is no conclusive proof. Mumford and Hutchinson made investigations in the mule-breeding district of Missouri, where mares often bear mule and later horse colts, but could find nothing to substantiate telegony. In the light of our present knowledge of inheritance, there is neither an experimental nor a theoretical basis for telegony.

Blood. In older animal husbandry literature and among breeders the term *blood* is often used more or less interchangeably with the word *inheritance*. Saying that an animal has three-fourths of the blood of a noted sire, or in the case of animals of mixed breeds, saying that a given animal is a half-blood, means in each case that these fractions of their hereditary material came from the source indicated. The use of this term is based in antiquity when it was believed that actual blood passed from mother to offspring during fetal life. We now know that this is not the case—rather the embryo is nourished by nutrients from the mother's blood, but the only thing she transmits directly to it is a sample half of her hereditary material. Since the term is based on a misconception, its use should be avoided.

SUMMARY

The occurrence of lethals and other genetic abnormalities was reviewed and examples given of the kinds of morphological variations sometimes observed. Guidelines were given for distinguishing between genetic abnormalities and environmentally induced conditions of similar appearance. Uses of progeny testing and other procedures were discussed for reducing frequencies of deleterious genes in a population. Basic information on heritable antigens, hemoglobin types, and transferrins was discussed in relation to possible correlations with economically important characters. Sex ratios and mechanisms of sex determination were considered in relation to farm animals. A number of false, outmoded, or frequently misused animal-breeding ideas were discussed.

REFERENCES

Anderson, David E. 1963. Genetic Aspects of Cancer with Special Reference to Cancer of the Eye in the Bovine, *Ann. N.Y. Acad. Sci.*, **108**:948–962.

Bogart, Ralph. 1959. "Improvement of Livestock," chaps. 7, 8, and 10, The Macmillan Company, New York.

Bovard, K. P. 1960. Hereditary Dwarfism in Beef Cattle, *Anim. Breed. Abstracts*, **28**:223–237.

Chu, Ernest H. Y. 1963. Mammalian Chromosome Cytology, pp. 3–14 in Recent Advances in Cytogenetics and Developmental Genetics, *Amer. Zoologist*, **3**:3–100.

DeSilva, P. L. G. 1965. Heterozygosity at Red Cell Antigen Locus *L* and Fertility in Chickens, *Genetics*, **51**:41–48.

Fishbein, Morris, ed. 1964. "Congenital Malformations: Papers and Discussions Presented at the Second International Conference," The International Medical Congress, Ltd., New York.

Gilmore, L. O. 1949. The Inheritance of Functional Causes of Reproductive Inefficiency: A Review, *Jour. Dairy Sci.*, **32**:71–91.

————. 1950. Inherited Non-Lethal Anatomical Characters in Cattle: A Review, *Jour. Dairy Sci.*, **33**:147–165.

Hetzer, H. O. 1945. Inheritance of Coat Color in Swine. I. General Survey of Major Color Variations in Swine, *Jour. Hered.*, **36**:121–128.

Hutt, Frederick B. 1958. "Genetic Resistance to Disease in Domestic Animals," Comstock Publishing Associates, Ithaca, N.Y.

Johansson, I. 1965. Hereditary Defects in Farm Animals, *World Rev. Anim. Production*, **3**:19–30.

Kiddy, Charles A. 1964. Inherited Differences in Specific Blood and Milk Proteins in Cattle: A Review, *Jour. Dairy Sci.*, **47**:510–515.

Marlowe, Thomas J. 1964. Evidence of Selection for the Snorter Dwarf Gene in Cattle, *Jour. Anim. Sci.*, **23**:454–460.

Mittwoch, Ursula. 1963. Sex Differences in Cells, *Scientific American*, **209**:54–62.

Morton, J. R., Gilmour, D. G., McDermid, E. M., and Ogden, A. L. 1965. Association of Blood-Group and Protein Polymorphisms with Embryonic Mortality in the Chicken, *Genetics*, **51**:97–107.

Rendel, Jan. 1961. Recent Studies on Relationships between Blood Groups and Production Characters in Farm Animals, *Ztschr. für Tierzücht. u. Züchtungsbiol.*, **75**:97–109.

Soller, M., Laor, M., Barnea, R., Weiss, Y., and Ayalon, N. 1963. Polledness and Infertility in Male Saanen Goats, *Jour. Hered.*, **54**:237–240.

Stansfield, W. D., Bradford, G. E., Stormont, C., and Blackwell, R. L. 1964. Blood Groups and Their Associations with Production and Reproduction in Sheep, *Genetics*, **50**:1357–1367.

Stormont, Clyde. 1958. Genetics and Disease, pp. 137–162 in "Advances in Veterinary Science," vol. 4, Academic Press Inc. (This article contains an extensive listing of hereditary defects of farm animals.)

——. 1962. Current Status of Blood Groups in Cattle, in Blood Groups in Infrahuman Species, *Ann. N.Y. Acad. Sci.*, **97**:251–268.

—— and Suzuki, Yoshiko. 1964. Genetic Systems of Blood Groups in Horses, *Genetics*, **50**:915–929.

Welshons, W. J. 1963. Cytological Contributions to Mammalian Genetics, pp. 15–23 in Recent Advances in Cytogenetics and Developmental Genetics, *Amer. Zoologist*, **3**:3–100.

CHAPTER SIX

Variation

Henry van Dyke is reported to have said that there is one thing in which all men are exactly alike and that is that they are all different. This statement holds true also for farm animals.

When we consider that probably no two blades of grass, no two calves, and no two humans have ever been exactly alike (though identical twins are essentially so), we get a glimpse of the resourcefulness of nature. It would seem that new plans and specifications must sometimes be depleted, yet life flows along in its thousands of varieties and millions of individuals, each different from all the rest of its kind. It is fortunate that this is so, because without this perpetuation of variability, progress would be impossible. Variation is the raw material which the breeder has available for herd or flock improvement. "Variation is at once the hope and despair of the breeder"; the hope because through it offspring better than their parents may be produced; and despair because after animals have been improved greatly, they may, and often do, vary again toward mediocrity.

Variation among animals in size, rate of growth, efficiency of feed utilization, carcass characteristics, disease resistance, milk production, speed, stamina, wool quality, and color has been observed and recorded numerous times. Of two steers, one gains 2 lb per day during a 140-day feeding period, the other gains 2.75 lb during the same period; of two cows, one produces 8,000 lb of milk in a year, the other produces 12,000 lb of milk in a year; of two litters of two sows, one litter gains 100 lb from 500 lb of feed, the other litter requires only 360 lb of the same feed for 100 lb of gain; of two sheep, one shears a fleece of 6 lb, and the other shears a 10-lb fleece.

Both heredity and environment are important in producing differences

Fig. 6-1. Variation within the genus *Ovis*—sheep. *Top left,* Bighorn. *Bottom left,* Punjab. *Top right,* Mouflon. *Bottom right,* Hampshire. (*Courtesy of New York Zoological Society.*)

among individual animals. In some instances the specific hereditary and environmental influences may produce variability which is not directly attributable to either but is a result of their joint action or interaction. By studying variation alone we are not able to determine which portion of the variation is certainly due to environment and which to heredity. Nevertheless, as will be evident in the succeeding chapters, the relative importance of hereditary and environmental influence on the variation for individual traits can be determined.

Environmental differences include the nongenetic variation resulting from managemental, nutritional, and climatic influences. Some animals may have been born in large litters and others in small litters, some may have had better care than others, some may have been born when the temperature was extremely hot or others when it was extremely cold. Certain animals may be infected with parasites, whereas others are not.

NATURE OF VARIATION

Discontinuous and Continuous Variation. Traits generally are grouped into those which show qualitative differences and those which show quantitative differences. In the former, the variations fall into a few clearly defined classes. This is usually due to the fact that these traits are under the control of one or a few pairs of genes whose final expression is not greatly influenced by external environmental factors. The polled or horned condition in cattle is an example of this type of trait. In practically all instances, an animal clearly displays one or the other of these two characteristics. Quantitative traits, on the other hand, show all manner of slight gradations from small to large as are found in milk production, wool clip, and rate of gain in the feed lot. The two sorts of variation associated with these traits are described as *discontinuous* and *continuous*. Qualitative suggests discontinuous variation (in sharply defined classes), and quantitative suggests continuous variation (many small gradations shading or intergrading almost imperceptibly into each other). Mendel dealt with one trait in peas which was quantitative (tall vs. dwarf) but which behaved in a qualitative manner; i.e., the peas were either tall or dwarf.

Students and livestock breeders often develop a feeling of frustration over the seeming inability to find clear-cut evidences of the known simple principles of heredity in farm animals. The basic reason that heredity in livestock cannot be resolved into simple terms is, of course, the fact that most of the economically important traits of our higher animals are highly complex. Few of them are determined by the action of only one or two pairs of genes, and very few of the qualities are immune to any and all environmental influences. With many genes involved and with qualities so subject to environmental influences, we naturally would expect to observe a graded series rather than sharply defined or discrete classes in most traits.

The number of genes present in the hereditary complex of farm animals is not known, but conservatively the number probably runs at least into the thousands. A goodly number of genes in any species are present in homozygous form; hence, they do not contribute to the observed variation in the species. Most quantitative traits such as size, growth rate, milk production, egg production, and prolificacy depend for their expression upon the interaction of a large number of heritable factors and environmental influences.

We do not know how many genes influence body size, but we do know that every organism begins as a single-celled zygote, and each organism is destined by its heredity to have its cells keep on dividing until a certain size is attained, provided suitable environment is available. The pituitary, thyroid, and thymus glands, perhaps also the adrenals and gonads, through their individual secretions and probably through complex interactions between them, have an influence on both the rate of growth and its extent. And how many individual genes and gene interactions function in this complicated process? Even in a relatively simple organism like the fruit fly it is known that as many as 30 genes scattered through the 4 pairs of chromosomes have a bearing on the color of a fly's eyes.

Other factors besides size play major roles in the ultimate performance of an animal. They include efficiency of digestion, absorption, and elimination; rate of blood flow, or basal metabolism in general; number and functional activity of the secreting cells in the udder; and the animal's disposition; as well as proper levels of feeding, presence or absence of good management practices, and freedom from various diseases. Enough has been said to indicate that the over-all physiologic functioning of any animal is highly complex, probably involving the actions and interactions of hundreds or thousands of genes.

Variation from Recombination.   Recombination of genes takes place as the chromosomes from the uniting gametes come together. In an organism with two pairs of chromosomes, there are four possible ways to recombine the intact chromosomes. In each resulting germ cell, one member of each pair of chromosomes is found. The possible number of different gametes with various numbers of pairs of chromosomes (n), where only a single pair is considered to be different for homologous chromosomes, is $2n$ as shown in Table 6-1.

If an *Aa* male is mated to an *Aa* female, two different types of gametes (*A* or *a*) can be produced by each parent, and offspring can be of three different genotypes (*AA, Aa,* or *aa*).

If a male of the genetic constitution *AaBb* (with genes *A* and *B* in different chromosomes) is mated to a female of the same genetic

Table 6-1. *Number of Different Kinds of Germ Cells Possible from Various Species When Each Pair of Chromosomes Is Heterozygous at Only One Locus*

Species	Number of pairs of chromosomes	Number of different kinds of germ cells
Ascaris	1	2^1 2
A. megalocephala	2	2^2 4
Fruit fly (D. melanogaster)	4	2^4 16
Corn (Zea mays)	10	2^{10} 1,024
Swine	19	2^{19} 524,288
Human	23	2^{23} 8,388,608
Sheep	27	2^{27} 134,217,728
Cattle, goat	30	2^{30} 1,073,741,824

make-up, four different types of gametes can be produced by each parent, and the offspring can be of nine different genotypes.

If two animals were of the genetic constitution shown below (loci on different chromosomes),

Sire: *AabbCCDd*
Dam: *AaBbccDD*

then it would be possible to have 12 different genotypes among the offspring:

$$3 \times 2 \times 1 \times 2 = 12$$

Thus from mating the above parents we could have 12 different sorts of full brothers or sisters as to genotype. All of these individuals would have identical pedigrees, but their genotypes might all be different, ranging from *AABbCcDD* to *aabbccDd*.

If the capital letters represented desirable traits and the small letters undesirable ones, we can see that there would be a vast difference between these full sisters or brothers. Identity of pedigree, then, does not mean identity of inheritance, except when the parents are homozygous for each pair of genes. Such a degree of homozygosity is almost never realized practically, but it is approached in crosses among highly inbred lines.

Our example has dealt with four pairs of genes, or eight in all. We do not know how many genes are present in the chromosomes of our farm animals. There are 60 chromosomes in some of these animals, and, if there were an average of 100 genes in each chromosome, there would be a total of 6,000 genes. Each parent would then be transmitting not 4 genes, as in our simple example, but 3,000. If two animals were heterozygous for all of these 3,000 pairs of genes, there could be $3^{3,000}$ different genotypes, assuming that all combinations were possible by means of independent assortment and crossing over. Such a number is, of course, beyond human comprehension. Even if linkage is complete in all chromo-

Table 6-2. Independently Segregating Gene Recommendations

Number of pairs of heterozygous genes	Number of different gametes	Number of different genotypes in F_2	Total number of combinations in F_2
1	2	3	4
2	4	9	16
3	8	27	64
n	2^n or $10^{0.301n}$	3^n or $10^{0.477n}$	4^n or $10^{0.602n}$

somes, we could still get 2^{30}, or 1,073,741,824, different recombinations of complete chromosomes, whereas there are probably less than 1 billion cattle now in the world. It is evident that the simple recombination of intact chromosomes in most of our farm animals can alone provide a tremendous amount of variability when considerable heterozygosity exists. The evidence available indicates that most of our animals are rather heterozygous.

Variation from Gene Mutations.   Gene mutations represent changes in the individual loci. The nature of mutational changes in genes is, of course, unknown, but it might conceivably be comparable to a change in the spatial arrangement of atoms in complex organic molecules.

Observable mutation rates under laboratory conditions are generally in the range of one mutation for each 100,000 to 1,000,000 loci. The rates of observable mutation vary for individual loci and can be increased under the influence of X ray, chemicals, ultraviolet light, and other extreme conditions.

Even though mutation rates can be increased by special treatment, they are still so rare as to impose a distinct limitation on the role of mutation in providing new variability. Furthermore, most of the mutations in our domestic animals are harmful, and many of them are recessive and deleterious or lethal when the gene is in homozygous condition. Some few mutations have not been harmful and have been preserved. The best-known mutation of this type probably is the one influencing polledness or hornlessness in cattle. It has been estimated that the mutation rate from horned to polled is about 1 in 20,000, much higher than the rate generally observed for most loci. The dominance of the polled gene was also of major importance in its detection and increase in frequency. The polled condition has been reported in practically all our breeds of cattle. This example also points up how the harmfulness or usefulness of a mutation may be determined by the environmental conditions under which it is expressed. Although we may consider polledness to be desirable under domestication, horns were undoubtedly of much value for protection for animals in the wild state.

Gene mutation is a reversible process, and loci which have mutated from the normal type are subject to mutation back to their original form. This can be summarized by a simple diagram:

$$A \underset{v}{\overset{u}{\rightleftharpoons}} a$$

in which A and a are alleles and u represents the mutation rate of locus A to a. The reverse mutation rate from a to A is v. These two rates of mutation, u and v, may differ greatly. With the generally recessive nature of most mutations, the mutation rate of a dominant to a recessive usually exceeds the reverse mutation rate.

The genetic variability with which a breeder works is the result of the accumulation of mutants which arose over the long span of evolutionary time. However, the rarity, recessiveness, and generally harmful nature of most gene mutations prevent them from being important in providing favorable variation for animal improvement within the lifetimes of even several generations of breeders. In fact, the breeder must exert at least a portion of his efforts toward purging the undesired mutants from his flock or herd as they appear.

Chromosomal Aberrations. In addition to gene or point mutations, a considerable variety of chromosomal aberrations involving various-sized blocks of genes, whole chromosomes, or whole sets of chromosomes have been discovered in recent years. Though such chromosomal changes in farm animals are chiefly of evolutionary significance, plant breeders in many instances have been able to utilize these chromosomal changes to produce new varieties or species. Changes of this nature are generally brought to light by discrepancies in breeding behavior resulting in disturbances of the normally expected Mendelian ratios, in the creation of new linkage relations, or in the appearance of variable offspring. When such things occur, cytological study generally has revealed certain changes in the number or arrangement of the loci within the whole chromosome complex or a change in the actual number of chromosomes. This close correspondence between the visible changes in the chromatin and changes in ratios or in individuals is, of course, the strongest evidence supporting the chromosomal theory of inheritance.

Inversion. One type of chromosomal mutation is called *inversion*. In this situation a portion of a chromosome has in some manner become inverted and the order of the genes is reversed. This might happen because a chromosome became looped upon itself with breakage of the chromosome and rehealing of the ends in reversed order (Fig. 6-2).

Duplication and deficiency. A chromosome may become deficient or lacking in certain genes by a portion of a chromosome becoming lost. This might occur through unequal crossing over between two homologous chromosomes. If this happens, the unequal interchange, besides producing a deficient chromosome, would also give rise to a chro-

Fig. 6-2. Diagrammatic representation of the process of inversion.

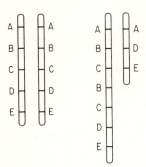

Fig. 6-3. Diagrammatic representation of duplication and deficiency following unequal crossing over.

mosome with some genes duplicated. This is illustrated in Fig. 6-3. One of the chromosomes is deficient for genes *B* and *C*, whereas the other has a double dose of these genes; i.e., they are duplicated.

Translocation. It sometimes happens that a chromosome becomes broken and one of the pieces is attached to some other chromosome. If the piece becomes attached to a nonhomologous chromosome, a new linkage relation is set up and a translocation is produced. If, for instance, a small piece of the X chromosome in *Drosophila* became attached to one of the two large pairs or to the small IV chromosome, then the genes which were contained in the broken piece of X chromosome and which had formerly shown sex-linked inheritance would no longer do so. Likewise, if a piece of one of the autosomes was broken off and became attached to an X chromosome, those autosomal genes that had formerly shown ordinary Mendelian inheritance would now show sex-linked inheritance.

Reciprocal translocation. During meiosis there might be an equal interchange of various-sized blocks of genes between non-sister, homologous chromatids. This process was called crossing over. So, too, there may be interchanges of blocks of genes between two non-sister, non-homologous chromatids or perhaps between nonhomologous chromosomes (illegitimate crossing over). This is called reciprocal translocation or segmental interchange. In ordinary translocation, there is no trading between the chromosomes or chromatids; a section of one chromosome simply goes over to another chromosome without any reciprocity on the part of the latter. In reciprocal translocation, there is reciprocity, and it is between nonhomologous chromatids or chromosomes. This type of translocation happens more frequently than does simple translocation.

Variation in Chromosome Number.[1]   A new organism generally arises from the union of an egg and a sperm of which the number of chromosomes has been reduced to half numbers at fertilization again restores

[1] Beatty, R. A., Heteroploidy in Mammals, *Anim. Breed. Abstracts*, 19:283–292, 1951.

the chromosomes to the diploid number $2n$. Sometimes a new organism gets one (rarely two) too many or too few chromosomes and is then $2n + 1$ or $2n - 1$. Some of these organisms are viable and have been studied both genetically and cytologically. These individuals are often less viable and less fertile than the normal type and do not breed true. Defects such as Mongolism and Klinefelter's syndrome in humans are due to such abnormalities.

The haploid (n) condition occurs in some species which reproduce parthenogenetically, e.g., bees, wasps, some moths, and rotifers. Triploids ($3n$) or tetraploids ($4n$) are rarely found in animals, but such polyploids are reasonably common among plants and certain lower animal forms including *Drosophila*. Actually the chromosomal behavior and composition of most of our farm animals have received little study.

Environmental Variation. The many changes in the genes and chromosomes have been pointed out as being responsible for genetic variation in animals. This genetic variation is expressed, however, only as the environmental stimuli and conditions permit its expression. The environment does not directly change the hereditary make-up of an individual, but the environmental circumstances do determine the extent to which the inherited tendency is expressed. Animals of a good strain may themselves be stunted and not developed to the limit of their inherent capacities, but these animals may still be capable of producing offspring of good genetic potential. The genes and chromosomes from the parents form the actual "bridge of inheritance" from parent to offspring, and are not affected directly by the stunting.

There is an abundance of examples showing that acquired environmental modifications have no influence on the hereditary make-up of the animal. When the tail of a sheep is docked, the sheep has acquired a character which is not passed on to descendants of this sheep. In succeeding generations it will be necessary to dock the lambs, notwithstanding the fact that docking has been practiced for hundreds of generations. In this category come also the dehorning of cattle, docking of horses' tails, clipping of dogs' ears and tails, etc. In none of these cases is the acquired character inherited by the offspring. In the human family, circumcision, which has been practiced for thousands of years, and the binding of feet to prevent their full development, as practiced by Chinese women, might be mentioned. No more than in the cases cited of farm animals is there evidence that these environmentally acquired characters are inherited in even the slightest degree.

Weismann, in order to test this hypothesis, cut off the tails of mice for 19 generations in succession and secured no shortening of the tails or absence of tails in any of the descendants. As Walter remarks, "it is a good thing that children of warriors do not inherit their parents' honorable scars of battle else we would have long since been a race of cripples." The same author also remarks in this regard that "evidently wooden legs are not inherited, but wooden heads are."

Even though environment does not directly influence the germ plasm, it is almost certain that in the evolutionary process the environment must have wielded a marked influence on life and caused gradual change in germ plasm. Examples from both the plant and animal kingdoms can be cited. Capsella, a wayside weed, has gradually climbed to a mountainous habitat. As it climbed it developed a dwarf character—at least it grows luxuriantly in the valley and is small, compact, and dwarfed in its mountain home. When taken back to the valley, it continues its dwarf characteristics. Presumably those plants genetically tending to be dwarfed had a better chance of surviving under the more rigorous conditions and decreased food supply that prevail to an ever-increasing degree as the mountains are ascended. The force bringing about the change was not the inheritance of acquired characters but *natural selection,* which no doubt always has and always will, provided man does not interfere, weed out those types and individuals poorly fitted to their environment and preserve and proliferate those better fitted to survive under the exising conditions.

Light, temperature, food, and moisture represent environmental stimuli which modify the development and expression of inherent qualities. Low temperatures stimulate the development of red flowers in the Chinese primrose, but high temperatures stimulate the formation of white flowers. The kind of food supplied to the larvae of bees determines whether the females shall be fertile (queen bees) or infertile (workers). The moisture supply influences the yearly growth in trees, as can be depicted from the width of the annual growth rings.

Rabbits of the genotype c^h/c^h show the Himalayan pattern and are white with black at the tips of the feet, ears, nose, and tail. The black pigment at the extremities results because the enzyme responsible for this pigment is inactivated at normal body temperatures. Only those parts normally cool develop color. This has been demonstrated experimentally by placing an ice pack on another portion of the rabbit with the resulting development of dark hair in this area.

We should always keep in mind the constant interplay between heredity and environment. Even though environmental variation is not transmitted, this should not deter breeders' efforts to provide as favorable an environment as is economically feasible to permit the attainment of the animals' inherent potentialities. In many cases our animals already have the inherent potential for a much higher level of performance than their environment permits them to express. A striking example of this comes from the report of a Danish dairy demonstration. Six cows which in the previous two years had produced an average of 370 lb of fat each in 365 days were taken from an average dairy herd. During the experimental demonstration the cows were fed four times a day and induced to eat as much as possible. They were given the best care that could be provided and were milked four times per day. At the end of the 365-day demonstration these cows had produced an average of 909 lb of fat per cow, an increase of 539 lb of fat over their average performance of the previous two years.

For most of our farm animals it is difficult to single out special environmental circumstances. The environment encompasses a complex of many measurable items as well as numerous seemingly intangible ones working in cooperation and competition. For most of our larger animals the herdsman or manager provides an important but often indeterminate part of the environment. With two groups of animals of comparable genetic worth and with comparable housing and feed available for each group, but with two different supervisors, wide differences in performance may be expected. The impossibility of providing managemental recommendations on a recipe or formula basis makes it difficult to standardize environmental conditions with large animals. Nevertheless it is in this special area that the art and experience of the skilled husbandman pay handsome dividends in increased performance.

The breeder's success depends ultimately on his ability to detect and breed from those animals in his herd and flock that show favorable variation which is of a genetic nature. To do this, he must keep records of performance on all his animals and make as reasonable an allowance as he can for the contribution that either good or bad environment has made to the merits or deficiencies which his animals exhibit. Only in this way can he put a firm genetic foundation under his attempts to breed better livestock.

MEASURING VARIATION

Probability. Variation within a population for qualitative traits can be summarized by assigning each individual to its appropriate discrete class. A frequency diagram showing the proportion of the population which possesses each of the qualities being studied provides a vivid picture of the population for the characteristic catalogued. In many situations the observations may be summarized as a ratio (such as 1:1 or 3:1) which may be expected on the basis of knowledge of the inheritance of the trait. Small samples may deviate widely from the expected ratios, but the true ratio will be closely approximated if the numbers are large.

An example of coin tossing provides an excellent analogy to certain genetic situations. Except for sex-linked genes, the genes in most farm animals are found in pairs. The composition of the gametes appears to be largely a matter of chance at segregation.

If we toss a coin, it has an even chance of falling heads or tails, and the probability or likelihood of occurrence of either is ½. If we toss two coins, the probability that one will fall heads is ½, and the probability that the second will fall heads is ½. The probability that both will fall heads is the product of the likelihood of each separate event: $(\frac{1}{2})(\frac{1}{2}) = \frac{1}{4}$. The same situation prevails for the likelihood that both coins will display tails. In addition, the two other possibilities, that the first coin displays heads and the second tails and that the first coin displays tails and the second heads, each have a probability of $(\frac{1}{2})(\frac{1}{2})$, or 1 chance in 8. If we toss three coins, the probability that all of them will fall heads is $(\frac{1}{2})(\frac{1}{2})(\frac{1}{2})$, or 1 chance in 8.

Referring again to the coin-tossing example, if we let p equal the probability of heads and q equal the probability of tails, the sum of these two probabilities for the toss of a single coin $(p+q)$ equals 1.0. The principle can be stated that the sum of probabilities of all possibilities of occurrence for a single or a series of independent events is 1.0. For the single coin toss,

$$p + q = \tfrac{1}{2} + \tfrac{1}{2} = 1.0$$

We can extend the use of this principle to cover the possibilities for the two independent coin tosses discussed above,

$$\tfrac{1}{4}(\text{HH}) + \tfrac{1}{4}(\text{HT}) + \tfrac{1}{4}(\text{TH}) + \tfrac{1}{4}(\text{TT}) = p^2 + pq + qp + q^2 = 1.0$$

A second important principle is that the likelihood of two or more of a series of independent events occurring together is the product of the probabilities of occurrence of the independent events. Thus the probability that three tosses of a coin would give three heads is

$$\tfrac{1}{2} \times \tfrac{1}{2} \times \tfrac{1}{2} = \tfrac{1}{8}$$

Binomial Distribution. Each of the probabilities mentioned can be derived from the expansion of the binomial $(p+q)^n$ where $p = q = \frac{1}{2}$ and n is the number of observations in the series. In tossing the coin three times, the probability of any one series of events can be determined from the expansion of the binomial

$$(p + q)^3 = p^3 + 3p^2q + 3pq^2 + q^3$$

The coefficient of each succeeding term in the binomial expansion can be obtained by multiplying the coefficient of the present term by the exponent of p and dividing by the number the present term represents in the series. In the second term the coefficient is 3 and the exponent of p is 2. Thus the coefficient of the third term is $(2 \times 3)/2 = 3$. The q^3 term represents the occurrence of three tails, and this is $\frac{1}{8}$. Of the eight possibilities that could arise from three tosses of a coin, the likelihood of getting three heads or three tails would be $\frac{1}{8}$. The probabilities of two heads and one tail or one head and two tails are each $\frac{3}{8}$. Again the sum of these probabilities add to 1.0 ($\frac{1}{8} + \frac{3}{8} + \frac{3}{8} + \frac{1}{8}$).

This procedure can be followed for more complex problems where the probabilities of the events may not each be $\frac{1}{2}$, such as $\frac{3}{4}$ or $\frac{1}{4}$, or where more than two alternatives may be possible, as with dice or multiple alleles at a specific locus. First find the probability of each separate event. These values are inserted in their proper place in the binomial (polynomial) expansion. In the case of three alternatives, one would have to expand the trinomial $(p + q + r)^n$ to examine the individual probabilities.

The binomial can be drawn upon most extensively in animal breeding. Where n represents the number of observations, the probability of a particular combination of two qualities can be obtained without needing to expand the complete binomial by using the following formula:

$$\frac{n!}{r!s!}\, p^r q^s$$

where factorial n = total number involved

factorial r = number of one alternative

factorial s = number of other alternative (and $r + s = n$)

p = probability or expectation of obtaining first alternative

q = probability or expectation of obtaining other alternative (and $p + q = 1$)

If we assume that the probability of a male or female from a particular birth was $\frac{1}{2}$ for most practical purposes, we can determine the probability that 4 females and 4 males would be born in 8 births. Following the formula above,

$$\frac{8!}{4!4!}\left(\frac{1}{2}\right)^4\left(\frac{1}{2}\right)^4 = \frac{8 \times 7 \times 6 \times 5 \times 4 \times 3 \times 2 \times 1}{(4 \times 3 \times 2 \times 1)(4 \times 3 \times 2 \times 1)}\left(\frac{1}{2}\right)^4\left(\frac{1}{2}\right)^4 = \frac{70}{256}$$
$$= 0.273$$

or about 1 chance in 4.

Chi-square. How well a given set of results fit those to be expected on the basis of a particular genetic hypothesis can be tested by chi-square (χ^2). This test is often termed *the test of goodness of fit*. First the expected ratio is determined on the basis of the hypothesis to be tested. Then the contribution which each of the classes in the ratio make to the chi-square value is computed by squaring the difference between the observed number and the expected number of observations in each class and dividing it by its expected number. The results for all classes are summed to give the chi-square value.

$$\Sigma \frac{(O - E)^2}{E} = \chi^2$$

Suppose there were 83 black animals and 37 red animals arising from crosses among heterozygous *Bb* parents. Since black is dominant, we expect a ratio of 3 black to 1 red among the F_2 offspring on the basis of a monohybrid-dominant hypothesis. We want to test whether the observations fit this hypothesis for this character or whether we must search for a more plausible theory. In short, what is the probability of getting as large or even larger deviations from the expected 3:1 ratio in the F_2? The computations required to provide the chi-square value are shown in Table 6-3.

Table 6-3. Computation of χ^2 for a 3:1 Expected Ratio

Class	Observed frequency O	Expected frequency E	Deviation O − E	$\dfrac{(O - E)^2}{E}$
Black	83	90	−7	0.59
Red	37	30	+7	1.63

$$\chi^2 = 2.22$$

We now consult the table of chi-square (Table 6-4) and read across the line for one degree of freedom. We find that our χ^2 value of 1.08 is very close to the value for 0.30. We interpret this to mean that the observed variation from the expected 3:1 ratio would be anticipated to occur by chance in about 30 per cent of such cases. This degree of departure would not usually be considered statistically significant, so we say that the monohybrid-dominance hypothesis can be accepted.

It is evident that the greater the deviation the larger χ^2 becomes. If it gets to the size of the figures in the 0.05 column of the chi-square table, we say the difference is significant (theory probably not true) because that large a deviation from expectation would occur only once in 20 times by chance alone. If it gets as large as or larger than the figure in the table under the column headed 0.01, we say the deviation from expectancy is highly significant; i.e., it would occur by chance

*Table 6-4. Table of Chi-square**

Degrees of freedom	0.99	0.95	0.70	0.50	0.30	0.05	0.01
1	0.00016	0.00393	0.148	0.455	1.074	3.841	6.635
2	0.0201	0.103	0.713	1.386	2.408	5.991	9.210
3	0.115	0.352	1.424	2.366	3.665	7.815	11.345
4	0.297	0.711	2.195	3.357	4.878	9.488	13.277
5	0.554	1.145	3.000	4.351	6.064	11.070	15.086
6	0.872	1.635	3.828	5.348	7.231	12.592	16.812
7	1.239	2.167	4.671	6.346	8.383	14.067	18.475
8	1.646	2.733	5.527	7.344	9.524	15.507	20.090
9	2.088	3.325	6.393	8.343	10.656	16.919	21.666
10	2.558	3.940	7.267	9.342	11.781	18.307	23.209

* Abridged from Table III of Fisher, R. A., "Statistical Methods for Research Workers," 1950, published by Oliver & Boyd, Ltd., Edinburgh and London, by permission of the author and publishers.

only once in 100 or more times. In using χ^2 always consult the line corresponding to one less than the number of F_2 classes involved in the theory. In our examples we expected two F_2 classes, so we used the figures in line 1 in interpreting χ^2. If we had been testing a dihybrid theory with its expected four F_2 classes, we would have used line 3, and so on.

Normal Curve and Mean. We have noted that the majority of the traits which are economically important probably are influenced by genes at a large number of loci and by many environmental factors. As a consequence, most of these traits are measured quantitatively, and they exhibit what we already have described as continuous variation. A number of special procedures and techniques have been devised to assist in characterizing and summarizing such continuous variation.

A single measurement describes the weight of a cow. A group of cows will vary in their weight, and a complete description of the weight of the entire group of cows requires an enumeration of the weight of each individual. Such a mass of data for a large herd of cows is unmanageable, and a statistical description of the population is desired. This description in its simplest form consists of a measure of the central tendency and a measure of the variability of the population. Such a description of the central tendency or the variation for the entire population of measurements is called a *parameter*. Ordinarily we do not have measurements on the entire population. By studying a sample of observa-

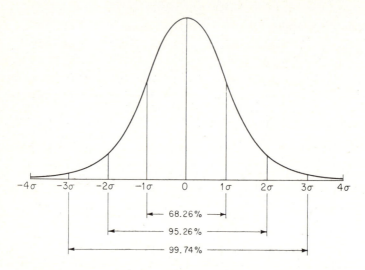

Fig. 6-4. A theoretical normal curve showing the area bracketed by one, two, three, and four standard deviations from the mean. Percentages of total area bracketed by one or more standard deviations above or below the mean are indicated.

tions from the population, statistics can be derived which are descriptive of the sample and which provide estimates of the corresponding population parameters. Statistics derived from a small sample may not be representative of the population parameters. A more reliable statistical description of the population usually can be obtained by increasing the size of the sample studied.

Our ability to describe a population of biological measurements in a simple yet meaningful manner is greatly aided by the fact that most of our biological measurements can be considered to be normally distributed. A distinctive feature of this normal distribution is that the values are clustered at a midpoint, thinning out symmetrically toward both extremes. Figure 6-4 shows a theoretical normal distribution. The height of the curve at a particular point represents the frequency of individuals having that particular value.

There are at least three common measures of central tendency: the median, representing the class value halfway between the two extreme values; the mode, representing the class with the highest frequency; and the mean, the average of all measurements in the population. For a true normal distribution, the median, the mode, and the mean will coincide. However, the mean is our most useful statistic in estimating the central tendency of most populations, since samples from a normally distributed population may show departures from normality.

The population mean is merely the arithmetic average of all the values included in the population. It is conventionally represented by the symbol $\bar{x}$. Obviously it is usually impossible to measure all individuals in the population; hence a sample chosen so as to be representative of the population is ordinarily used to estimate the population mean. The symbol $\bar{x}$ is conveniently used to represent the sample mean. Individual measurements in the sample can be symbolized by X_1, X_2, X_3, . . . , X_n, and the mean can be computed according to the formula

$$\bar{x} = \frac{X_1 + X_2 + X_3 + \cdots X_n}{n}$$

where n represents the number of observations included in the sample.

Table 6-5 illustrates the use of the above formula in showing the sums and the means for heart-girth measurement of 75 in., and the mean weight of 1,260 lb provides a base to which the individual cows in the sample can be compared.

Variance. The degree of dispersion or variation exhibited by a population can be expressed as the average deviation or difference from the mean, ignoring signs. The range between the extremes of the population also provides some indication of variability. However, each of these measures lacks flexibility, and complete reliance upon them soon blocks the detailed analysis of the observed variation. Variance s^2, which is the average squared deviation of the individual measurements from the population mean, is the most useful measure of variation for studying the variability of populations. Actually we deal with a sample from the population and compute estimates of the population variance according to the formula

$$s^2 = \frac{(X_1 - \bar{x})^2 + (X_2 - \bar{x})^2 + \cdots + (X_n - \bar{x})^2}{n - 1}$$

Since the deviations are squared, variance is a positive value with zero as a lower limit. The use of this formula is shown in Table 6-5, where the sample variance for heart girth is 13.4 and that for weight is 19,575. The sum of the squared deviations from the mean is divided by $n - 1$ rather than n to compute the average squared deviation, since a limited sample ordinarily does not encompass the entire range of the population. As a consequence, division by n consistently provides an underestimate of the population variance, but divisions by $n - 1$ provides an unbiased estimate. Unbiased estimates are ones which show no consistent tendency to be either above or below the population parameter.

Table 6-5. *Heart-girth Measurements and Weights of 25 Holstein Cows Taken Approximately One Month after Calving, Showing Calculations for Determining Linear Correlation and Regression Coefficients*

Cow no.	Heart girth, in. X	Weight, lb Y	Deviations* from means x	Deviations* from means y	Squares and products of deviations x^2	Squares and products of deviations xy	Squares and products of deviations y^2
1	82	1,490	7	230	49	1,610	52,900
2	76	1,330	1	70	1	70	4,900
3	77	1,400	2	140	4	280	19,600
4	75	1,300	0	40	0	0	1,600
5	84	1,570	9	310	81	2,790	96,100
6	74	1,240	−1	−20	1	20	400
7	76	1,360	1	100	1	100	10,000
8	81	1,480	6	220	36	1,320	48,400
9	75	1,200	0	−60	0	0	3,600
10	76	1,250	1	−10	1	−10	100
11	75	1,340	0	80	0	0	6,400
12	73	1,240	−2	−20	4	40	400
13	76	1,230	1	−30	1	−30	900
14	72	1,140	−3	120	9	360	14,400
15	76	1,300	1	40	1	40	1,600
16	74	1,130	−1	−130	1	130	16,900
17	72	1,150	−3	−90	9	270	8,100
18	70	1,020	−5	−240	25	1,200	57,600
19	72	1,120	−3	−140	9	420	19,600
20	72	1,130	−3	−130	9	390	16,900
21	70	1,130	−5	−130	25	650	16,900
22	74	1,200	−1	−60	1	60	36,000
23	80	1,440	5	180	25	900	32,400
24	73	1,240	−2	−20	4	40	400
25	70	1,070	−5	190	25	950	36,100
Sums (Σ)	1,875	31,500	0	0	322	11,600	469,800

$\bar{x} = 75$ $\bar{y} = 1,260$ $s_x^2 = 13.4$ $s_y^2 = 19,575$
                                        $s_x = 3.66$ $s_y = 139.9$

*x = deviation of X's from their mean $\bar{x}$
 y = deviation of Y's from their mean $\bar{y}$
 xy = product of a paired set of deviations

Table 6-6 shows how the variance due to different environmental and genetic causes can be subdivided. This study by Plum represents one of the first comprehensive analyses of the total variance in a particular trait. Variance is a most useful mathematical concept. Its mastery requires much thought and experience.

Table 6-6. *Relative Importance of Causes of Variation in Fat Production**

Source of variation		Per cent of total variance
Breed		2
Herd:		
Feeding policy	12	
Other (genetic or environmental)	21	
		33
Cow (mostly genetic)		26
Residual (year-to-year variations):		
Feeding variation within the herd	6	
Length of dry period	1	
Season of calving	3	
Other year-to-year differences	1	
Other factors	28	39
Total		100

* Plum, M., *Jour. Dairy Sci.*, **38**:824, 1935.

Standard Deviation. The population standard deviation which is the square root of the variance is symbolized by σ. The estimate of the standard deviation from a sample is the square root of the sample variance s^2. Although the units of variance are pounds or inches squared, the units for the standard deviation are pounds or inches, just as the original items were measured. Since the standard deviation is the square root of the variance, it can be computed as follows:

$$s = \sqrt{\frac{(X_1 - \bar{x})^2 + (X_2 - \bar{x})^2 + (X_3 - \bar{x})^2 + \cdots + (X_n - \bar{x})^2}{n - 1}}$$

This is illustrated in Table 6-5, where the standard deviation for heart girth is 3.66 inches and the standard deviation for weight is 139.9 lb.

When reliable estimates of the mean and standard deviation are available for a normally distributed population, the expected proportion of the population which will fall within a designated area of the normal distribution can be computed from specially prepared tables. It has been possible to develop such tables because a consistent relationship between the mean and the standard deviation exists for a normal distribution. Figure 6-4 shows this relationship, with 68.26 per cent of the observations being included in the area bounded by one standard deviation on either side of mean. Similarly, 95.46 per cent of the population is included in the area bracketed by two standard deviations from the mean, and 99.74 per cent of the population is included in the area bracketed by three standard deviations on either side of the mean.

As an example of the application of this concept, studies show that the standard deviation of average breeding (genetic) values for sires, as appraised from a study of their daughters' lactation fat records, is about 40 lb. In a population of 1,000 bulls whose breeding values are normally distributed, 161 would be expected to have breeding values which are at least one standard deviation (40 lb) above the population mean. Only 23 would be expected to be at least two standard deviations (80 lb) above the mean, and only 1 would be expected to be at least three standard deviations above the mean (120 lb). Rather frequently when a sire's daughters are compared with their dams, increases of from 80 to 100 lb of fat are found. Since the bull transmits only a sample half of his genetic make-up, a bull that is genetically 120 lb above the population average would be expected to have daughters that were 60 lb above average dams of the population. When the daughters of a bull average 80 to 100 lb of fat more than their dams, an environmental contribution to this increase should be suspected. Bulls capable of genetically increasing their daughters to this extent are extremely rare.

Coefficient of Variation. It is sometimes desirable to compare the variability of traits measured in different units. Generally, large things vary much and small things little, making it convenient to express the standard deviation as a percentage of the mean to make such comparisons. The standard deviation as a percentage of the mean is termed the *coefficient of variation C*, and the formula for computing it is

$$C = \frac{s \times 100}{\bar{x}}$$

For example, the variance of heart girth in the example in Table 6-5 is 13.4 and that for weight is 19,575. However, the coefficients of variability are 18 and 11 per cent respectively. Certain traits for a class of animals have characteristic coefficients of variation, a knowledge of which is valuable for planning and evaluating experiments.

Correlation. We are often interested in learning whether or not one trait in an animal is associated with another trait. For example, is the height at the withers in beef cattle associated with body length? Is the weight of a dairy cow related to the amount of milk she will produce? Is there an association between back-fat thickness and the percentage of lean cuts in swine carcasses?

The *correlation coefficient* r measures the degree of association between two traits or variables. It ranges in value from −1.0 to +1.0
and is an abstract value possessing no common units. A correlation of
+1.0 indicates that for each unit increase in one variable there is a
unit increase in the correlated trait. The correlation coefficient may be
anywhere between these two extremes, with a value of 0 indicating
no association between the two variables.

Heart-girth measurements and weights for 25 Holstein cows were
given in Table 6-5. The high correlation between heart girth and body
size has long been recognized and has provided the basis for measuring
tapes used to estimate body weight. Table 6-5 also shows the calculations
necessary to compute the correlation coefficient.

The correlation coefficient r is then computed according to the following formula:

$$r = \frac{\Sigma xy}{\sqrt{(\Sigma x^2)(\Sigma y^2)}} = \frac{11,600}{\sqrt{(322)(469,800)}} = \frac{11,600}{12,299} = 0.943$$

Correlation coefficients are subject to sampling fluctuations, and the
value 0.943 obtained from this sample of 25 cows will be limited in
its representation of the association between these two traits, say, in
the population of Holstein cows in the United States. Methods have
been developed to ascertain the probability that a correlation of the
magnitude found in the sample actually exists in the population. A significant correlation, from the statistical standpoint, means that there is
a high probability that there is an association between the traits examined of the magnitude indicated by the sample value. With a large
volume of data, statistical significance may be realized, but logical judgment is necessary to determine whether or not the correlation is large
enough to be practically useful in prediction and culling.

Correlation and variance are related and in many respects represent
two different ways of viewing variation. The square of the correlation
coefficient r^2 measures the portion of the variance in one variable, say
y, that can be accounted for by variation in a related variable x. The
square of the correlation 0.943 shows that 89 per cent of the variance
in body weight is associated with variation in heart-girth measurements.
The remaining 11 per cent $(1 − r^2)$ is associated with variation in things
other than heart girth.

Too often, when a correlation is found between two variables, a cause-
and-effect relationship is assumed. One must be most cautious in arriving
at a conclusion regarding cause and effect. The correlation coefficient
provides no evidence of itself as to which variable is the cause and
which is the effect. Such evidence must come from a specific investigation of the biological relations between the traits.

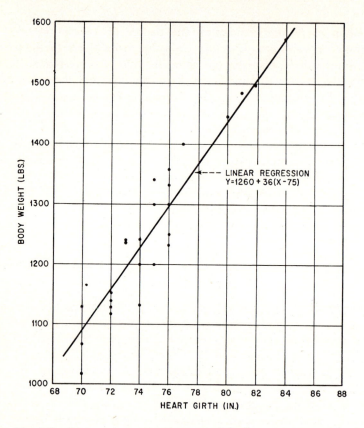

Fig. 6-5. Diagrammatic representation of the linear regression of body weight on heart-girth measurements for the data in Table 6-5.

Regression. Whereas the correlation coefficient measures the degree of association between two variables, the *regression coefficient b* measures the amount of change in one variable associated with a unit change in the second variable. For example, from the data in Table 6-5 we might ask how many pounds change in body weight on the average are associated with each inch change in heart girth.

The information in Table 6-5 has been plotted in Fig. 6-5 to demonstrate visually the association between heart girth and body weight. The regression line has been drawn to provide the best linear fit to the series of paired observations so as to minimize the sum of the squared deviations from this line. Again referring to Table 6-5, the regression coefficient which represents the slope of the line is computed as follows:

$$b_{yx} = \frac{\Sigma xy}{\Sigma x^2} = \frac{11,600}{322} = 36 \text{ lb}$$

In this example the regression coefficient indicates that for each change of 1 in. in heart girth an average change of 36 lb in body weight is expected.

The regression coefficient and the correlation coefficient are related. If the correlation between heart girth and body weight were perfect, all the plotted points in Fig. 6-5 would be on the regression line. Under these circumstances we would be able to predict the weight of a cow exactly by knowing her heart-girth measurement. The relationship between the correlation coefficient and the regression coefficient is pointed out further, since, if r is known, b can be obtained as follows:

$$b_{yx} = r\frac{s_y}{s_x} = 0.943 \left(\frac{139.9}{3.66}\right) = 36$$

The regression coefficient finds most use in predicting or estimating one variable, provided the other variable is known. Presume we know the heart-girth measurement of an animal, and on the basis of the analysis of the sample of data in Table 6-5 we wish to predict her weight. Symbolically the equation for predicting a value of $Y(\bar{Y})$ is

$$\hat{Y} = \bar{Y} + \frac{\Sigma xy}{\Sigma x^2}(X - \bar{X}) = \bar{Y} + b_{yx}(X - \bar{X})$$

If a cow's heart girth measured 72 inches, the predicted body weight for the cow would be

$$\hat{Y} = 1{,}260 + 36(72 - 75) = 1{,}260 - 108 = 1{,}152 \text{ lb}$$

Note that this and other predicted values will lie on the regression line.

Regression as originally introduced by Sir Francis Galton had a slightly different, but related, meaning. In studying human stature, Galton found that the progeny of tall parents usually are not as tall as their parents and the progeny of short parents are not as short as their parents. In each case the progeny were nearer the population mean, or, as Galton expressed it, the progeny regressed toward the population average.

Average parents tend to produce average offspring; parents above average tend to produce progeny above average; parents below average tend to produce progeny below average. However, the progeny of extreme parents, above or below the average, exhibit their parental characteristics to a less-marked degree than these traits were manifest in the parents themselves.

This regression in the Galtonian sense is universal in material where the correlation between the two variables is not perfect. In the case of stature, in Galton's studies the correlation between the average of the two parents and their offspring approached 1.0, and though regression was evident, the progeny retained a large proportion of the parental superiority. The high parent-offspring correlation indicated that most of the variation in stature was genetically conditioned, hence transmissible from parent to offspring. For traits where the parent-offspring correlation is low, nongenetic variation is comparatively more influential. Only the parental superiority resulting from superior genetic potential is transmissible, and a more marked regression, in the Galtonian sense, toward the population average is evidenced. It should be obvious that this law of regression can be applied with confidence only to the average of large numbers of observations.

A more comprehensive assessment of the variation in a trait can be obtained from the statistical technique commonly described as the *analysis of variance*. In our earlier illustration of the computation of variance, standard deviation, and coefficient of variation, only the total variance was considered. However, in Table 6-6 the concept that variance can be subdivided according to its various sources was suggested. Such a subdivision of the total variance into its genetical and environmental components is most useful to analyze problems in animal breeding. Detailed procedures involved in the analysis of variance and the estimation of components of variance are described in statistical texts listed at the end of this chapter. The example to be discussed here is presented to illustrate the biological interpretation of components of variance.

We shall consider data from litters of swine analyzed so that farrowing season and breed differences do not influence the variation. The thickness of the back fat of each of the pigs was obtained from back-fat probes at 154 days of age. An analysis of variance was computed for the measurements as shown in Table 6-7. A new term has been introduced, *degrees of freedom*. This can be taken to represent the number of independent items in a particular grouping, and generally it will be one less than the total number of items in a set or group ($n - 1$).

The sums of squares represent the sum of the squared deviations of the respective values from the general mean. The total sum of squares is the sum of the squared deviations of the measurements for the individual pigs from the mean back-fat measurement for the data (2.79 cm). Likewise, the sum of squares among sire families can be represented as the sum of the squared deviations of the sire family means (paternal half-sib averages) from the over-all mean.[1]

---

[1] This is precisely true only when there are the same number of offspring per sire. However, the concept holds even when the family size varies.

Table 6-7. Analysis of Variance for Back-fat Thickness (cm) in Swine at 154 Days of Age*

Source of variation	Degrees of freedom	Sums of squares	Mean squares	Components of mean squares
Total	7,590	1,906.2		
Among sire families	342	395.4	1.1561	$\sigma^2\omega + 7.07\,\sigma^2 f$ $+ 18.16\,\sigma^2 s$
Among dam families, in sire	734	552.6	0.7529	$\sigma^2\omega + 6.59\,\sigma^2 f$
Among full-sibs (littermates)	6,514	958.2	0.1471	$\sigma^2\omega$

* Data adapted from Cox, D. F., *Jour. Anim. Sci.*, **23**:447–450, 1964.

When the sums of squares are divided by the degrees of freedom, we obtain the mean squares. These represent variances on a per item basis. A further subdivision of these mean squares into components of variance can be made to assign the variance to its source.

The components included in the mean squares for the last three lines of the analysis of variance in Table 6-7 are given in Table 6-8. We shall begin with the simplest variance component $\sigma^2\omega$ and then consider the added sources of variance. Numerical values for the components are obtained by setting the mean squares equal to their components. The value for $\sigma^2\omega$ is obtained directly and the value for $\sigma^2 f$ is

$$\frac{0.7529 - 0.1471}{6.59} = 0.0919$$

Of course, $\sigma^2 s$ is obtained by further substitution after $\sigma^2\omega$ and $\sigma^2 f$ are available.

$$\frac{1.1561 - 7.07(0.0919) - 0.1471}{18.16} = 0.0204$$

The relative importance of the components of variance for back-fat thickness is shown by the percentage values in Table 6-8. Such percentages vary considerably for different traits depending upon the genetic and environmental influences.

The variance component $\sigma^2\omega$ arises from differences among pigs from the same litter, i.e., full sibs. Average differences in the measurements due to the sex of the pigs were adjusted before analysis. Although the pigs of the same litter have the same parents, they do not necessarily have the same genes since each received independent samples of genes from each parent. They also occupied different positions in their dam's uterus and undoubtedly experienced slightly different conditions after

Component	Magnitude	Per cent
$\sigma^2 s$	0.0204	8
$\sigma^2 f$	0.0919	36
$\sigma^2 \omega$	0.1471	56
Total	0.2594	100

birth. One might have contracted a disease while nursing, but the other remained free. The magnitude of these effects are reflected in $\sigma^2 \omega^2$.

Measurements of two pigs, each by a different dam but by the same sire, also may vary due to the differing genetic contributions of the two mothers. Prenatal and postnatal influences which were alike for littermates may differ from one litter to another. The contribution of these factors to the differences among pigs are measured by the component of variance $\sigma^2 f$. Maternal environment, including the milk available from the mother, may be most influential on early growth and development, whereas in later life these effects diminish in importance.

Additional variation usually is evident when two pigs have different sires. This would arise because of the different samples of genes contributed by the gametes from their sires. If there is much genetic variation in the trait, $\sigma^2 s$ should be relatively large, since it is associated with the genetic contribution of the sire.

SUMMARY

Variation is the raw material the breeder has available for herd and flock improvement. The variation which is observed arises from both genetic and environmental influences. Genetic variation results from variation in the genes provided by genetic recombination and less frequently by gene mutations and chromosomal changes. Traits are usually grouped into those which show qualitative or discontinuous variation and those which show quantitative or continuous variation. Specific ratios can be expected for qualitative traits, depending on the number of loci influencing them, the degree of linkage, and the degree of dominance. The observed ratio can be tested against the expected theoretical ratio using chi square. Quantitative traits cannot be characterized so easily. The mean or average of the observations and a measure of the variability about the mean are needed to describe the variation for a particular trait. Correlation and regression coefficients are useful to express relationships between two quantitative traits. The correlation coefficient expresses the degree of association between two variables, being 1.0 or —1.0 when there is complete association. It is zero when the two traits are independent and not associated. The regression coefficient expresses the amount of change in one trait that is expected for a specified change in the second trait.

REFERENCES

Fisher, R. A. 1950. "Statistical Methods for Research Workers," Oliver & Boyd Ltd., Edinburgh and London.

Mather, K. 1947. "Statistical Analysis in Biology," Interscience Publishers, New York.

Sinnott, E. W., Dunn, L. C., and Dobzhansky, T. 1950. "Principles of Genetics," McGraw-Hill Book Company, New York.

Snedecor, G. W. 1946. "Statistical Methods," The Iowa State University Press, Ames, Iowa.

Srb, A. M., and Owen, R. D. 1953. "General Genetics," W. H. Freeman and Company, San Francisco.

Steel, R. G. D., and Torrie, J. H. 1960. "Principles and Procedures of Statistics," McGraw-Hill Book Company, New York.

CHAPTER SEVEN

Population Genetics

Thus far we have been concerned chiefly with a specific mating and the genetics of an individual or individuals resulting from this mating. The discovery of many important genetic principles has resulted from such studies. However, a somewhat different complexion is given to several of these principles when the heredity of a population or a group of interbreeding individuals is considered. *Population genetics* is a field of inquiry in which genetics as related to a group or population is considered in contrast to the genetics of individuals. This is sometimes termed *statistical genetics,* but the term population genetics is much to be preferred. Statistics has entered the picture to characterize and to aid in providing a rational explanation of the dynamics of the population. The need to examine populations from the statistical viewpoint is also necessitated because most of the economically important traits are influenced by a large number of segregating loci. As was pointed out in the previous discussion of quantitative inheritance, the action of this large number of genes on individual traits precludes the separation of the population into discretely distinct genotypes.

The desire to unravel the mystery of evolution has served as a great stimulus to the growth of population genetics, especially the work of Fisher and Wright. With most of the economically important traits in animals, our efforts toward improvement really represent attempts to accelerate the action of evolutionary forces. Before man realized he could modify his plants and animals in accordance with the purposes they were to fulfill, natural selection was actively developing populations better adapted to existing environmental conditions. The underlying biological circumstances which produced inherent change from generation to generation was the same for natural selection as it is for artificial selection. Whereas evolutionary change is measured against a base of centuries, breeders try to operate in animal improvement against a time base of years or a few generations.

GENERAL CONSIDERATIONS

Genetic improvement of a population results from underlying changes in gene frequency and/or changes in the way in which the mating system in the population permits the genes to unite as the zygotes are produced. Although the frequencies of most of the genes influencing quantitative traits are not known, an understanding of gene frequency is essential to appreciate and understand the genetic dynamics of populations.

Gene Frequency. The frequency of a gene, in its most general sense, means the degree to which it is rare or abundant. More precisely, the frequency of a gene is the proportion of the loci of a given allelic series occupied by a particular gene. When only two members of an allelic series, say A and a, are involved, and there are n individuals in the population, the gene frequency can assume any one of $2n + 1$ values. The frequency of a gene is conventionally designated by the letter q and that of its allele by $1 - q$. For example,

$$q = \frac{\Sigma A \text{ loci}}{\Sigma A \text{ loci} + a \text{ loci}} \quad \text{and} \quad 1 - q = \frac{\Sigma a \text{ loci}}{\Sigma A \text{ loci} + \Sigma a \text{ loci}}$$

When several members of an allelic series are involved,

$$q_{A_1} = \frac{\Sigma A_1}{\Sigma A_1 + \Sigma A_2 + \cdots + \Sigma a}$$

Gene frequency as defined takes on a range of values between 0 and 1.0.

For a numerical example consider the case of coat color in Shorthorn cattle, in which the genetic situation is reasonably well-explained on the basis of a single locus where an RR animal is red, an Rr animal is roan, and an rr animal is white. In a herd of 100 Shorthorns there are 47 red, 44 roan, and 9 white animals. There are 200 loci in the population, and 138 of the 200 loci are occupied by the R gene (two in each of the 47 red animals plus one in each of the 44 roan animals), giving it a frequency of 0.69. The r gene occupies 62 of the 200 loci in the population, giving it a frequency of 0.31.

$$q_R = \frac{2 \times 47 + 44}{200} = 0.69 \qquad (1 - q)r = \frac{2 \times 9 + 44}{200} = 0.31$$

For loci where dominance is exhibited, it is not possible to detect the heterozygotes directly and determine the gene frequency. However, the Hardy-Weinburg rule, proposed independently by Hardy, a British mathematician, and Weinburg, a German physician, in 1908, has provided a basis for deducing the gene frequency in many situations where all of the individual genotypes are not distinguishable. This underlying idea of gene distribution postulates that in the absence of selection the relative frequencies of each allele in a population mating randomly tends to remain constant from generation to generation.

Random Mating.   To explore this idea in more detail, we must understand what is meant by random mating. The situation of random mating exists when the probability of the mating of a given female to a particular male is in direct proportion to the different types of males in the population. Random mating with respect to a trait exists when an individual which possesses this trait is no more or no less likely to mate with another possessing the trait than would be expected from the frequency of individuals of the opposite sex possessing that trait in the population. Using the example of human blood groups, data that have been collected indicate that mating is essentially random with respect to the AB, A, B, and O blood types. A person with blood-grouping AB is not influenced in choosing a mate by the prospective mate's blood type. This means that in a population the proportion of AB individuals which will marry O-type individuals can be predicted from the proportion of O-type individuals of the opposite sex in that population.

Some may prefer the term *panmixia* on the grounds that the term *random mating* may suggest carelessness. Such is not the case, however. It should also be pointed out that the method of selecting animals may in no way be random, although the manner in which these selected individuals are combined may be random. The method of choosing the individuals and the mating system employed generally will not be contingent on one another.

The concept of random mating may be further illustrated by again referring to information on the ABO blood types. We will let d, a, b, and o represent the relative frequencies of the AB, A, B, and O phenotypes respectively $(d + a + b + o = 1.0)$. Then the expected fre-

Table 7-1. Summary of the Symbolic Frequencies of Various Types of Matings for the ABO Blood-group System Assuming Random Mating

Female blood types	*Male blood types*			
	AB	*A*	*B*	*O*
AB	*d²*	*da*	*db*	*do*
A	*da*	*a²*	*ab*	*ao*
B	*db*	*ab*	*b²*	*bo*
O	*do*	*ao*	*bo*	*o²*

quencies of the various types of mating can be determined by squaring the phenotypic frequencies $(d + a + b + o)^2$. The proportion of the various types of matings can then be catalogued as in Table 7-1. The relative frequencies of the various mating types are directly proportional to the relative frequencies of the various genotypes in the population.

Hardy-Weinberg Law. In a population satisfying the Hardy-Weinburg law, the proportion of the different types of gametes produced in a population is directly proportional to their respective gene frequencies. Thus, in our previous example with Shorthorns, were q was 0.69 and $1 - q$ was 0.31, the gametic frequencies in that population would be $0.69R + 0.31r$. In a large population mating randomly with no selection, mutation, or migration, the frequency of each allele tends to remain constant, and the square of the population's gametic array gives the zygotic distribution. With only two alleles, the gametic array would be $[qA + (1 - q)a]$. The zygotic array under random mating would be the square of this expression, or

$$[qA + (1 - q)a]^2 = q^2A + 2q(1 - q)Aa + (1 - q)^2aa$$

This expression will permit the extension of the checkerboard method for determining the expected proportion of the different kinds of genotypes from a particular mating to include the entire range of gene frequencies. Most of the previous examples in basic genetics were special cases in which q was equal to 0.5, as in F_1's and F_2's.

The Hardy-Weinberg formula permits an indirect estimation of gene frequencies when we are working with populations in which:

1. The reproductive and survival rates of individuals carrying *AA, Aa,* or *aa* genotypes are equal.
2. Mating is random.
3. Mutation is infrequent.
4. The population is large enough to make accidents of sampling inconsequential.

When complete dominance is exhibited, the heterozygotes are indistinguishable from the homozygous dominants. Nevertheless, in a population that satisfies reasonably well the requirements enumerated above, an estimate of gene frequency can be obtained from an accurate count of the proportion of recessives in the population. Although recent counts are not available, earlier information has indicated that in Holsteins about 1 of 200 calves born in red and white. Red is recessive to black, and from the Hardy-Weinburg rule the recessives should make up $(1 - q)^2$ of the population. Hence,

$$(1 - q)^2 = \tfrac{1}{200} = 0.005 \quad 1 - q = \sqrt{0.005} = 0.07 \quad \text{and} \quad q = 0.93$$

For sex-linked traits, the zygotic and gene frequencies in the heterogametic sex are the same. If complete dominance is expressed, the gene frequency in the homogamic sex could be determined in the same manner as with autosomic inheritance. Red-green color blindness in man is sex-linked. If 5 per cent of the males are color-blind, the frequency of the recessive gene would be $1 - q = 0.05$. Assuming the same gene

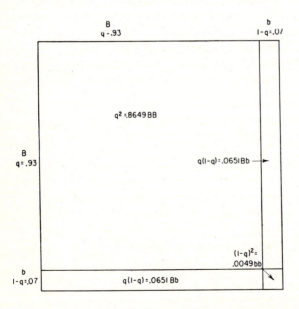

Fig. 7-1. An example of the Hardy-Weinburg law where *BB* and *Bb* individuals have black coat color and *bb* individuals have red coat color. Rather than a ratio of 3 black to 1 red which would be expected when gene frequency (q) was 0.5, the ratio here is approximately 199 black to 1 red.

frequency in the females, the incidence of such color blindness would be $(0.05)^2 = 0.0025$, or 0.25 per cent. Red-green color blindness would be about 20 times as common in males as in females under random mating.

CHANGES IN GENE FREQUENCY

Since changes in the mean of a population under random mating are dependent on gene-frequency change, the forces influencing gene frequency are found to be in various states of balance from time to time. The gene frequency at a particular instant is a result of the equilibrium between selection, migration, chance, and mutation. Wright has summarized the effects of these forces on evolutionary change and provided a basis for interpreting the effect of these forces on gene frequencies in livestock populations.

Mutation. The source of genetic variability was discussed in the preceding chapter. Mutation proposes new genes, and selection dictates whether these genes are retained or rogued from the population. Mutations are generally much less important in changing gene frequency than De Vries originally presumed in 1901 after studying *Oenothera*. In contrast to Darwin's belief that small variations were responsible for eventual species formation, De Vries looked upon mutation as providing the means for drastic changes.

It has already been pointed out that several things limit the role of mutation in providing usable variation. These same conditions impose limitations to its influence in changing gene frequency. The quantitative change in gene frequency by mutation can be expressed by letting u be the mutation rate from A to a and by letting v be the reverse mutation rate from a to A. The change in q, symbolized as Δq, per generation can be expressed

$$\Delta q = -uq + v(1 - q)$$

The balance between the two mutation rates in the absence of other disturbing factors makes for a decrease, increase, or a stable gene frequency in the population. Under circumstances where the mutation rates are permitted to seek their own equilibrium, $\Delta q = 0$. At this point

$$q = \frac{v}{u + v}$$

Mutation must be recognized for the important part it plays in providing new variation over long periods of time, permitting the population to adapt to changing environmental stress. It is not a potent force in changing gene frequency in most economically important species, because most mutation rates are low and most mutations are harmful. Their general recessive nature makes it difficult to increase their frequency in initial selections.

Migration. The introduction of new genes into a population can change gene frequency. Widespread introductions are utilized in crop breeding, but in purebred livestock populations it is not usually of widespread importance. The pedigree barrier imposed by the closed registration in most of the pure breeds restricts the introduction of new genes. Any change which migration invokes is dependent on the frequency of the gene in the immigrant population and the extent to which the immigrants are allowed to propagate, i.e., the proportion of gametes they are permitted to furnish.

Several examples are available in which importation of animals has been made to introduce genes for specific traits. Indian cattle were introduced into the beef populations in Texas and the Gulf states to provide resistance to tick fever and the ability to gain on the poor quality native roughage. More recently Red Sindhi have been used in experimental crosses to introduce genes for heat tolerance in the dairy cattle of this region. The introduction of the Danish Landrace hog to provide genes to improve bacon qualities of our American breeds of swine is another interesting example.

In a more limited sense, certain seedstock herds or flocks may serve as centers of radiation. Purchase of animals from these herds for use in commercial herds and flocks might be looked upon as selection for the breed as a whole. However, for the individual herd or flock it is comparable to migration, with possible new genes being introduced from the seedstock herds or flocks.

Chance. Changes in gene frequency also can result just from chance due to Mendelian sampling which takes place in providing the gametes which represent the gene pool for the next generation. To the extent that these sampling changes do occur, they should be random and cancel one another in a large population. However, with a small population the continuous chain of sampling to provide the gametes for each succeeding generation may become a potent force in changing gene frequency.

If the sampling from each parental population were entirely random, the knowledge of the characteristics of the binomial distribution permits the examination of the chance variation expected in a given-sized population. We shall consider two alleles, A and a, where $q = 0.5$ in two populations, one consisting of 100,000 individuals and the other consisting of 50 individuals. The larger population's 200,000 gametes arise from a pool consisting of equal numbers of *A* and *a* gametes. Exactly equal numbers of the two gametes will not necessarily enter into the formation of the next generation. By chance there might be more or less than 100,000 of each kind of gamete. According to the sampling of a binomial distribution, the expected number of *A* or *a* gametes in the large population would be

$$100,000 \pm \sqrt{\frac{100,000 \times 100,000}{200,000}} = 100,000 \pm 223$$

Likewise in the small population of 50 individuals the expected number of *A* or *a* gametes would be

$$50 \pm \sqrt{\frac{50 \times 50}{100}} = 50 \pm 5$$

The sampling or standard deviation of 223 for the large population is much larger than the 5 for the small population. Comparatively, the standard deviation for the large population is only 0.223 per cent of the expected number of gametes, whereas in the small population it is 10 per cent of the expected number. From this, one can see that in the absence of selection, mutation, and migration the proportion of gametes, and hence the gene frequency, can remain seemingly unchanged from generation to generation in a large population. With small populations, much variation in gametic proportions is expected, and the extinction or fixation of a given locus is not uncommon. Intermediate between extinction and fixation, the gene frequencies may drift in one direction or another, providing for a change which has been referred to by Wright as *genetic drift*.

The decrease in heterozygosity accompanying inbreeding is a manifestation of the above property. Under inbreeding, the population number is restricted since matings take place within a closed herd or flock. The smaller the population, the more drastic may be the genetic drift or possible change due to chance, just as the smaller the intermating population, the more intense the rate of inbreeding.

Selection. Selection occurs whenever forces permit some individuals to produce more offspring than others. Individuals which leave no offspring are genetically culled from the population. Selection creates no new genes, but it permits the possessors of some genes and gene combinations to have more offspring than individuals lacking these genes or gene combinations. The underlying influence of selection is to change gene frequency, and the consequences of selection depend on the magni-

Table 7-2. The Relative Frequencies of Three Zygotes Arising from One Locus under Random Mating with Their Relative Reproductive Rates

Parental genotypes	Relative frequencies	Reproductive rates
BB	q^2	1.0
Bb	$2q(1 - q)$	$1 - hs$
bb	$(1 - q)^2$	$1 - s$

tude of the change in q. The apparent creativeness of selection in producing, after continuous selection, types not present in the original population results from the accumulation of genes and gene combinations over many generations of selection.

Although actual selection involves saving or rejecting the individual and its full complex of genes, for illustrative purposes we shall simplify the situation to involve a single pair or two allelic genes. Even in this simplified situation, selection cannot be directly for individual genes, rather individuals (zygotes) must be saved or rejected.

The amount of change in gene frequency that is to be expected from a generation of selection can be determined if we know the relative number of offspring that parents with each genotype will produce. Wright has developed the background for measuring this change in gene frequency when a constant selection pressure is applied. We shall assume that the BB, Bb, and bb genotypes arising from segregation at a single locus reproduce in the ratio $1:1 - hs:1 - s$.

The symbol s is the selection coefficient representing the intensity of selection against the bb zygotes, assuming the BB zygotes are most desirable. The value of s represents the proportion of the bb types which are rejected. Thus in the case of the red genotypes for registered Angus and Holsteins, s would be 1.0, since red animals are not accepted in the regular registry. Likewise for a recessive lethal condition, s equals 1.0.

The symbol h represents the intensity of selection against Bb types in terms of selection against the bb types. With zygotic selection and complete dominance ($h = 0$), the heterozygote is indistinguishable from the desired homozygote (BB); consequently, no selection pressure can be applied to the heterozygote. With no dominance ($h = \frac{1}{2}$), the heterozygote is intermediate in breeding value between BB and bb. The value of $h = \frac{1}{2}$ means that selection against Bb types is one-half as intense as it is against bb types. When the recessive type bb is desired ($h = 1$), the BB and Bb types would be equally restricted in their reproductive rates.

On the basis of the above definitions, Wright has shown that the expected change in gene frequency Δq for one generation of selection in which dominance is complete ($h = 0$) is

$$\Delta q = \frac{sq(1 - q)^2}{1 - s(1 - q)^2}$$

For no dominance where $h = \frac{1}{2}$,

$$\Delta q = \frac{sq(1 - q)}{2[1 - s(1 - q)]}$$

When the recessive gene is the desired one and $h = 1$,

$$\Delta q = \frac{sq^2(1 - q)}{1 - s(1 - q^2)}$$

The effectiveness of selection for genes exhibiting these degrees of dominance can be seen in Fig. 7-3, which was originally presented by Wright. The heights of the curved lines represent the rate of change in gene frequency to be expected from selection when the original gene frequency in the population ranges from 0 to 1.0. Fluctuation in the height of the curves makes it clearly evident that Δq is dependent on the existing gene frequency in the population when selection is being practiced.

For the gene showing no dominance, Δq is largest when $q = 0.5$. Beginning on the left of the figure, where $q = 0.1$, progress will become more rapid with each successive generation of selection until q reaches 0.5. From this point the rate at which q approaches 1.0 will slow down more and more as the actual value of q in the selected population increases.

Dominance is an aid to increasing the frequency of a dominant gene when q is low. So long as the undesired recessive is more abundant than the desired dominant, even the retention of heterozygous individuals serves to increase the frequency of the dominant gene. Nevertheless, after the desired dominant gene has reached a frequency of 0.5, saving the heterozygotes does not contribute to increasing q. At this stage the complete dominance which assisted the change in q when the frequency of the desired gene was low now actually hinders progress.

In Fig. 7-2 the curve shows that efforts to increase the frequency of a rare gene are not so rewarding when the gene is recessive. This is another way of pointing up the difficulty experienced in establishing a recessive mutation. At the early stages even the saving of the heterozygotes would aid in increasing the frequency of the desired gene.

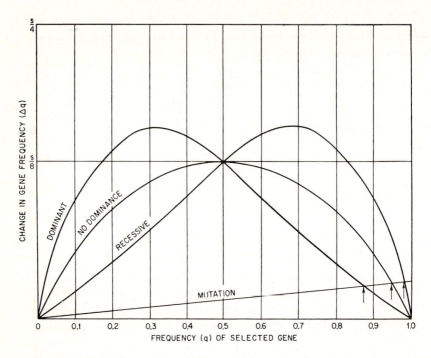

Fig. 7-2. Rate of change of gene frequency under selection and mutation only. Equilibrium frequencies are shown at the junction of the mutation line and the curve for the respective type of desired gene. Mutation rate (u) assumed equal to $s/32$. (*From Wright, 1931.*)

However, it cannot be recognized and differentiated from the even less desired homozygote.

Mutation and Selection. As has been pointed out, the gene frequency at a given time is the result of the equilibrium between selection, mutation, chance, and migration. In exploring the explanation for evolutionary changes, considerable progress has been made in expressing the relationship among these forces and in predicting the ultimate equilibrium values of gene frequency for a specific population.

Illustratively, we shall consider the special case in which only selection and mutation have an important influence on gene frequency. We shall further consider that reverse mutation is negligible; hence the net change in gene frequency due to mutation would be uq. At equilibrium between mutation and selection, Δq would equal zero. The pressure which selection exerts to change gene frequency would be counterbalanced by the change due to mutation, and Δq would equal zero. When selection is for a dominant gene and the population is at equilibrium between mutation and selection,

$$uq = sq(1 - q)^2$$

Solving this expression for q gives the gene frequency reached at equilibrium, which for the above expression is

$$q = 1 - \sqrt{\frac{u}{s}}$$

For the other two cases discussed in the preceding section, the equilibrium value of q is $1 - 2u/s$ when selection is for a gene showing no dominance and $1 - u/s$ when selection is for a recessive. The equilibrium points for each of these special cases are shown in Fig. 7-2.

For an illustration of the equilibrium frequency, consider the problem of eliminating the red gene from either Holstein or Angus cattle. In these situations the double recessives are eliminated, but the heterozygotes are allowed to breed. The selection coefficient s equals 1 and h is 0, since the desired gene is dominant. Information on the actual mutation rate u for this locus is not readily available. Assuming a mutation rate of 1 for each 20,000 loci (0.00005), which is a value suggested for the polled locus in cattle, an equilibrium gene frequency would be obtained when

$$q = 1 - \sqrt{\frac{u}{s}} = 1 - \sqrt{\frac{0.00005}{1}} = 0.993$$

Earlier in this chapter it was pointed out that the frequency of the black gene was probably near 0.93. The above equilibrium value indicates that mass selection can still increase the frequency of the black gene, but it will be a time-consuming process, as illustrated in the curve of Δq for a dominant gene in Fig. 7-2. If breeders considered eliminating the red gene extremely important and were willing to go to the trouble and expense to test all males by breeding each of them to 5 or 6 red cows, progress toward eliminating the red gene would be accelerated. Under such a plan, selection against the red gene would be more intense, since most of the heterozygous males would be detected. If such a plan were kept active for a great many generations, a new equilibrium value for gene frequency would be reached somewhat nearer 1.0 than the 0.993 of the above example.

HEREDITARY VARIANCE

Both hereditary and environmental influences are responsible for the observed variation in records of performance. The hereditary variation results from the action of the genes and gene combinations in response to the environmental conditions provided the individual members of the population. Only genes and chromosomes conditioning the hereditary

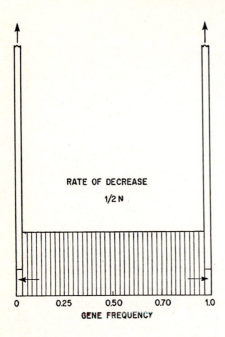

RATE OF DECREASE

1/2 N

GENE FREQUENCY

Fig. 7-3. Distribution of gene frequencies in a closed population where only chance is operating to move the genes toward extinction $(q = 0)$ or fixation $(q = 1.0)$. Mutation, migration, and selection are assumed to be inactive. The size of the population (N) determines the rapidity with which the intermediate gene frequencies move to zero or 1.0. (*From Wright, 1931.*)

variability can be transmitted from parent to offspring or selected for in breeding programs.

Most important quantitative traits are usually influenced by an unknown number of genes. Moreover, their mode of action, either chemically or descriptively, such as additive, multiplicative, etc., is also rarely known. Fortunately this, in most cases, is not an insurmountable handicap to animal breeders. While somewhat more accurate choices of animals might be possible if all the details as to the genes present and their mode of action were known for the simpler inherited traits, what the breeder would do corresponds closely to what he can do when the individual performance, the mean, and the relative importance of the several sources of variability are known.

Subdivision of Hereditary Variance. The fundamental genetic theory which provides the basis for the logical direction of applied breeding programs as developed largely by Fisher and Wright has been interpreted in the light of animal-breeding applications by Lush and his associates. Its main thesis relies on the partitioning of the total hereditary variance into three components based on a statistically convenient description of gene effects. These three component parts of the hereditary variance $\sigma^2 h$ are the additively genetic variance $\sigma^2 g$, the variance due to dominance deviations from the additive description $\sigma^2 d$, and the residual genotypic variance $\sigma^2 i$ termed *epistatic variance*.

The simplest way for genes to be combined is for a gene to produce a certain plus or minus change in the trait or quality being observed (Fig. 7-4). Some gene effects presumably do combine in this way, and the effects on phenotypic variability of most other genes can be explained statistically by an additive description or model. The variance associated with the additive or average effects summed up for all loci influencing the trait is termed the *additively genetic variance* $\sigma^2 g$. The sum of the additive or average effects for all loci influencing a trait represents what is commonly referred to as the average breeding value for an individual.

In some instances the full effect of a gene on the phenotype cannot be adequately represented by the additive description. For many of such instances the effect of a gene may be dependent on the other genes present in the genotype. In the case of dominance, the substitution of *A* for *a* in an *aa* genotype has a more marked effect on the phenotype than when *A* is substituted for *a* in an *Aa* genotype. Dominance represents the condition of nonadditivity where the phenotype of the heterozygote is not exactly midway between the two homozygotes. The effect of a gene substitution of *A* for *a* is not additive for all genotypes in the population.

Fig. 7-4. Phenotypic values for a single locus where additive effects and varying degrees of dominance are expressed.

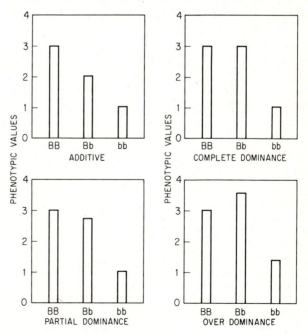

Even with complete dominance, much of the relationship between phenotypes and genotypes in the population can be accounted for by an additive relationship. The portion of variability which cannot be thus accounted for is termed *dominance deviations* from the additive description, and the variance contributed by these deviations is the *dominance variance* $\sigma^2 d$.

Joint influences (interactions) of nonallelic genes also may influence a trait. A classical example is that of purple and white flower color in sweet peas reported by Bateson. Two independent pairs of genes (*Cc, Pp*) each showing dominance were found to be responsible. White flowers result when either one or both of the dominant loci are absent. At least one of each of the dominant genes at the two loci, however, are required to produce the alternative purple color. Hence the presence of these genes, each at different loci, is required for the expression of purple color. Though dominance may be regarded as interaction among allelic genes, this example represents interaction also among nonallelic genes and is referred to as *epistatis*. Hereditary variance not included in the additive or dominance portions is termed *epistatic variance* $\sigma^2 i$.

These types of genetic effects are shown in Table 7-3 for a hypothetical example for two loci. Phenotypic values have been assigned the genotypes to illustrate additive effects, dominance, and epistatic deviations. The values given in the table represent deviations from the mean, and the relative frequencies are those of a simple dihybrid F_2 ratio. Phenotypic values have been arrived at by assuming complete dominance at the two loci with $BB = Bb = CC = Cc = 2$ and $bb = cc = -6$ expressed as deviations from the mean. Then an interallelic effect was

Table 7-3. Illustration of Phenotypic Values As Composed of Additive, Dominance, and Epistatic Effects

Genotype	Relative frequency	Phenotypic value	Sum of additive effects	Sum of dom. dev.	Sum of epistatic dev.
BBCC	1/16	6.0	11.0	−5.5	0.5
BBCc	2/16	6.0	5.5	0.0	0.5
BBcc	1/16	−7.0	0.0	−5.5	−1.5
BbCC	2/16	6.0	5.5	0.0	0.5
BbCc	4/16	6.0	0.0	5.5	0.5
Bbcc	2/16	−7.0	−5.5	0.0	−1.5
bbCC	1/16	−7.0	0.0	−5.5	−1.5
bbCc	2/16	−7.0	−5.5	0.0	−1.5
bbcc	1/16	−12.0	−11.0	−5.5	4.5

Table 7-4. Summary of the Components of Hereditary Variances for the Genetic Model of Table 7-3

Component	Variance	Percentage
$\sigma^2 g$	30.250	63.5
$\sigma^2 d$	15.125	31.8
$\sigma^2 i$	2.250	4.7
$\sigma^2 h$	47.625	100.0

introduced by allowing $B - C -$ combinations to add 2 to the phenotypic values, $B - cc$ and $bbC -$ to add -3 to the phenotype values, and $bbcc$ to add zero.

These phenotypic values were then analyzed to partition the values into the additive effects at the two loci plus the dominance and epistatic deviations.[1] No environmental influences have been added to the phenotypic values; hence, all of the variance will be genotypic in the broad sense $(\sigma^2 h)$. As was suggested earlier, the total hereditary variance can be subdivided into its additive $\sigma^2 g$, dominance $\sigma^2 d$, and epistatic $\sigma^2 i$ components: $(\sigma^2 h = \sigma^2 g + \sigma^2 d + \sigma^2 i)$.

The components of variance have been computed for this example, but the procedures are too complex to be given in detail in this text. The magnitudes and the relative contribution which each type of variance makes to the total variance are given in Table 7-4.

Note that the assignment of numerical values to the various genotypes and effects provides no explanation of the physiological basis of the gene effects. It has permitted a statistical description of the variation, which can be used to predict genetic changes in large populations. It should be noted that the gene frequency for both alleles in the above example was 0.5. If the gene frequencies changed, the relative frequencies of the various genotypes would change, as well as the relative importance of the three types of hereditary variance.

The usefulness of this concept of partitioning hereditary variance will be pointed out numerous times. In partial explanation, the gene effects which can be described in an additive manner, and which are responsible for the additively genetic variance, represent those influences which are subject to molding and change by selection. Those effects expressed as dominance and epistatic deviations are not amenable to mass selection

[1] Additive effects, dominance, and epistatic deviations, and components of variance computed according to Kempthorne, O., "An Introduction to Genetic Statistics," John Wiley and Sons, Inc., New York, 1957, p. 428.

pressure. They represent specific effects which provide the basis for heterosis in inbred-line and breed crosses.

Heritability and Repeatability. Two important concepts which tie together the statistical methodology and the principles of population genetics are *repeatability* and *heritability*. Their introduction into animal-breeding literature can be credited to Lush.

Heritability is the fraction of the observed phenotypic variance which results from differences in heredity—among the genes and gene combinations of the individual genotypes as a unit. This is the broad concept of heritability in which the hereditary variance is considered as the sum of the additively genetic, the dominance, and the epistatic, variances.

A more restrictive or narrower definition of heritability is more useful in most aspects of animal improvement. The narrower definition of heritability represents the fraction of the observed phenotypic variance which is additively genetic or which is associated with differences in average breeding values. This is expressed as

$$h^2 = \frac{\sigma^2 g}{\sigma^2 g + \sigma^2 d + \sigma^2 i + \sigma^2 e}$$

where $\sigma^2 e$ represents all other nongenetic variance.

Theoretically, heritability can range from 0 to 1.0, but these extreme values are rarely encountered in practice. A particular heritability value is descriptive of a trait in a particular population at a given time. Since it is a fraction, its value can be varied by changes in the additively genetic variance of the numerator or by changes in any one or all of the components of variance in the denominator. The additively genetic variance is closely associated with the gene frequency of the genes influencing the trait. For most situations $\sigma^2 g$ is largest when q for the genes influencing the trait is near 0.5.

Repeatability is a concept closely allied to heritability and is useful for those traits which are expressed several times during an animal's lifetime, such as lactation milk yield for dairy cows, number farrowed and litter weight for swine, and weaned weight of lamb or calf for sheep or beef cattle. Since neither the genes nor gene combinations influencing the successive expressions of a trait change, repeatability should be at least as large as heritability in a broad sense. It may be larger, since certain permanent environmental influences may be included in the numerator of the repeatability fraction, but they, of course, would be nongenetic.

Repeatability can be computed as the regression of future performance on past performance. In addition to being computed as a regression it may be derived from an analysis of variance as an intraceless correlation among records or observations of traits on the same individual. Repeatability can be expressed as

$$\frac{\sigma^2 g + \sigma^2 d + \sigma^2 i + \sigma^2 pe}{\sigma^2 g + \sigma^2 d + \sigma^2 i + \sigma^2 pe + \sigma^2}$$

where $\sigma^2 pe$ represents variance associated with possible permanent environmental influences which make for differences in the expression of a particular trait for the several individuals in the population. For example, a cow may accidentally but permanently lose one quarter, and this would influence milk yield in all future lactations. The feeding and care of young animals may be such as to stunt them, and the influence of this poor feeding and care would appear again and again in the subsequent expressions of a trait.

Some data of Lush et al. illustrate both the concepts of heritability and repeatability. They studied 676 daughter-dam comparisons used to prove 103 sires in Iowa dairy herd improvement associations. The lactation fat yields for each female were expressed on a standardized mature basis. When the mates of a bull had only one record, the data for her and her daughter were discarded. The mates (dams) were then divided into a low and a high half on the basis of the first available record on each dam. All available later records were then averaged to provide a single value which was used in representing the later records of these cows. Each dam was then represented by only one value in the average of first and later records. The summary of these data is given in Fig. 7-5.

If a bull had an even number of daughter-dam comparisons, all were used. When a bull had an odd number of such comparisons, the mate having a median first record was discarded along with the information on her daughter. In this way, each of the 103 sires had exactly the same number of mates in the high group as in the low group. Thus, differences in herd averages or in the genetic merit of the sires would not affect the differences between the high and low groups.

As shown in Fig. 7-5, the difference X between the mates' first records was 102.1 lb of fat. This would be representative of the total variability between these two groups of cows. The difference Y between the later records of these groups of cows was 43.6 lb. This is representative of the real difference in producing ability among these cows if successive records of these cows in the same herds were obtained. When the mates for each sire were divided into a low and a high half, the low half, for example was represented by cows with lower inherent ability, and, in addition, by selecting on the basis of the size of the record, poorer-

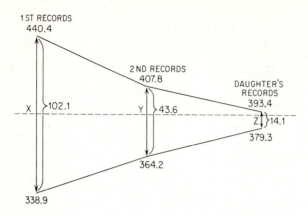

Fig. 7-5. Regression of later records and daughter's records on the dam's selected record. Dotted line represents midpoint between selected high and selected low records. (*From Lush et al., J. Dairy Sci.,* **24:**699, 1941.)

than-average environmental circumstances were represented also. In the successive records, a new sample of herd circumstances, now presumably representative for both low and high groups, was included. Hence, only the repeatable or real difference between these groups of cows is exhibited in the contrast of the later records. The above procedure provides an approximation to the regression of future performance on present performance, and as such is an estimate of repeatability. The repeatability of differences in single lactation fat yields in these data is 43.6/102.1 = 0.43.

The averages for the daughter lactations for these two groups of cows are shown in Fig. 7-5. The daughters from the high groups of dams averaged 393.4 lb and those from the low group averaged 379.3 lb. Each sire was represented by an equal number of daughters in the low and the high groups, and sire differences do not contribute to the 14.1-lb difference between the daughters of these two groups of dams. However, since the daughters received only a sample half of the dam's genotype, the 14.1-lb difference represents only one-half of what might have been expected if the sires as well as their mates (dams) could have been divided into comparable low and high groups.

On the basis of the first available record, the difference between the low and the high group of cows was 102.1 lb. However, on the basis of future records, only 43.6 lb represent real or repeatable differences among cows. The daughters of these two groups of cows point out that of the 102.1-lb initial difference in the cows, one-half of the average difference in breeding value among these cows was 14.1 lb. Hence the heritability of differences in the lactation fat records is estimated to be $2(14.1/102.1) = 0.28$. Multiplication by 2 is necessary since the effect of the sires is canceled out in this example.

The concept of regression in the Galtonian sense as referred to in the preceding chapter is aptly illustrated in the present example (Fig. 7-5) with the regression of the average of the first records and the average of the daughters toward the mean of X. With perfect repeatability, the second records also should have averaged 440.4, and no regression would have been evident. With complete heritability, the difference between the daughters of these two groups of dams should have been 51.5 rather than the 14.1 actually observed.

Determining Heritability. Numerous determinations of heritability have been made for economically important traits in farm livestock. Due to statistical sampling and variation, rather large bodies of information are required to provide meaningful estimates. Average heritability values for individual traits are given in Chaps. 11, 12, 13, and 14 for each specific class of livestock.

Numerous procedures for computing heritability are used in research studies. The underlying basis for determining heritability depends on evaluating how much more alike individuals with similar genotypes are than unrelated or less closely related individuals. The parent-offspring and the paternal half-sib relationships are most widely used to determine heritability in farm animals.

Parent-offspring. The regression of offspring on parent is most commonly used. The procedure begins by obtaining measurements of the traits on both the parents and their offspring. These measurements are corrected for such influences as age or sex when they are likely to influence the measurements. The sum of products (Σxy as in Table 6-5) for the paired observations is obtained. When this· is divided by the degrees of freedom $(n-1)$, an estimate of the covariance between parent and offspring is obtained. The offspring receive a sample half of the genes of the parent, and the covariance between parent and offspring is expected to include one-half of the additively genetic variance for the trait. The variance of the parents' measurements (x) is obtained, and the regression of offspring on parent represents one-half of heritability.

$$b_{op} = \frac{\text{cov } op}{\sigma^2 p} = \frac{\sigma^2 g/2}{\sigma^2 p}$$

Since relationship to only one of the two parents is involved, the regression must be multiplied by 2 to compute heritability.

An example of the results from using this procedure to compute heritability for the percentage of nonfat milk solids in dairy cattle is given in

Table 7-5. Heritability of the Percentage of Nonfat Milk Solids from Off-spring-Parent Regression

Source	Degrees of freedom	Σx	Σxy	Variance	Covariance
Total	813	55.54	17.87		
Among herds and years	257	22.16	7.34		
Within herd and year	556	33.38	10.53	0.0600	0.0189

$$b_{op} = \frac{0.0189}{0.0600} = 0.315 \qquad h^2 = 2b_{op} = 2(0.315) = 0.63$$

Table 7-5. The regression has been computed on a within-herd-and-year basis to avoid the influence of herd or yearly effects on the regression. Note that the variance for the parents and the covariance of parents and offspring for the regression are on a within-herd-and-year basis. Often the intra-sire regression of daughter on dam is used, since the progeny of a sire are nearly contemporary and perform under similar managemental conditions. Furthermore, if specific mates are chosen for the different sires, the intra-sire regression does not permit the peculiarities of the mating system to contribute to the daughter-dam regression. For the above example, the regression of offspring of parent is 0.315 and heritability of the percentage of nonfat solids is 0.63.

Paternal half sibs. Half sibs have on the average only one-quarter of their genes in common rather than one-half as with parent and offspring. Usually paternal half sibs are used since there are few maternal half sibs in a set, and their similarity may be influenced by maternal effects. Heritability is determined from an analysis of variance, such as was illustrated in Table 6-8. The intra-class correlation t between paternal half sibs is computed as a ratio of the components of variance.

$$t = \frac{\sigma^2 s}{\sigma^2 s + \sigma^2 f + \sigma^2}$$

Since half sibs have only one-fourth of their genes in common, the covariance $\sigma^2 s$ among them represents $\sigma^2 g/4$. Thus heritability is computed by multiplying t by 4. The heritability value for back-fat thickness (Table 6-8) would be 0.32 on this basis ($t = 0.0204/0.0204 + 0.0919$ 0.1471 = 0.08).

In cattle, monozygotic twins have been used to determine the relative influence of heredity and environment on various traits. These one-egg twins have the same genetic composition. Differences between them

Fig. 7-6. A pair of monozygotic twins. Such twins arise from the same fertilized ovum and carry identical inheritance. Approximately 5 per cent of all twins in dairy cattle are monozygotic. (*Courtesy A. E. Freeman, Iowa State University.*)

should be of an environmental nature. Heritability values obtained from identical-twin studies, almost without exception, have been larger than those computed from field data. They usually approximate heritability in the broad sense, since all of the dominance and epistatic effects contribute to the similarity among monozygotic twins. Possible maternal and contemporary environmental factors also could serve to inflate these heritability values. As a consequence, heritability estimates from identical twin data are usually high, and they do not provide a true indication of the progress that can be expected from selection for a particular trait.

The parent-offspring and paternal half-sib resemblances are intended to provide an estimate of the narrow definition of heritability. Nevertheless, practically all of the estimates of heritability include some unavoidable environmental contribution to the resemblance among related animals. In addition to possible environmental contributions to the resemblance, some epistatic influences may contribute to the observed likeness. Generally this contribution is believed to be small, and we shall consider the estimates of heritability using parent-offspring and half-sib relationships as being indicative of the importance of the additively genetic fraction of the phenotypic variance.

Why Estimate Heritability? The usefulness of the concept of heritability will be illustrated in the succeeding chapters. Some of the reasons why it is an important concept can be profitably considered at this point.

When the breeding program consists of individual selection, with the most desirable animals being chosen as parents, the anticipated progress or change is the inheritable fraction (heritability in the narrow sense) of the superiority of the parents over the average of the population from which they were chosen. The intensity of selection determines the superiority of the parents, but only the portion of the superiority resulting from additively genetic differences is to be recovered on the average in the progeny. This can be represented as

$$\Delta M = h^2 SD$$

where ΔM is the expected change in the mean and SD is the superiority of the parents. Much more consideration of this point will be given in Chap. 10.

A reliable estimate of heritability is also needed to decide which breeding plan is likely to be most effective. When heritability is high for the desired characteristic, individual selection with little emphasis on pedigree, family performance, or progeny testing may permit most rapid improvement. If heritability is low, progress from straight selection may be slow enough to warrant emphasis on pedigree, family selection, or even the progeny test.

When estimates of heritability in both the broad and narrow concepts are available, the relative importance of additively genetic variance in comparison of epistatic and dominance variance can be appraised. If there is much epistatic and dominance variance, breeding systems which will capitalize on the specific genetic combinations are to be recommended. This may call for inbred-line formation if crosses among already existing breeds do not give practically satisfactory recovery of the potential hybrid vigor.

Partitioning the Hereditary Variance. In the consideration of hereditary variance, it was mentioned that the deviation of the phenotypic values from the additive representation was responsible for dominance deviations and dominance variance. The following partitioning of the variance at a single locus may be of interest to those desiring a mathematical expression of the additive and dominance variance. The relationship between the average breeding value and the phenotypic value can be expressed as the regression of phenotypic value on the number of the

Table 7-6. Genotypic Frequencies and Phenotypic Values in Relation to the Number of the Most Desired Alleles in the Genotype

Genotype	Frequency	Phenotypic values, Y	No. of desired genes, X
BB	q^2	a	2
Bb	$2q(1-q)$	d	1
bb	$(1-q)^2$	$-a$	0

$a = \dfrac{BB - bb}{2}$ one-half the difference between the two homozygotes.

d = phenotypic value for heterozygote, $d = a$ for complete dominance, and $d > a$ if overdominance is expressed.

most desired alleles in the genotype. Only additive and dominance effects and variances can be illustrated with this example involving a single locus.

The gene designated B is taken to be the desired one. The linear regression of phenotypic values Y on the number of desired genes X represents the average effect of substituting gene B for b. It can be derived as follows from the information in Table 7-6:

$$b_{YX} = \frac{\Sigma XY - (\Sigma X)(\Sigma Y)/n}{\Sigma X^2 - (\Sigma X)^2/n} = \frac{\Sigma xy}{\Sigma x^2}$$

where n is equal to 1.0 since the frequencies of the genotypes are expressed as proportions.

$$\Sigma XY = 2aq^2 + 2q(1-q)d$$
$$\Sigma X = 2q^2 + 2q(1-q) = 2q$$
$$\Sigma Y = aq^2 + 2q(1-q)d - a(1-q)^2 = a(2q-1) + 2q(1-q)d$$
$$\Sigma X^2 = 4q^2 + 2q(1-q)$$
$$\Sigma xy = 2q(1-q)a + 2q(1-q)d - 4q^2(1-q)d$$
$$\Sigma x^2 = 2q(1-q)$$
$$\Sigma y^2 = 2q(1-q)a + d - 2qd^2 + 4q^2(1-q)^2d^2$$

Then the regression coefficient

$$b_{YX} = \frac{2q(1-q)a + 2q(1-q)d - [2q(1-q)d]^2}{2q(1-q)}$$
$$= a + d(1-2q) = \alpha$$

The variance in phenotypic values Y accounted for by the linear regression of phenotypic values on the number of desired genes represents the additively genetic variance $\sigma^2 g$.

$$V_{reg} = \frac{(\Sigma xy)^2}{\Sigma x^2} = \frac{[2q(1 - q)(a + d - 2qd)]^2}{2q(1 - q)}$$
$$= 2q(1 - q)\alpha^2 = \sigma^2 g$$

The additional variance is termed the *dominance variance,* and it results because the phenotypic value of the heterozygote does not lie exactly intermediate between the values for the two homozygotes. This dominance variance turns out to be

$$\sigma^2 d = [2q(1 - q)d]^2$$

Since no environmental variance is included in this simple model,

$$\sigma^2 h = \sigma^2 g + \sigma^2 d = 2q(1 - q)\alpha^2 + 2q(1 - q)d^2$$

When many loci influence a trait, the contributions from the several individual loci are summed to give the total variance. This example involves only additive and dominance effects and no epistatic effects, since only a single locus is involved.

Drawing upon the example for black and red pigmentation in Holsteins (Fig. 7-7), a single locus representation can be given. Since dominance is complete, $a = d$ for the definition of phenotypic values in Table 7-6. Arbitrarily we will say that black and white individuals are 2 units

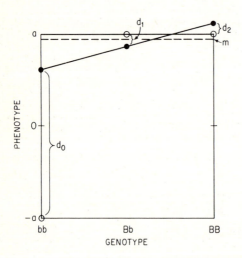

Fig. 7-7. Diagrammatic representation of the regression of phenotype on genotype for a single locus ($q = 0.93$).

Table 7-7. *Fractions of the Hereditary Variance Represented by the Additively Genetic and Dominance Variance for Varying Values of Gene Frequency for Complete Dominance*

Ratio of variances	Gene frequency q				
	0.10	0.30	0.50	0.70	0.90
$\sigma^2 g / \sigma^2 h$	0.95	0.82	0.67	0.46	0.18
$\sigma^2 d / \sigma^2 h$	0.05	0.18	0.33	0.54	0.82

superior to red and white individuals. Thus $a = d = 1.0$. The regression of Y on X then is

$$b_{YX} = a + d(1 - 2q) = 0.14 = \alpha$$

since q is 0.93. The total variance for this locus can be computed from the above expressions for $\sigma^2 g$ and $\sigma^2 h$, remembering that α is 0.14 and q is 0.93.

$$\sigma^2 h = \sigma^2 g + \sigma^2 d = 0.0026 + 0.0173 = 0.0199$$

For this gene frequency the dominance variance is larger than the additively genetic variance. The proportion of the total variance which the additively genetic and the dominance variances represent of the total hereditary variance differs as values of gene frequency change. In Table 7-7 this change in these proportions is illustrated for several values for gene frequency.

When dominance is complete and the frequency of the desired allele is high, less of the hereditary variance is additively genetic. This situation reflects the difficulty that would be encountered in making further progress once a high frequency of the desired gene(s) had been attained (Fig. 7-2).

SUMMARY

Population genetics seeks to relate genetics to a group or population in contrast to an individual. The genetics of a population is dependent upon the frequencies of the various genes in the population. Gene frequencies are influenced by mutation, migration, chance, and selection. Selection is the most important force available to the breeder for changing gene frequencies. The expected genetic change for a trait in a population is closely related to the nature of the hereditary variation. The total hereditary variance $\sigma^2 h$ has been partitioned into three portions, the additively genetic variance $\sigma^2 g$, the variance due to dominance deviations $\sigma^2 d$, and the variance due to epistatic deviations. The expected progress from selection for improvement of a trait is closely related to the proportion or ratio of the total phenotypic variance which is additively genetic. This ratio is termed *heritability*, and it also can be defined as the fraction of the difference between the parents and their generation mean that is expected to be transmitted to their progeny. Heritability for a trait can be estimated by several methods. Those which utilize the parent-offspring and paternal half-sib relationships are most widely used in animal populations. Variances due to dominance and epistatic deviations are much more difficult to estimate. When these kinds of hereditary variance are proportionately important, breeding systems requiring the crossing of various lines or strains are needed to obtain the specific genetic combinations which are most desirable.

REFERENCES

Books

Falconer, D. S. 1960. "Introduction to Quantitative Genetics," The Ronald Press Company, New York.

Fisher, R. A. 1930. "The Genetical Theory of Natural Selection," Oxford University Press, Fair Lawn, N.J.

Huxley, Julian. 1943. "Evolution: The Modern Synthesis," Harper & Row, Publishers, Incorporated, New York.

Lerner, I. M. 1950. "Population Genetics and Animal Improvement," Cambridge University Press, London.

Li, C. C. 1955. "Population Genetics," The University of Chicago Press, Chicago.

Lush, Jay L. 1945. "Animal Breeding Plans," 3d ed., The Iowa State University Press, Ames, Iowa.

Articles

Fisher, R. A. 1918. The Correlation between Relatives on the Supposition of Mendelian Inheritance, *Trans. Roy. Soc. Edinburgh*, **52**:399–433.

Lush, J. L. 1949. Heritability of Quantitative Characters in Farm Animals, *Proc. Eighth Intern. Congr. Genetics,* Hereditas, supplementary vol., pp. 356–375.

Wright, S. 1921. Systems of Mating, *Genetics*, **6**:111–178.

Wright, S. 1931. Evolution in Mendelian Populations, *Genetics*, **16**:97–159.

CHAPTER EIGHT

Inbreeding and Relationship

The forces which can produce changes in the frequency of the genes in a herd or population—selection, mutation, migration, and chance—were enumerated in our consideration of population genetics. Selection is the most important of these forces as a breeder seeks to develop animals to meet his goal. Some control over the distribution of the zygotes in a population can be exercised by the breeder's choice of a mating system. Mating individuals which are alike because of pedigree or phenotypic similarity tends to increase homozygosity. Conversely, mating individuals which are unlike, either from a pedigree or phenotypic standpoint, increases heterozygosity. In both cases, choosing mates on a pedigree basis has a stronger influence on the zygotic make-up.

When the breeding practices of most of the outstanding early breeders have been examined, it will be seen that some close matings, or inbreeding, have been practiced, although generally the inbreeding was mild. The genetic consequences of inbreeding were not fully understood at that time, but practical men gained an appreciation for it through their experiences.

INBREEDING

Inbreeding is a system of mating whereby the mates are more closely related than are average members of the breed or population being intermated. Previously we considered random mating, in which after selections are completed, no special effort is made to assign the particular mates. With inbreeding, the mates are chosen to have a common ancestral (pedigree) background or relationship. For very mild inbreeding, the parents may be related only as second cousins, but intense inbreeding may involve the mating of full brothers and sisters.

When related individuals are mated, the offspring tend to become more homozygous. This increase in homozygosity and accompanying decrease in heterozygosity is the underlying reason for the genotypic and phenotypic changes which are associated with inbreeding.

The increase in homozygosity can be most vividly illustrated by what happens in self-fertilization, the mating of an individual with itself. Many species of plants normally reproduce by self-fertilization, and many normally cross-fertilized species, such as corn, bear both male and female gametes on the same plant. These can be self-fertilized (selfed) by artificially putting pollen grains (bearers of male germ cells) on the female portion of the flower.

The expected increase in homozygosity for a single pair of genes when an individual which is heterozygous for this pair of genes is selfed is shown in Table 8-1. Selfing of a *Bb* individual is equivalent to a *Bb* × *Bb* mating and produces on the average ¼ *BB*, ½ *Bb*, and ¼ *bb* offspring. In subsequent generations, selfing of *BB* individuals (*BB* × *BB*) will produce only *BB* offspring. Selfing *bb* individuals will likewise produce only *bb* individuals. Continued self-fertilization removes genes from the *Bb* group and does not return any. In a few generations, the population is virtually all of the *BB* or *bb* types.

This same process occurs with inbreeding in animals. However, the increase in homozygosity takes place much more slowly, since inbreeding cannot be as intense as selfing. Even with full-brother-sister matings, the most intense type of inbreeding possible with animals, the increase in homozygosity is less than half as rapid as with self-fertilization in plants.

Table 8-1. Increase in Homozygosity in the Descendants Produced by Continued Self-fertilization from a Single Heterozygous Individual

	Proportion of each genotype				
Generation	*BB*	*Bb*	*bb*	*Per cent homozygous*	*Inbreeding coefficient*
1	1	2	1	50.0	0
2	3	2	3	75.0	0.500
3	7	2	7	87.5	0.750
4	15	2	15	93.8	0.875
5	31	2	31	96.9	0.938
6	63	2	63	98.4	0.969
n	$2^n - 1$	2	$2^n - 1$	$1 - (\frac{1}{2})^n$	$1 - (\frac{1}{2})^{n-1}$

DEGREE OF INBREEDING

Although certain of the consequences of inbreeding were realized during the early period of breed formation, a serious handicap existed in that there was no realistic method for expressing the degree or intensity of inbreeding. In laboratory animals, such as mice, the amount of in-breeding was expressed in terms of the number of generations of full-brother-sister mating. In plant species, inbreeding could be reckoned on the basis of the number of generations of selfing that had occurred.

Although many attempts were made to provide a measure for the intensity of inbreeding, the coefficient of inbreeding F proposed by Wright[1] in 1921 continues in general usage. Even before considering the definition of the inbreeding coefficient, we should reemphasize that the underlying genetic consequence of inbreeding is to increase homozygosity. As related individuals are mated, genes which trace to ancestors common to each parent are recombined in the resulting offspring. Bringing together the two samples of genes (gametes) which arise at least in part from an earlier common gene pool increases the opportunity for loci in the offspring to become homozygous. Thus it appears logical that the measure of the intensity of inbreeding should reflect this probable increase in homozygosity.

Coefficient of Inbreeding. The coefficient of inbreeding represents the probable increase of homozygosity resulting from the mating of individuals more closely related than the average for the population. It may range in value from 0 to 1.0. It also is often referred to as a percentage ranging from 0 to 100. The intensity of inbreeding as measured by the inbreeding coefficient is relative to a particular breed or population at a specified time. The inbreeding coefficient then represents the probable increase in homozygosity which has occurred as a result of the mating of related individuals in the population since the reference date. Conversely, as the value of F increases, the relative proportion of heterozygous, again measured from the specific base, declines a proportionate amount equal to $1 - F$.

This increase in homozygosity expressed by the inbreeding coefficient is a most probable result. Many loci would be involved, and by chance either a larger or smaller proportion of the loci might have become homozygous than the computed value of F indicates. Even though two individuals from a line may have the same inbreeding coefficients, they would not necessarily have become homozygous for the same loci. The expected proportion of the loci that would have become homozygous

[1] Wright, S., *Amer. Nat.*, **56**:330–338, 1921.

would be the same for the two individuals. There are two alternatives at each locus, and F represents the proportion of loci originally heterozygous that have become homozygous. The second alternative is that the loci are still heterozygous, and $1 - F$ represents the probable proportion of such loci. Thus the standard deviation for a computed value of F would be $\sqrt{F(1 - F)/n}$, where n would be the effective number of independently segregating loci in the population at the base time.

When viewed from the standpoint of an individual locus, the inbreeding coefficient represents the probability that a random locus originally heterozygous has become homozygous following inbreeding. Reasoning in terms of probability leads to a very useful concept, and it is the basis of the definition of inbreeding that was introduced by Malecot.[1] In an inbred population, an individual may possess two genes at a locus that are the same (homozygous). The homozygosity which is exhibited may basically arise from two sources. In one situation, homozygosity may be present because the two genes are *alike in state;* that is, they may be randomly drawn from the same population and both happen to be either B or b. Such would be the case in random mating where $q = 0.5$. In accord with the Hardy-Weinburg law, the probability that these genes are alike in state would be 0.50, or $0.25AA + 0.25aa$. With inbreeding there is an additional probability of homozygosity because the two genes arise from the direct replication of the same gene from a common ancestor in an earlier generation. Two such genes are said to be *identical by descent.* The additional homozygosity contributed by gene pairs which are identical by descent represents the impact of inbreeding. Thus the inbreeding coefficient also can be defined as the probability that the two alleles at a locus in an individual are identical by descent.

Calculation of Inbreeding Coefficient. In order to illustrate the calculation of the inbreeding coefficient, consider the pedigree in Fig. 8-1. We are interested in determining the inbreeding coefficient of individual X. We desire to determine the probability that X received genes that were identical by descent from D, transmitted through both B and C. In other words, what is the probable proportion of loci in X that are homozygous because X received two replicates of a gene at these loci from D? Consider B and C. The probability that they received replicates of the same gene at a given locus from D is $\frac{1}{2}$, and the probability that they received different genes is also $\frac{1}{2}$ when D is not inbred. If D is inbred, the probability that both B and C will receive replicates of identical genes from D becomes $\frac{1}{2}(1 + F_D)$ rather than

[1] Malecot, G., "Les Mathematiques de l'heredité," Masson et Cie, Paris, 1948.

Table 8-2. Exponential Values of ½

$(\frac{1}{2})^2 = 0.250000$	$(\frac{1}{2})^6 = 0.015625$
$(\frac{1}{2})^3 = 0.125000$	$(\frac{1}{2})^7 = 0.007812$
$(\frac{1}{2})^4 = 0.062500$	$(\frac{1}{2})^8 = 0.003906$
$(\frac{1}{2})^5 = 0.031250$	$(\frac{1}{2})^9 = 0.001953$

merely ½. Considering the remainder of the paths, the probability that B passes the gene it got from D to X is ½, and the probability that C passes the gene it got from D on to X is also ½. Obtaining the product of these probabilities (paths), we find that the probability that X received replicates of a specific gene from D is

$$(\tfrac{1}{2})(\tfrac{1}{2})(\tfrac{1}{2})(1 + F_D) = (\tfrac{1}{2})^{n+1}(1 + F_D) = F_X$$

when n represents the number of segregations included in the chain of inheritance from the two parents of X traced through the common ancestor.[1,2]

In the above example, $n = 2$ (1 for D to B and 1 for D to C). If the common ancestor is non-inbred ($F = 0$), then $F_X = (\frac{1}{2})^3 = 0.125$. This can be interpreted as meaning that one-eighth of the loci in individual X are probably homozygous because X received replicates of genes from D. The pedigree in Fig. 8-1 represents the mating of paternal half sibs; hence, only one common ancestor and only one path connects B and C. The complete expression for the inbreeding coefficient must take into account the possibility that there may be more than one common ancestor and that more than a single path may connect the sire and the dam of the individual X whose inbreeding coefficient is being determined. This complete formula can be expressed

$$F_X = \Sigma(\tfrac{1}{2})^{n+1}(1 + F_A)$$

where F_X = inbreeding coefficient of individual X

Σ = summation of all independent paths of inheritance which connect sire and dam of X

n = number of segregations in a specific path between sire and dam of X

F_A = inbreeding coefficient of common ancestor for each path

---

[1] One also can consider n to represent the sum of the number of generations from the sire to the common ancestor plus the number of generations from the dam of the individual to the common ancestor.

[2] The word *common* as used in this connection merely means *shared* and in no sense implies that the ancestor was ordinary or inferior.

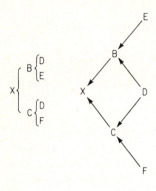

Fig. 8-1. Bracket- and arrow-style pedigrees of an animal (X) produced by mating of a half-brother and half-sister.

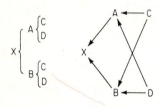

Fig. 8-2. Bracket- and arrow-style pedigrees of an animal resulting from a mating of full brother and full sister.

In the more complicated pedigrees, it is often convenient to set up the pedigree in the *arrow* style rather than in the conventional bracket form if relationships are being studied. In the arrow style, each common ancestor is included only once, with lines drawn to each of his (or her) offspring in the pedigree. These lines represent the paths of inheritance by which genes are transmitted.

Full Brothers and Sisters. Matings of full brothers and sisters represent a more intense system of inbreeding than does the half-sib mating in Fig. 8-1. The pedigree in both bracket and arrow styles of an individual resulting from a full-brother-sister mating is given in Fig. 8-2. Here we have two common ancestors and two independent paths of inheritance to evaluate. The calculations required to compute F, assuming C and D are non-inbred, are summarized in Table 8-3.

Table 8-3. Calculation of Inbreeding of Individual X in Fig. 8-2

Common ancestor	n	Contributions
C	2	$(\tfrac{1}{2})^{2+1} = (\tfrac{1}{2})^3 = 0.125$
C	2	$(\tfrac{1}{2})^{2+1} = (\tfrac{1}{2})^3 = 0.125$
		Sum $= \overline{0.25}$*

*$F_X = 25$ per cent.

Table 8-4. Inbreeding of R in Fig. 8-3

Common ancestor	n	Contribution
T	1	$(\frac{1}{2})^{1+1} = 0.25*$

* $F_R = 25$ per cent.

This is as intense a form of inbreeding as is possible in mammals. The inbreeding coefficient of X is 0.25. Parent-offspring matings are equal to full-brother-sister matings in intensity of inbreeding. This is illustrated in Fig. 8-3 and Table 8-4. The only common ancestor is T, and $F_R = 0.25$.

Further Example of Inbreeding. The above examples were kept very simple in order to illustrate the principles involved in computing inbreeding coefficients. The principles apply equally well to more complicated pedigrees. As we scan Fig. 8-4 we see that animals B, E, and F occur on both sides (top and bottom) of it, and we also note that B is himself $12\frac{1}{2}$ per cent inbred, since he is the result of a half-brother-sister mating. Table 8-5 shows the calculations necessary to compute the inbreeding of A as 35.9 per cent.

Fig. 8-3. Bracket- and arrow-style pedigrees of an animal resulting from the mating of parent and offspring.

Fig. 8-4. Bracket- and arrow-style pedigrees of an inbred animal also having an inbred common ancestor.

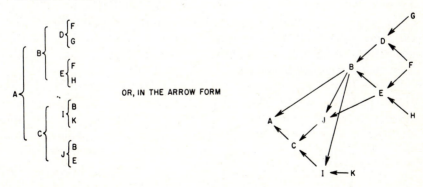

OR, IN THE ARROW FORM

Table 8-5. Calculation of Inbreeding of A in Fig. 8-4

Common ancestor	n	$1 + F_B$	Contributions
B (as sire of A and I)	2	1.125	$(\frac{1}{2})^3(1.125) = 0.1406$
B (as sire of A and J)	2	1.125	$(\frac{1}{2})^3(1.125) = 0.1406$
E (as dam of B and J)	3	. . . . .	$(\frac{1}{2})^4 = 0.0625$
F (as sire of D and E)	5	. . . . .	$(\frac{1}{2})^6 = 0.0156$
			Sum $= 0.3593^*$

$^* F_A = 35.9$ per cent.

In Table 8-4, B is listed twice because the same gene or genes from B might have been transmitted to A directly and also indirectly by either of two routes—through J or through I. Segregation of allelic genes occurs in the formation of each germ cell, so these two routes represent independent chances for A to receive B's genes in homozygous form.

RELATIONSHIP

Being related means that two individuals have one or more common ancestors. Actually any two animals in a breed are usually related in this sense. If the pedigrees of a pair of Holsteins, Herefords, Hampshires, or Hackneys were traced far enough back, each pair would have some common ancestors.

With regard to the human family, we might be surprised at the duplications in each of our pedigrees if we could trace them back to the time of Christ. Allowing an average of 28 years per generation, we find that we are more than 70 generations removed from that historic period. Since the number of ancestors doubles at each generation, there would be a total of 2^{70}, or about 1,200,000,000,000,000,000, individuals in each of our pedigrees in the single generation around 1 A.D. It is obvious that that many people were not alive at that time (there are more people in the world now than ever before, nearly 3 billion). Thus, the same individuals must have occurred many times in that and other generations of each of our pedigrees. All our pedigrees would doubtless show many ancestors in common.

Obviously, too, in a relatively few generations the number of ancestors in the pedigree of any farm animal is greater than the total number of animals which were alive in the breed at that time. For example, the twentieth generation of a pedigree contains over 2 million individuals. Thus, in a broad sense, all the animals in a breed are related. In farm animals, however, we use the term *related* in a more restricted sense to mean that the animals mated are more closely related than average animals of their breed. This usually means that there are common ancestors at least in the first four to six generations of their pedigrees. If two animals had an ancestor in common in the tenth generation, that common ancestor's inheritance would have been halved ten times in getting down through the ten generations to each of the animals in question. Obviously after ten halvings of that remote ancestor's heredity, the two animals would have little genetic relationship because of the common remote ancestor. But if, for example, the shared ancestor is only two generations removed, his inheritance has been halved only twice in getting to each of the two related individuals.

Measurement of Relationship. The coefficient for expressing the degree of relationship between two individuals also was developed by Wright. It measures the probable proportion of genes that are the same for two individuals due to their common ancestry, over and above that in the base population.

Parent-offspring relationships are the simplest. They are fundamental to all other degrees of relationships as these represent combinations of several parent-offspring relationships. Since half the genes of any animal come from his sire and half from his dam, any offspring is 50 per cent related to each parent. Since each parent in turn received half his genes from his sire and half from his dam, and since a sample half is transmitted to each offspring, on the average 25 per cent of the genes of any animal originally came from each grandparent. Thus, on the average, an animal is 25 per cent related to each of his grandparents. Again it should be kept in mind that the relationship between two individuals is the extra similarity in the genes they possess due to their common ancestry. Many of their genes will already be alike because of the high frequency of these genes in the population (breed).

The key to measuring relationship is the number of generations between the two animals being studied and their common ancestor or ancestors. The first step in computing a relationship coefficient consists of counting the number of generations intervening between some common ancestor and the two descendants in question.

In Fig. 8-5, animal C is a grandsire of both A and Z. In other words, A and Z are more closely related than average animals of their breed because they have an ancestor in common close-up in their pedigrees. To find their degree of relationship (because of common grandsire C), we count the generations from C to A, which is 2, and from C to Z, also 2. Heredity is a halving process, so we see that C's inheritance has been halved twice in getting to A and twice in getting to Z. In short, it has been halved a total of 4 times. So the relationship of A

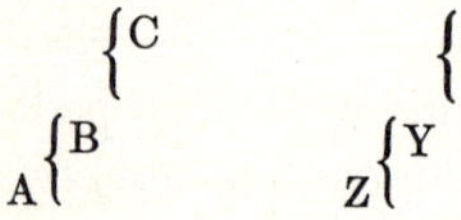

$$\text{A}\begin{cases} \text{B}\begin{cases} \text{C} \\ \\ \end{cases} \end{cases} \qquad \text{Z}\begin{cases} \text{Y}\begin{cases} \text{C} \\ \\ \end{cases} \end{cases}$$

Fig. 8-5. Pedigrees of animals with common grandsire.

and Z is C's inheritance halved 4 times, or $(\frac{1}{2})^4$, which is $(\frac{1}{2})$ $(\frac{1}{2})$ $(\frac{1}{2})$ $(\frac{1}{2})$, or $\frac{1}{16}$, or $6\frac{1}{4}$ per cent. The relationship between A and Z is $6\frac{1}{4}$ per cent. This simply means that about 6 per cent more of A's and Z's genes are the same than would be the case with average animals of their breed. This is true since they both received genes from the same grandsire C. This is expressed more concisely as follows:

$$R_{\text{AZ}} = (\tfrac{1}{2})^4$$

The foregoing example involved the relationship between animals that are related because they are descendants of some of the same animals. A and Z were single first cousins, since they had one grandparent in common. The relationship of animals (or people) with two grandparents in common is $12\frac{1}{2}$ per cent. These animals are termed *single first cousins*. Such relationships are called *collateral*. The other possible type of relationship occurs between individuals when one is a descendant of the other. This is called *direct relationship* as in parent and offspring.

Calculation of the relationship of X and Y (Fig. 8-6) is somewhat more complicated than in the previous examples because X and Y have two common ancestors, S and D. When there are two or more common ancestors, the contributions of each are added to arrive at the complete coefficient of relationship. In such cases it is convenient to set up the calculations as in Table 8-6.

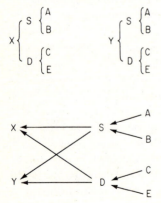

Fig. 8-6. Bracket- and arrow-style pedigrees for two full sibs (X and Y).

Table 8-6. Relationship of X and Y from Fig. 8-6

Common ancestor	n	Contributions
S	2	$(\frac{1}{2})^2 = (\frac{1}{2})^2 = 0.25$
D	2	$(\frac{1}{2})^2 = (\frac{1}{2})^2 = 0.25$
		Sum $= 0.50$

Thus X and Y have 0.5 of their genes in common and are related 50 per cent. You can readily calculate the relationship of half brothers as 25 per cent. In the above example you may wonder why A, B, C, and E were not considered as common ancestors of X and Y. The answer is that they make their contributions only through S or D, not through both. For example, the only way A is an ancestor of both X and Y is through S, and by including S as a common ancestor, we automatically take care of A's contributions. If A had been an ancestor of both S and D, we would have included him as a common ancestor, since he could have contributed the same genes to X and Y through both S and D.

Coefficient of Relationship. Because inbreeding increases homozygosity, an inbred animal will transmit similar genes to each of his offspring more frequently than will a non-inbred individual. If an inbred animal is the common ancestor of two related individuals, they will therefore have more genes in common and thus be more highly related than if the common ancestor had not been inbred. To take care of this, the contribution of each inbred common ancestor must be multiplied by $1 + F_A$.

Inbreeding also makes a population more variable by producing separate inbred strains. Inbred descendants of any animal will be homozygous in a greater percentage of their gene pairs than if they were not inbred, but they may be homozygous for different alleles of the same gene pair and thus less related than if they were not inbred. The denominator for the relationship formula takes this into account, making the complete formula:

$$R_{XY} = \frac{\Sigma(\frac{1}{2})^n (1 + F_A)}{\sqrt{(1 + F_X)(1 + F_Y)}}$$

This correction makes the relationship coefficient a measure of the degree to which the genotypes of related animals are similar rather than leave it in terms of the proportion of genes from a common source.

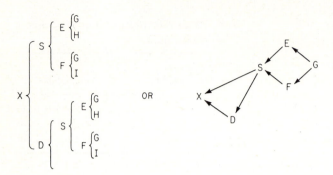

Fig. 8-7. Pedigree where inbreeding of the two related individuals must be taken into account.

As an example of the use of this formula, consider the pedigree shown in Fig. 8-7 of an animal produced by mating an inbred sire to his daughter. S is 12½ per cent inbred since he is the product of a half-brother-sister mating.

From Table 8-8 the value of the numerator of the relationship coefficient is 0.84375. The complete relationship is

$$R_{XS} = \frac{0.84375}{\sqrt{(1 + F_X)(1 + F_S)}}$$

$$R_{XS} = \frac{0.84375}{\sqrt{1.28125)(1.12500)}}$$

$$R_{XS} = 0.7028$$

In the above parent-offspring mating, it is most likely that 75 per cent of the genes of X came from S. Since X is more highly inbred than S, however, X will probably have some homozygous gene pairs that were heterozygous in S, thus reducing the similarity of their genotypes.

There has been so little inbreeding in most of our farm animals that the denominator of the relationship coefficient seldom is very much larger than 1 and can therefore usually be omitted.

In the original derivation of the relationship coefficient by Wright, the path-coefficient approach was used. Accordingly the relationship

Table 8-7. *Calculation of Inbreeding of X in Fig. 8-7*

Common ancestor	n	n'	$1 + F_A$	Contribution
S	0	1	1.125	$(\frac{1}{2})^2(1.125) = 0.28125$

Table 8-8. Calculation of Relationship of X to S

Common ancestor	n	n'	$1 + F_A$	Contributions
S	0	1	1.125	$(\frac{1}{2})(1.125) = 0.5625$
S	0	2	1.125	$(\frac{1}{2})^2(1.125) = 0.28125$
				Sum $= 0.84375$

coefficient was derived as the correlation between the breeding values of two individuals. This concept is still expressed in the literature. A higher relationship (correlation) between two individuals reflects a higher likelihood that the two individuals have the same genes.

IRREGULAR INBREEDING SYSTEMS

The rate of increase in homozygosity with inbreeding is dependent on the closeness of the relationship of the individuals which are mated. Previously it was stated that full-brother-sister matings are the closest that can be made in mammals. The mating of sire to daughter or son to dam gives the same rate of increase in homozygosis. However, it is much more difficult to carry out these latter schemes than the full-brother-sister mating for an extended time. Expressions have been developed to compute the inbreeding coefficient for such irregular inbreeding systems. The inbreeding coefficients for successive generations of full-brother-sister mating are given in Table 8-9. Note that the increase in the inbreeding coefficients is much less rapid than for self-fertilization. Actually about six generations of brother-sister mating is required to attain the degree of inbreeding attained in two generations of selfing.

A plan of inbreeding that has been useful in the formation of families with a high relationship has been to mate a sire to his daughters, as shown in Fig. 8-8. Such a mating plan would start first with the choice of a desirable male and a number of his female progeny. With frozen semen now available in cattle, this plan for inbreeding could be continued for several generations before a new sire would need to be chosen. The main advantage of such a plan is that the inbreeding can be increased rapidly to above 0.40 in three generations. Following these early generations, the system becomes more like a one-sire herd or line. This is especially so when a new foundation sire (son) must be chosen to head up the line.

Such extreme inbreeding is largely relegated to experimental studies. In practical breeding with livestock, the breeding plan cannot usually

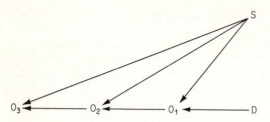

Fig. 8-8. Diagram of inbreeding resulting from the mating of a sire back to his offspring in successive generations. This mating plan allows for a comparatively rapid increase in inbreeding in cattle. Frozen semen would extend the serviceable line of the sire (S).

be specified regularly over an extended period. More commonly a herd may be closed to outside introductions and a limited number of sires placed in service. Often the breeding population is described in terms of a one-, two-, or three-sire herd. On occasion a breeder may desire to close his herd, yet he would like to keep the loss of heterozygosity to what might be termed a "reasonable level."

Wright[1] has shown that approximately $\frac{1}{8}N_m + \frac{1}{8}N_f$ of the heterozygosity present in a closed population would be lost each generation. The number of males used each generation is represented by N_m, and N_f represents the number of females in the breeding herd. In a herd of 3 sires and 100 females, $\frac{1}{24} + \frac{1}{800}$ of the loci that are heterozygous would be expected to have become homozygous each generation. Since the value for N_f will almost invariably be many times larger than N_m, the $\frac{1}{8}N_f$ term can be neglected with little error. Using the expression $\frac{1}{8}N_m$, the loss of heterozygosis (gain in homozygosis) for one-sire, two-sire, and three-sire herds is approximately 12, 6, and 4 per cent respectively per generation. The expected increase in the inbreeding for successive generations in such populations is given in Table 8-9.

The above values are based on random mating within the herd. With a one-sire herd, inbreeding cannot be avoided, but when several sires are used, the inbreeding could be reduced some by mating females from one sire line to males from another sire line. When the mating plan is devised to give the maximum avoidance of inbreeding, the rate of increase in homozygosity is reduced to approximately one-half that given by the formula. Consequently, the closing of a herd does not automatically mean that a high level of inbreeding will develop. With several sires, a herd can be closed for many generations before the average inbreeding reaches a high level.

[1] Wright, Sewall, *Genetics*, 16:107–111, 1931.

Table 8-9. Inbreeding Coefficients in Various Generations of Different Intensities of Inbreeding

Generation	Selfing	Full-sibbing	Sire back to off-spring*	One-sire herd	Two-sire herd	Three-sire herd
1	0.500	0.250	0.250	0.125	0.066	0.042
2	0.750	0.375	0.375	0.218	0.128	0.082
3	0.875	0.500	0.438	0.304	0.185	0.119
4	0.938	0.594	0.469	0.380	0.239	0.155
5	0.969	0.672	0.484	0.448	0.289	0.189

* See Fig. 8-8.

Robert Bakewell, the great pioneer English breeder of Longhorn cattle, Leicester sheep, and Shire horses, is reported to have inbred intensively with apparent good results. Culley, writing in 1794, said:

Mr. Bakewell has not had a cross for upwards of 20 years, his best stock has been inbred by the nearest affinities, yet they have not decreased in size, neither are they less hardy, or more liable to disorder, but, on the contrary, they have kept in a progressive state of improvement.

Fig. 8-9. Comet (155). A Shorthorn bull bred by Thomas Bates and reported to be the first bull to have sold for as much as $5,000. His pedigree (Fig. 8-10) indicates that he was highly inbred. (*From Sanders, Shorthorn Cattle, The Breeder's Gazette.*)

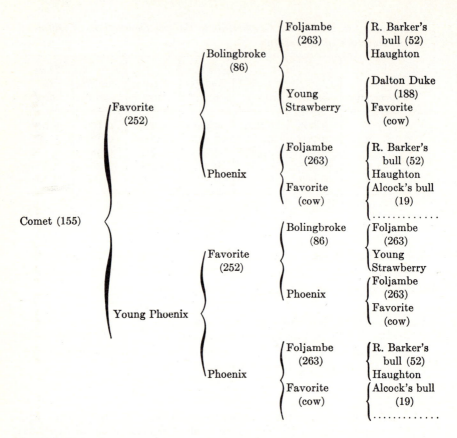

Fig. 8-10. Pedigree of the Shorthorn bull Comet. His inbreeding coefficient is 46.87, and that of his sire (Favorite 252) is 18.75.

Although many highly inbred animals undoubtedly were developed by Bakewell, he is also reported to have loaned out many sires for possible later use in his herd. Thus even though he had not "had a cross for upwards of 20 years," the average inbreeding coefficients need not have become excessively high.

GENETIC EFFECTS OF INBREEDING

Inbreeding increases the proportion of gene pairs that are homozygous and decreases the proportion that are heterozygous. Inbreeding itself does not change gene frequency, but in small populations some alleles may be lost as inbreeding progresses. In a small population the gene frequency would fluctuate rather extremely and by chance certain genes would be lost as others become fixed (homozygosis). In large populations the fluctuations of gene frequency would be less extreme and few of the alleles would be lost. As inbreeding progressed without selection, the large population would become a series of subpopulations, each likely to be homozygous for a number of different alleles.

Assumptions. Certain conditions were accepted in the development of the coefficients of inbreeding and relationship that we should clarify. The expressions were derived on the assumptions that mutation and selection were not important in the population. Such an assumption for mutation may be practically satisfactory over the few generations normally involved in the compilation of the inbreeding coefficients of farm animals. Available evidence suggests that mutation rates are low; hence, few loci expected to be homozygous from inbreeding would be rendered heterozygous by mutation.

Acceptance of the assumption that selection does not measurably influence the degree of homozygosity expressed by the inbreeding coefficient may be more difficult. Natural selection is at work at all times in all populations. Certain of the weaker, less-vigorous animals die in every herd or flock.

Just what effect selection has on the degree of homozygosity in a specific situation is problematical. A number of situations have been reported where heterozygosity in inbred lines is apparently greater than would be expected from the computed inbreeding coefficients. These cases point up the possibility that selection during inbreeding favored the heterozygotes over either homozygote.[1] How frequently heterozygote superiority is manifested is not known. It would likely be more frequent for traits influencing reproductive capacity.

Since inbreeding on the average does tend to depress performance, and since there may be heterozygote superiority, it is probable that selection for performance traits during inbreeding favors animals more heterozygous than average. This selection process, continued over a period of several generations, might result in substantially more heterozygosity in an inbred line than inbreeding coefficients would indicate if (1) gene pairs with major effects exhibited heterozygote superiority and (2) the inbreeding system was a mild one, permitting selection on traits other than reproductive fitness.

Basis for Inbreeding Effects. What accounts for the depression of performance which accompanies inbreeding? If all the gene effects are additive, inbreeding would not be expected to produce a reduction in mean performance. Alleles are made homozygous without regard to their effects, and it seems that the more desirable and the less desirable gene

---

[1] Shultz, F. T., and Briles, W. E., *Genetics*, **38**:34–50, 1953.

effects would tend to balance. On the contrary, if some degree of dominance is expressed, an increase in homozygosity would produce a decline in average merit, since the proportion of homozygous recessive loci would increase. Thus, the hypothesis of dominance of favorable genes was advanced early in the consideration of inbreeding.

Another possible genetic reason for the harmful effects of inbreeding is that, in at least some cases, the heterozygote may be better than either homozygote. For example, a pair of genes *B* and *b* might have effects such that *BB* is better than *bb* but that *Bb* is better than either. Lerner[1] reviewed the experimental evidence and made a strong case for the existence of genetic mechanisms such that animals heterozygous for either single gene pairs or gene complexes have superior fitness and performance characteristics. The term *overdominance* has been applied to describe this phenomenon.

If an appreciable fraction of the gene pairs have this effect, the inbreds lack vigor and productiveness because of homozygosity itself. Development of highly productive inbreds would therefore be impossible, and the ultimate in productive ability could occur only in crosses or hybrids produced by breeding systems giving maximum heterozygosity.

Either of the above theories would fit observed results to date reasonably well. The most probable situation would appear to be that both hypotheses are partially correct. Certainly inbreeding brings many recessives to light. In many of these there has never been any suggestion of the heterozygote having a selective advantage. This indicates that the dominance hypothesis is important. Neither can the apparent evidences of overdominance in some cases be overlooked. Further work will doubtless clarify the situation. Obviously the genetic basis for inbreeding depression and heterosis are closely related, since the same loci are involved. Those traits which show a marked inbreeding depression also exhibit heterosis when genetically diverse stocks are crossed.

PHENOTYPIC EFFECTS OF INBREEDING

The outward effects of inbreeding were noted and recorded long before the experimental approach was undertaken. Darwin's statement, "The consequences of close interbreeding carried on for too long a time are, as is generally believed, loss of size, constitutional vigor and fertility sometimes accompanied by a tendency to malformation"[2] is pertinent

[1] Lerner, I. Michael, "Genetic Homeostasis," John Wiley & Sons, Inc., New York, 1954.
[2] Darwin, C., "The Variation of Plants and Animals under Domestication," London, 1868.

today. For many years the belief persisted that the detrimental effects of inbreeding resulted from inbreeding per se. This viewpoint was corrected when the Mendelian nature of inheritance was elaborated. Detrimental effects were a consequence of inbreeding, but inbreeding itself was not the culprit. It merely uncovered the latent tendencies and permitted them to express themselves.

During the past 20 to 25 years, a large amount of experimental work has been reported on inbreeding effects in swine and poultry. Lesser amounts have been reported in cattle and sheep. Evidence from these species is reinforced by a large amount of experimental work involving the inbreeding of laboratory mice, rats, and guinea pigs.

Effects on Growth Rates. Although the growth rates of certain laboratory species are apparently little affected by inbreeding, the average effect (and this is what a breeder must assume he will obtain in embarking on a particular breeding system) is for inbreeding to depress growth rate. The average reduction is remarkably similar for the various classes of farm mammals. The reductions are definitely large enough to discourage the routine use of even moderate inbreeding in commercial herds and flocks where the margin of profit may be small and dependent in large part on live-weight production per animal unit.

Effects on Reproductive Performance. With the exception of a few cases in laboratory animals and a few inbred lines of swine, the results of inbreeding experiments are unanimous in indicating a reduction in reproductive efficiency with inbreeding. As a matter of fact, an extremely high percentage of the inbred lines started have been discarded because of low reproductive performance.

Efforts to find the physiologic bases for reduced reproductive efficiency have suggested that in at least some lines inbreeding slows testicular development, delays puberty in both sexes, reduces the number of ova shed by the females, and increases early embryonic death rates. In spite of the rather drastic average effects of inbreeding on reproductive performance certain inbred lines have performed reasonably well.

Effects of Inbreeding on Vigor. Vigor is difficult to define in quantitative terms, but observation and objective evidence indicate that inbreeding often lowers vitality. Some inbred lines showed major increases in death rates with increased inbreeding. Death losses in other lines were little affected. The observation of most people who have worked with inbred farm animals is that they are much more susceptible to critical environmental conditions than are non-inbred animals.

In general terms, it appears that reductions in reproductive performance together with increased mortality are perhaps the most striking effects of inbreeding. These effects are by no means universal in inbreds, but they are consistent enough to make the continuation of many inbred lines difficult, impossible, or at least economically impracticable, even if animals of the line produce superior progeny when used in outcrosses. The average effects of inbreeding on reproductive performance indicate strongly that the commercial producer cannot afford to follow an inbreeding program.

Appearance of Hereditary Lethals or Other Abnormalities with Inbreeding. In many inbreeding experiments, hereditary abnormalities or lethal genes have appeared. Such traits are almost always recessive in inheritance. Genes for them may be present in low frequencies in outbred herds but remain hidden and unsuspected, usually or always being covered by their dominant alleles. When inbreeding occurs, the probability of these genes appearing in homozygous form is increased, just as the probability of all genes being homozygous is increased. Inbreeding does not create such factors; it merely permits them to be expressed and identified.

USEFULNESS OF INBREEDING

In spite of the occasional good results obtained from inbreeding, it is becoming increasingly apparent that the development of highly productive inbreds in our farm animals is not likely within the foreseeable future. Rather, inbreeding must find its usefulness as an aid in producing seedstock which can be used with predictable results as parents for outbred or crossbred commercial animals. The use of inbreds for commercial production is being exploited more widely in poultry than with other farm animals.

Linebreeding. Most breeders, as well as commercial producers, are wary of intense inbreeding, but they are often willing to undertake mild inbreeding to maintain a high relationship to a supposedly outstanding ancestor. Although linebreeding is inbreeding in the fundamental sense, its primary purpose is not to increase homozygosity but to retain a goodly proportion of the genes of the designated individual. In addition to trying to retain desired genes and gene combinations, linebreeding is often used when there is a high likelihood of reducing the merit of the herd when outside sires are introduced. Thus only breeders with superior herds can really justify linebreeding.

The structure of a linebreeding program can take a variety of forms.

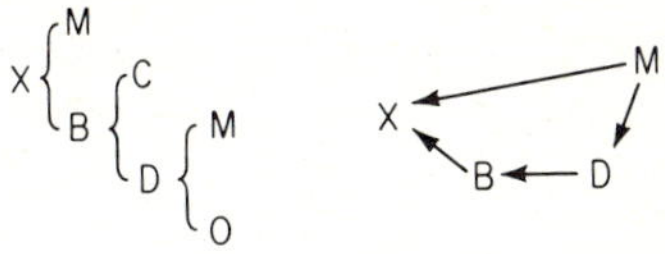

Fig. 8-11. Bracket- and arrow-style pedigrees of a common linebred mating, sire to grand-daughter.

A plan based on maintaining a high relationship to an outstanding sire is practically most workable. Males can leave many more offspring than females, and the use of artificial insemination and frozen semen can extend their useful life. Occasional examples of successful linebreeding to females have been recorded. Genetically their influence has been carried forward through their sons and grandsons.

Disagreement over terminology arises because some choose to define linebreeding merely as the mating of distantly related animals, regardless of whether it is directed toward a single favored ancestor or not. Actually closebreeding and linebreeding are varieties of inbreeding and will be so considered here. Some matings cannot be satisfactorily classified. Mating of sire to daughter (or conversely, son to dam) is one of the closest forms of inbreeding and is therefore classed as closebreeding. It is also the most effective type of mating for concentrating the hereditary material of the sire and might logically be called *linebreeding,* or perhaps *close linebreeding.* A pedigree which illustrates a common type of linebreeding is shown in Fig. 8-11.

Some of the key points regarding linebreeding are:

1. Linebreeding should be practiced only in better-than-average herds, in which the quality is high enough to make it difficult or impossible to obtain sires outside the herd with a reasonable likelihood that they will be as good as home-raised sires. The use of home-raised sires often has the additional advantage of having the prospective sires' performance evaluated in the same herd environment his later progeny will experience.
2. Linebreeding is justified only in herds whose owner is well informed regarding both its possibilities and pitfalls.
3. Linebreeding is usually not feasible unless the herd uses at least two sires. Otherwise inbreeding will rise to dangerous levels in a few years.
4. Linebreeding with inferior animals merely as an economy measure should be strongly discouraged.

Elimination of Undesirable Recessives. At one stage in corn improvement it was thought that inbreeding could be used to remove deleterious genes from the breeding stocks. Then the stocks could be recombined and selection could progress without the hindrance of the deleterious genes. Such an approach did not gain acceptance, but it is true that stocks which have undergone inbreeding should have less likelihood of transmitting these undesirable alleles. The practice of mating a sire to 15 to 20 of his daughters is used in some situations to test for the presence of deleterious genes which are expressed early in life. Suspected carriers of lethal or semilethal genes must be tested in some manner before being used extensively. When the sire is mated to his daughters rather than to specific tester animals, all of the deleterious genes rather than only those which the special tester carries have a chance to be expressed. This is a rather drastic testing procedure since the inbred offspring ($F = 0.25$) could be a distinct burden even though no lethals or sublethals were detected.

Development of Families.   Inbreeding tends to develop distinct lines or families as the breeding plan continues. If family relationships in excess of 0.50 are to be developed, some inbreeding must be undertaken. Such family formation makes possible effective family selection for traits such as mortality and carcass merit. The mating of a sire to his daughters, as discussed earlier in this chapter, is the most extreme approach to forming distinct families by inbreeding.

Homozygosity and Prepotency.   Those animals which perform even reasonably well in an inbreeding program should be homozygous for a larger-than-average number of desirable genes. Their increased homozygosity means that the composition of their gametes will be more uniform than those of outbred animals. Their offspring should be more uniform.

Since desirable genes are often dominant, good inbred animals are often prepotent; that is, they stamp their own characteristics on the offspring to the exclusion of those of the other parent. Inbreeding is the only known method of increasing prepotency. Prepotency depends upon the homozygosity of dominant genes.

We would also expect increased uniformity of productive characters, such as growth rates and litter size in inbreds as compared with outbreds. Actual data on this point are limited, but as far as they go they seem to indicate no increase in uniformity. Perhaps the increased genetic uniformity is offset by an increased susceptibility to adverse environmental factors that affect some individuals more than others, and thus increases environmental variation. However, even if there were no increase in variation due to environmental factors, the decrease in variation in quantitative characters with inbreeding would be relatively small and might well remain undetected except at high levels of inbreeding. For characters only 20 to 30 per cent heritable, a reduction of 20 or even 40 per cent in the heritable portion of the variation would have little measurable effect on the total variance.

INBREEDING EXPERIMENTS

Before the turn of the nineteenth century, Crampe[1] and Ritzema-Bos[2] demonstrated the adverse effect of inbreeding on fertility and litter size in rats. Numerous abnormalities also appeared as inbreeding progressed. Crampe's rats were inbred for about 17 generations and those of Ritzema-Bos about 30 generations; however, since the inbreeding was irregular, a reliable estimate of the inbreeding coefficients could not be made.

At the Wistar Institute, rats have been inbred by brother-sister mating for more than 125 generations. King's[3] results were different from those of the earlier workers. By rigorous selection along with inbreeding, neither fertility nor constitutional vigor were reduced by inbreeding. Her foundation rats came from a previously interbred colony, and special care was given to the nutritional requirements of the inbreds. Wistar rats continue to be one of the most widely used strains of laboratory animals in the United States.

Extensive inbreeding research with guinea pigs initiated by G. M. Rommel of the U.S. Department of Agriculture in 1906 has resulted in major contributions to our understanding of the consequence of inbreeding in livestock. Later, upon joining the Department, Sewall Wright became responsible for this research. A brief quotation from Wright's summary of the inbreeding studies with guinea pigs shows the remarkable general agreement between these results and those since reported in farm animals.

There has been an average decline in vigor in all characteristics during the course of 13 years of inbreeding of guinea pigs, brother with sister. The decline is most marked in the frequency and size of litter, in which it is so great that it would have to be accounted for even though the decline in other respects was assumed to be due wholly to a deterioration in the environmental conditions. The decline is greater in the gains after birth than in the birth weight, and greater in the percentage raised of the young born alive than in the percentage born alive. The ability to raise larger litters has fallen off much more than ability to raise small litters.

A comparison of the inbred guinea pigs with a control stock, raised under identical conditions without inbreeding, and derived in the main from the same linebred stock as the inbred families, indicates that the inbreds have suffered a genetic decline in vigor in all characteristics. The decline in fertility is again shown to be marked. Experimental inoculation with tuberculosis has shown that the inbreds were inferior, on the average, to the controls in disease

[1] Crampe, H., *Landw. Jahrb.*, **12**:389–458, 1883.
[2] Ritzema-Bos, J., *Biol. Cent.*, **16**:75–81, 1894.
[3] King, H. D., *Jour. Expt. Zool.*, **26**:1–98, 1918.

resistance. A study of sex ratio yields results in marked contrast to those obtained in connection with the other characters. There are no significant fluctuations from year to year, no contrast between inbreds and controls, and no indications of change due to inbreeding.[1]

Inbreeding experiments with cattle and even the smaller farm animals are expensive and time consuming. Nevertheless, results of experiments have been most valuable in corroborating results from laboratory mammals. Although additional studies are needed, available results have provided guides to the decline in performance that might be expected when selection is accompanied by mild inbreeding in farm animals.

SIMILARITY BETWEEN INBREEDING AND RELATIONSHIP

The inbreeding and relationship coefficients can both be considered as probability statements. Inbreeding is the probability that two genes at a given locus in an individual are alike by descent, and relationship is the probability that two individuals related by descent possess more of the same genes than unrelated individuals from the population.

Inspection of the formula for the inbreeding coefficient and the complete expression for the coefficient of relationship reveals a further similarity. The inbreeding coefficient is one-half the numerator of the relationship coefficient or one-half the relationship coefficient when the two related individuals are not inbred.

$$F = \Sigma(\tfrac{1}{2})^n(1 + F_A)(\tfrac{1}{2})$$

When the related individuals are the sire and dam, the inbreeding of the offspring can be expressed as one-half the numerator of the relationship coefficient between the sire and dam.

$$F \text{ (offspring)} = \tfrac{1}{2} \text{ (numerator relationship for sire and dam)}$$

In the original derivation of the relationship coefficient by Wright, it was viewed as a genetic correlation expressing the correlation between two related genotypes. As a corrollary to the simple correlation coefficient the numerator relationship has been termed the *covariance*. The above relations have been drawn upon to reduce the labor of computing inbreeding and relationship coefficients in closed herds or lines. Convenient reference tables of covariance values (numerator relationships) are available for all potential pairs of breeding individuals in the closed herd or inbred line. Such calculation of covariance tables has been a

[1] Wright, Sewall, *U.S. Dept. Agr. Bul.* 1090, 1922.

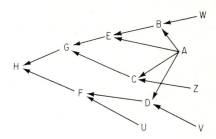

Fig. 8-12. Arrow-style pedigree of a closed line showing the inbreeding buildup of H.

routine procedure with Lush and co-workers since the 1930s. The table of covariances permits making least related or most closely related matings, whichever may be desired. It also allows for easy computation of all necessary inbreeding coefficients.

In essence, the procedure of covariance is practically workable because covariance values among the parents can be used to compute the values for the offspring. The tedium of counting independent paths, with its concomitant errors, is reduced to a simple addition of covariances. Once one becomes familiar with the process it generally is preferable to compute inbreeding coefficients for complex extended pedigrees using this procedure rather than by tracing out the independent paths.

The pedigree of individual H (Fig. 8-12) can be used to illustrate the calculation of covariance tables. The principle of averaging covariances is most helpful and necessary. From the above pedigree, the covariance between F and G is needed to compute the inbreeding of H ($F = $ cov FG $\times \frac{1}{2}$). Several relationships can be drawn upon to compute the covariance between F and G. Two forms are given below.

$$\text{cov FG} = \frac{1}{4} (\text{cov ED} + \text{cov EU} + \text{cov CD} + \text{cov CU})$$
$$= \frac{1}{2} (\text{cov FE} + \text{cov EC}) = \frac{1}{2} (\text{cov GD} + \text{cov GU})$$

When the covariances are compiled by hand, generally the second of the above two relationships is used. In situations where both parents may not be in the same generation, it is important to take the average of the two covariances between the older animal with the parents of the younger. Consider the pedigree below to illustrate the need to apply this rule to avoid duplicating contributions.

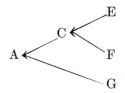

cov AC $= \frac{1}{2}$ (cov CC + cov CG) but cov AC $\neq$
$$\frac{1}{2} (\text{cov AE} + \text{cov AF})$$

Table 8-10. Covariance Table for Individuals in the Pedigree of H

	A	B	C	D	E	F	G	H
A	1.000							
B	0.500	1.000						
C	0.500	0.250	1.000					
D	0.500	0.250	0.250	1.000				
E	0.750	0.750	0.375	0.375	1.250			
F	0.500	0.250	0.625	0.625	0.375	1.125		
G	0.625	0.500	0.688	0.312	0.813	0.500	1.187	
H	0.562	0.375	0.656	0.469	0.594	0.812	0.844	1.250

The following sample calculations will be helpful to illustrate the technique.

$$\text{cov } AB = \tfrac{1}{2} \text{ (cov } AA + \text{cov } AW) = \tfrac{1}{2} \text{ } (1.000 + 0) = 0.500$$
$$\text{cov } AC = \tfrac{1}{2} \text{ (cov } AA + \text{cov } AZ) = \tfrac{1}{2} \text{ } (1.000 + 0) = 0.500$$
$$\text{cov } AD = \tfrac{1}{2} \text{ (cov } AA + \text{cov } AV) = \tfrac{1}{2} \text{ } (1.000 + 0) = 0.500$$
$$\text{cov } AE = \tfrac{1}{2} \text{ (cov } AA + \text{cov } AB) = \tfrac{1}{2} \text{ } (1.000 + 0.500) = 0.750$$
$$\text{cov } BC = \tfrac{1}{2} \text{ (cov } BA + \text{cov } BZ) = \tfrac{1}{2} \text{ } (0.500 + 0) = 0.250$$
$$\text{cov } BD = \tfrac{1}{2} \text{ (cov } BA + \text{cov } BV) = \tfrac{1}{2} \text{ } (0.500 + 0) = 0.250$$
$$\text{cov } BE = \tfrac{1}{2} \text{ (cov } BB + \text{cov } BA) = \tfrac{1}{2} \text{ } (1.000 + 0.500) = 0.750$$
$$\text{cov } CD = \tfrac{1}{2} \text{ (cov } DA + \text{cov } DZ) = \tfrac{1}{2} \text{ } (0.500 + 0) = 0.250$$
$$\text{cov } CE = \tfrac{1}{2} \text{ (cov } BC + \text{cov } AC) = \tfrac{1}{2} \text{ } (0.250 + 0.500) = 0.375$$
$$\text{cov } DE = \tfrac{1}{2} \text{ (cov } BD + \text{cov } AD) = \tfrac{1}{2} \text{ } (0.250 + 0.500) = 0.375$$

Since the relationship of an individual with itself is by definition 1.00, the numerator relationship of covariance of an individual with itself (X) is $1 + F_X$.

$$R_{XX} = \frac{1 + F_X}{\sqrt{1 + F_X} \sqrt{1 + F_X}}$$

Thus the diagonal values in Table 8-10 are 1.000 plus the inbreeding coefficient of the individual. The inbreeding coefficient of the individual is one-half the covariance between its parents. Individual E's inbreeding coefficient is

$$F_E = \tfrac{1}{2} \text{ (cov } AB) = \tfrac{1}{2} \text{ } (0.500) = 0.250$$

The remaining values in Table 8-10 can be verified by the reader to gain additional experience with this approach.

Relationship coefficients between any two individuals included in the table also can be computed:

$$R_{\mathrm{DE}} = \frac{0.375}{\sqrt{1.000}\ \sqrt{1.250}} = \frac{0.375}{\sqrt{1.250}}$$

When the two individuals are not inbred, the relationship can be read directly from the covariance table.

SUMMARY

Inbreeding is the mating of individuals which are more closely related than are average members of the breed. Inbreeding increases homozygosity, and the genetic consequence of inbreeding results directly from the increased homozygosity. The inbreeding coefficient measures the proportion of the loci which was heterozygous in the base population, which probably has become homozygous due to inbreeding. Uncovering deleterious recessives and reducing the proportion of heterozygotes are deleterious responsible for the reduced fertility and the loss in vigor and productivity associated with inbreeding. Lower performance with inbreeding make it unwise to follow an inbreeding program in a commercial herd. Nevertheless it may be useful in breeding herds in which it is desired to linebreed or to develop distinct families for family selection. The increased homozygosity of breeding animals for dominant genes would increase prepotency but any increase in uniformity for most quantitative traits would be difficult to detect.

Relationship is based on the probability that two individuals are alike in more of their genes than are random individuals from the breed or population. The coefficient of relationship measures the probable proportion of genes that are alike for two individuals due to their common ancestry. Since higher relationships do reflect a higher likelihood that two individuals have the same genes, it is useful in weighting evidence from relatives when appraising an individual's merit.

REFERENCES

Books

East, E. M., and Jones, D. F. 1919. "Inbreeding and Outbreeding," J. B. Lippincott Company, Philadelphia.

Falconer, D. S. 1960. "Introduction to Quantitative Genetics," The Ronald Press Company, New York.

Gowen, John W., ed. 1952. "Heterosis," The Iowa State University Press, Ames, Iowa.

Lerner, I. Michael. 1954. "Genetic Homeostasis," John Wiley & Sons, Inc., New York.

Lush, Jay L. 1945. "Animal Breeding Plans," 3d ed., The Iowa State University Press, Ames, Iowa.

Winters, L. M. 1954. "Animal Breeding," 5th ed., John Wiley & Sons, Inc., New York.

Articles

Carter, R. C. 1950. A Genetic History of Hampshire Sheep, *Jour. Hered.*, 31:89–93.

Dickson, W. F., and Lush, Jay L. 1933. Inbreeding and the Genetic History of the Rambouillet Sheep in America, *Jour. Hered.*, **24**:19–33.

Fowler, A. B. 1932. The Ayrshire Breed: A Genetic Study, *Jour. Dairy Res.*, 4:11–27.

Lush, Jay L. 1946. Chance as a Cause of Changes in Gene Frequency within Pure Breeds of Livestock, *Amer. Nat.*, **80**:318–342.

———, Holbert, J. C., and Willham, O. S. 1936. Genetic History of the Holstein-Friesian Cattle in the United States, *Jour. Hered.*, 26:205–215.

——— and Anderson, A. L. 1939. A Genetic History of Poland-China Swine, *Jour. Hered.*, **30**:149–156 and 219–224.

McPhee, H. G., and Wright, S. 1925. Mendelian Analysis of the Pure Breeds of Livestock. III. The Shorthorns, *Jour. Hered.*, **16**:205–215.

Robertson, A., and Asker, A. A. 1951. The Genetic History and Breed Structure of British-Friesian Cattle, *Empire Jour. Expt. Agr.*, **19**:113.

Stonaker, H. H. 1943. The Breeding Structure of the Aberdeen-Angus Breed, *Jour. Hered.*, **34**:322–328.

Willham, O. S. 1937. A Genetic History of Hereford Cattle in the United States, *Jour. Hered.*, **28**:283–294.

Wright, S. 1921. Systems of Mating, *Genetics*, **6**:111–178.

———. 1922. Coefficients of Inbreeding and Relationship, *Amer. Nat.*, **56**:330–338.

———. 1922a. The Effects of Inbreeding and Crossbreeding on Guinea Pigs. I. Decline in Vigor. II. Differentiation among Inbred Families, *U.S. Dept. Agr. Bul.* 1090.

———. 1922b. The Effects of Inbreeding and Crossbreeding on Guinea Pigs. III. Crosses between Highly Inbred Families, *U.S. Dept. Agr. Bul.* 1121.

Yoder, D. M., and Lush, Jay L. 1937. A Genetic History of Brown Swiss Cattle in the United States, *Jour. Hered.*, **28**:154–160.

Young, G. B., and Purser, A. F. 1962. Breed Structure and Genetic Analysis of Border Leicester Sheep, *Anim. Prod.*, **4**:379–389.

CHAPTER NINE

Outbreeding

The term *outbreeding* means following a breeding system in which matings are between animals less closely related than the average of the
group to which they belong. For practical purposes we usually consider
matings to be of an outbred type if the animals mated have no ancestors
in common in the first four to six generations of their pedigrees.

Outbreeding includes mating of unrelated animals within breeds,
crossbreeding, grading, the crossing of inbred lines, and extreme crosses
between animals belonging to different species, such as the cross of
the ass and the horse to produce the mule. Our purpose in this chapter
is to discuss the usefulness and limitations of these various types of
outbreeding.

OUTCROSSING

Outcrossing is the term applied to mating unrelated animals within the
same pure breed. It is the most common breeding system in use by
American breeders of seedstock herds. The common practice of continually selecting the best available, but unrelated, sires for use on the
females of a herd is an outcrossing system. Although unspectacular and
in many ways less effective than some other breeding systems, it has
nevertheless been responsible for a high percentage of the changes in
livestock since breeds evolved.

The pool of genes in any breed has, in general, been elastic enough
to permit outbreeding systems to bring about marked changes in type
and other characteristics. Indeed, selection to change the proportions

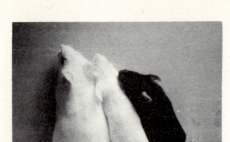

Fig. 9-1. 60-day-old male mice illustrating differences in size produced under outbreeding systems. *Left,* a representative of Goodale's Large White. *Center,* an albino bred without selection for either large or small size. *Right,* a representative of MacArthur's small line. (*Courtesy of Dr. W. H. Kyle, Animal Husbandry Research Division, U.S. Department of Agriculture.*)

of the various genes and inbreeding to increase homozygosity are the only two breeding tools available to the breeders of most of our modern breeds, because breed-society rules prohibit the introduction of outside genes into a breed.

A rather prevalent but erroneous opinion is that inbreeding is increasing rapidly in our breeds and that they are truly "pure strains." A good deal of inbreeding did occur in the formative stages of many breeds. As one of the most extreme examples, average animals of the Shorthorn breed in 1920 were found to be about 26 per cent inbred, mainly because intensive inbreeding was practiced by a few pioneer breeders. Studies of many breeds, largely by Sewall Wright and Jay L. Lush and their associates, show that in general the present increase in average inbreeding coefficient is about 0.5 per cent per generation. Thus, it will take the average breed about 25 generations (around 35 to 40 years in swine and 110 to 125 in cattle) to increase in homozygosity as much as would result from one generation of half-brother-sister mating. Even this much increase in homozygosity is likely to be reduced somewhat by occasional registration of a grade or crossbred through mistake or fraud.

Another factor minimizing increases in homozygosity within pure breeds is the probable tendency to select animals with less than average homozygosity for breeding purposes. This follows from the fact that inbreeding, with resultant homozygosity, tends to depress performance. Thus, if selections are made purely on the basis of performance (phenotype) there will be a tendency to select those less homozygous than average because *on the average* their performance will tend to be superior.

From the foregoing, the conclusion seems inescapable that outcrossing combined with selection has been responsible for most of the changes and improvements in modern pure breeds of livestock. Many past accomplishments can be used to illustrate the power of this breeding system.

Perhaps the most marked changes during the past half century or so have occurred in swine type. They are discussed further in the chapter on Swine. Probably if a group of breeders of the small, fine-boned, "chuffy" hogs in 1900 had been shown a picture of the "race horse" type which evolved by 1930 and told that their job was to create such a type, their first reaction would have been, "It can't be done." However, such a change was made, and doubtless our present-day intermediate-type hogs have enough hidden variability to permit the re-creation of either extreme type should it ever be desirable to do so.

Other changes, such as the smoothing-up of dairy cows, the increased ability of beef types to fatten at young ages, and the divergence of the Shorthorn breed into beef and milking types could be used as equally good examples of changes which have taken place.

Genetic Aspects of Outcrossing. For characters that are largely under the control of genes with additive effects and are highly hereditary, a system of selection and outcrossing is very effective. High heritability indicates a high correlation between genotype and phenotype. Individual, or phenotypic, selection is therefore reasonably accurate in locating those animals with greater-than-average numbers of desired genes. The mating of selected animals in outcrosses results in relatively few undesired genes being fixed in homozygous form. In other words, a breeding system of this kind brings about immediate improvement and at the same time does not shut the door on future improvement, as an intense inbreeding program might do through the fixation of deleterious genes.

Growth rate in beef cattle is an example of a highly hereditary trait for which a breeding system of this kind should be most effective. In fact, the system will be effective to a degree for traits with much lower heritabilities, there being no definite lower limit at which we could suddenly say some other system was preferable.

Experiments involving selection for rather highly hereditary quantitative traits under outcrossing systems with laboratory animals including mice, rats, the fruit fly *Drosophila melanogaster,* and the flour moth *Tribolium castaneum* have often shown rapid responses for 10 to 30 generations. Improvements from selection have usually started slowing at some point beyond about 10 to 15 generations and in some cases have ceased after around 20 to 30 generations of selection. The level at which failure to respond occurs is often spoken of as a plateau, and such populations are said to have *plateaued.* The reasons for plateauing are not fully understood but can include the following:

1. *Exhaustion of genetic variability.* This is another way of saying that the strain has become homozygous for all genes affecting the trait under selection. This could occur if the number of gene pairs were small, but is unlikely if the number of gene pairs affecting a given trait is in the 100s or 1,000s as seems probable. Often, detailed analysis of data from plateaued populations show that additive genetic variability is still present, thus proving that in at least some such populations, exhaustion of genetic variability is not wholly responsible for failure to respond to further selection.

2. *Negative genetic relations between reproductive ability or viability and the trait or traits under selection.* Relations of this kind would make it impossible to select intensely for the desired trait or traits and in extreme cases could result in negative selection. Simultaneous selection for two or more negatively related, desired traits could reduce or eliminate selection response for both.

3. *Traits are under influence of genes with dominance, overdominance, or epistatic effects.* Outcrossing with selection is an ineffective system for fixing genetic factors having these types of action because maximum expression of the traits depends upon heterozygosity and is thus not fixable or because phenotype and genotype are not closely related. Selection thus becomes more and more inaccurate. This is another way of saying that heritability (thought of in the narrow sense as hereditary variation due to additive gene effects) will be reduced.

It is not certain whether any economically important traits in large farm animals have plateaued, or reached the point where selection is not effective. The usefulness of outcrossing depends almost wholly on the *effectiveness of selection.* There is no doubt that selection has been effective in the past, but the important question of whether farm animals may be reaching the point where selection will decrease in effectiveness is only beginning to be studied intensively. Current work will be reviewed in chapters on improving each of the classes of farm animal. Although there is evidence that selection should still be effective for most characters, indications are that for at least certain traits (e.g., litter size in swine) a ceiling may have been approached and that future selection cannot be expected to be very effective. Further improvement in these characters will depend on use of other breeding techniques.

It appears that outcrossing should be used with selection in most seedstock herds until it has carried a breeder to the point where further progress is difficult, or when selection is for characters initially low in heritability. The breeder should then turn to inbeeding of linebreeding to fix the desirable characters and permit more effective selection against undesirable combinations.

The sequence envisioned here is perhaps best illustrated by the breeding history of Indian corn, or maize. In the relatively short span of years from the discovery of America to 1900 and probably mostly during the nineteenth century, the white man, largely by selection among cross progeny of existing diverse strains, increased the yielding capacity of corn manyfold. An equilibrium had apparently been reached, however, so that selection with the techniques then used seemed powerless to effect further improvement. Plant breeders turned to inbreeding and the crossing of selected lines. The resulting *hybrid* corn had a higher yielding capacity than selection had been able to achieve. Its use virtually revolutionized corn growing.

Most livestock breeders whose herds have truly exerted an influence on their breeds have practiced intensive selection. Many of them have also practiced inbreeding or linebreeding with a resulting improved breeding value of their animals. Nevertheless, almost invariably these master breeders at times have used outcrosses to regain lost vigor or to introduce new genes. Only when used in the latter ways with a specific aim in view is outcrossing in an outstanding herd to be recommended. In herds of less-than-average merit it is usually the best breeding method, at least according to our present knowledge and our current practice of merchandising breeding stock on the basis of individual merit.

Outcrossing is a useful procedure in cases in which it is desired to change the type of herd rather drastically. The necessity for doing this might arise because of changes in market demands, changes in fashion, or because of previous unrealistic selection standards in a herd. If sires of the desired or popular type are available in the breed, they can be used in such herds to change the type rather quickly. Sometimes sires more extreme in the desired direction than would be desirable in a true breeding strain can be used in the expectation that the "corrective matings" will result in intermediate offspring of approximately the desired type.

GRADING

Grading is the practice of breeding purebred sires of a given breed to nondescript, scrub, or native females and their female offspring generation after generation.

The first generation offspring carry 50 per cent of the inheritance of the pure breed and, depending upon the quality of the original females, can usually be expected to be a considerable improvement over their dams. The next generation results in offspring with 75 per cent of the hereditary material of the pure breed. In subsequent generations the proportion of inheritance remaining from the original scrub females

Fig. 9-2. Results of grading in dairy cattle. (*From Iowa Agr. Expt. Sta. Bul.* 188.)

Scrub cow No. 60. Average production 3,313.2 lb of milk and 178.47 lb of fat.

Half-blood Jersey No. 241 out of Scrub No. 60. Average production 6,126.4 lb of milk and 348.98 lb of fat.

Three-quarter-blood Jersey No. 348, out of half-blood Jersey No. 241.

is halved with each cross. After five or six crosses of purebred sires, the grade animals carry 96.9 and 98.3 per cent, respectively, of the hereditary material of the pure breed. From a genetic standpoint they are essentially the same as purebreds.

Grading is thus a process by which a few purebred sires can rather quickly transform a nondescript population into a group resembling the pure breed. The quality and productivity of the resulting grades depends entirely upon the genetic quality of the purebred sires used— the grading process is not a creative one.

Many experiments have shown the value of grading. The second-generation daughters of good dairy sires in an Iowa experiment many years ago produced 116 per cent more milk than their scrub granddams. Essentially similar results could be quoted from experiments with beef cattle, sheep, and hogs.

Limitations on Usefulness of Grading. A word of caution should be introduced against the use of purebreds in grading programs in areas where the particular purebreds used are not adapted. Purebred dairy and beef cattle from temperate zones often degenerate when used in tropical or subtropical areas, and their offspring do not have the vigor and constitution for high production in the warmer climate. Sometimes, in such cases, the first crosses, still having 50 per cent native genes plus the genes for high production from the introduced purebreds, will be a distinct improvement over their native dams. Second and later crosses having 75 per cent or more of purebred ancestry may deteriorate rapidly, however. An example of this type of thing is shown in the accompanying table. These data, accumulated under variable conditions in several parts of India, indicate that animals with ½ to ⅝ imported background produced more milk than those with more imported hereditary material, even though the latter animals presumably had more genes for high milk production. Lack of adaptability apparently put a limit on the production of animals with high proportions of imported ancestry.

For grading to be successful, it is not sufficient to use purebreds indiscriminately. The purebreds used must have evidenced the ability to perform well under the conditions which their offspring will have to meet.

Table 9-1. Average Production of Cattle in India Carrying Various Proportions of Imported Ancestry, Mostly Holstein-Friesian°

Breeding of cows	Number of records	Av. amounts of milk, lb
⅛ imported ancestry	21	4,839
¼ imported ancestry	175	5,982
½ imported ancestry	589	6,977
⅝ imported ancestry	204	6,985
¾ imported ancestry	396	6,664
⅞ imported ancestry	86	6,180

* Adapted from *United Nations, FAO Agr. Studies* 1, 1949.

Future Usefulness of Grading. The so-called *purebred era* in the United States is generally thought of as having begun about 1875. Since that time, genetic improvement of commercial livestock has for the most part depended upon the purebred breeder. Early in the purebred era it was generally true that most purebreds were superior enough to most nonpurebred or commercial stock that use of purebred sires could be expected to bring about improvement in conformation, market quality, or productivity.

With the passage of time, many high-grade herds and flocks have been developed. In some cases careful selection of purebred sires for use in these herds generation after generation has resulted in herds with apparent genetic merit as great and perhaps greater than that of many purebred herds. Two questions can be raised with respect to such herds: (1) Should the owner of such a herd continue to purchase purebred sires from outside the herd or should he use high-grade sires produced in his own or similar herds? and (2) Should the high-grade animals of superior quality be accepted as purebred and registered in the appropriate breed herdbook?

The answer to the first of these questions depends solely upon the decision of the individual producer. Whether continued use of purebred sires will result in further improvement depends upon whether the pure breed is improving and whether selection programs of breeders within the breed are directed toward the same characters which are important to further improvement of a given commercial herd or flock. Thus, to maintain its position as a supplier of sires for commercial production, a pure breed must continuously advance in merit and maintain selection criteria related to needs of commercial producers.

Registration of High Grades. The question of the registration of high grades must be answered by each breed association. From a genetic standpoint, the registration of high grades would have the advantages of

1. Increasing numbers in the breed and thus broadening the genetic base for future selection
2. Increasing the frequency of desirable genes for quantitative characters if the grades registered were higher in merit than the average of the pure breed
3. Permitting the introduction of genes not already present in the breed

Increasing numbers in a breed is especially important to breeds which are numerically small. Increasing numbers would permit more intensive selection of sires and wider use of outstanding sires without increasing inbreeding levels at unduly rapid rates.

The introduction of specific genes into a breed through the registration of high grades can be useful primarily for traits under the control of only one or a few gene pairs. Introduction of a gene for polled into a horned breed would be an example. The process would necessarily be slow if four or more top crosses were required for registration.

The only genetic danger of registering high-grade animals is the possibility of inadvertently introducing an undesirable recessive gene into a breed not already having it. The possibilities of this could be reduced to a negligible level by appropriate safeguards.

Thus, from a genetic standpoint, provision for registration of high grades would seem to be desirable. It would be especially desirable for numerically small breeds. For breeds with large numbers, the option would probably be exercised infrequently and only when there was a real need of gene introduction.

Registration of high grades might be either desirable or undesirable when looked at from the point of view of the business interests of persons already owning and breeding animals of a given breed. In the case of a breed with limited numbers, expansion of these numbers more rapidly than would otherwise be possible could facilitate its becoming of regional or national importance. If the breed were truly valuable, this would likely result in building a demand from commercial producers which would benefit all breeders. For a breed already present in a country in large numbers, creating additional purebreds could oversupply the market, result in lower average prices, and be detrimental to the financial interests of existing individual breeders.

Almost all breeds of livestock now considered pure had their beginnings in one herd or in the herds of a few breeders in a given locality. In some cases the ultimate formation of a breed had been planned in advance. This has been true of many of the breeds formed in recent years. In other cases the establishment of a breed occurred as an outgrowth of the development of a useful type of livestock by a group of commercial producers.

Once a group of animals has reached a stage of development at which the founder, founders, or others owning representatives of the new strain believe designation as a breed is justified, the typical procedure has been for those owning the animals to band together into a breed society or association and to form a herdbook or registry.

Regardless of the mode of origin of a breed, decisions have to be made on (1) which animals to admit to the original herdbook, and (2) procedures for determining which animals will be admitted in the future and thus be considered as purebred.

The first decision has usually been based on many factors, typically including herd of origin, as much information as available on ancestry, and adherence to color, conformation, performance, or other standards which may be established.

Regulations governing which animals will be admitted to registry after the original foundation stocks have been agreed upon have varied widely between breeds and between countries.

European breeds have quite often, but not in all cases, maintained provision for registration of animals with four or five top crosses of registered sires. In many European breeds, registration of any animal, regardless of whether it descends from registered ancestors in all lines of its pedigree or is the result of a grading process, is dependent upon adherence to breed standards as determined by a physical inspection. In some countries, maintenance of herdbooks is a government function and the inspectors are government officials.

Registration is usually on an individual pedigree basis, but in some European breeds so-called *flock* or *herd registration* is used in which females are not registered individually. In this case the entire flock or herd owned by a breeder is considered registered if established standards are met.

Eligibility for registration in the United States has typically been based on the *closed herdbook* principle. It is required that descent in all lines of the pedigree be from registered animals. This has been true both for European breeds imported to the United States and for breeds developed in this country during the nineteenth century. Except for exclusion from registry of animals with specified defects or deviations from breed color or horn standards as established by each association, any animal with registered parents is eligible for registration regardless of conformation or producing ability. In the United States, establishment of standards is entirely a function of the individual associations. There is no governmental regulation.

Several of the newer breeds or those of which only limited numbers have been imported from the country of origin currently have procedures which deviate from the foregoing. Some of these include the registration of animals with four or five top crosses of registered sires by the American Brahman Breeder's Association and the American-International Charolais Association. The Santa Gertrudis Breeder's International certifies animals with four top crosses, recognizes the produce of multiple sire herds (i.e., does not require individual pedigrees), and requires an inspection by an approved classifier before certification. The Red Angus Association requires that animals meet minimum performance standards.

The foregoing is not intended to be exhaustive. Rules of all associations can change from time to time. Persons interested in a particular breed should familiarize themselves with current rules of that breed.

HYBRID VIGOR OR HETEROSIS

Crosses of animals from different strains or lines of the same breed, from different breeds, or from different species, often result in offspring whose level of production is *above that of the average* of the parental types and often *above the level of either* parental type. The increased productivity may be due to improvements in one or more of such things as increased fertility, increased pre- and post-natal viability, faster and more efficient growth, improved mothering ability of cross-type females, etc. The increased level of performance as compared to the average of the parental types is known as *heterosis* or *hybrid vigor*.

Possible Genetic Bases for Heterosis. The genetic reason for heterosis is not known with certainty. As a species, breed, strain, or line develops, it becomes homozygous for some favorable dominant genes, and probably also for some unfavorable recessive ones. Another breed or species would probably not be homozygous for the same undesirable recessive genes. Thus, when two breeds were crossed, the cross for some allelic pairs (two pairs are assumed here purely for illustrative purposes) might be *AAbb* × *aaBB*, with the offspring being *AaBb*. Since it carries one dominant of each pair, the offspring would be expected to be more vigorous or productive than either parental type. This explanation of heterosis assumes that it is largely a result of covering up the effects of recessive genes.

A second explanation sometimes thought to account for heterosis, at least to a degree, is that of interallelic interactions, or epistasis. Going back to our previous simplified illustration, this possible basis for heterosis would occur if the combination *AaBb* results in an interaction such that the presence of *both A* and *B* gives a phenotype larger or in other ways more desirable than would be expected from average phenotypes of *AAbb* and *aaBB*. A classic illustration of this theory comes from plants in which it was observed that one dominant gene resulted in *long* internodes and another resulted in *many* internodes. The net effect of the presence of both dominant genes tends to be multiplicative and results in offspring taller than would be expected from averaging the sizes of parental types having many but short internodes with those having few but long ones.

Many other types of interallelic interactions which might result in heterosis can be visualized. Some might at least theoretically result in *negative heterosis*. As a hypothetical example, suppose one line of animal carried a dominant gene resulting in excessive production of some body chemical but with the excessive production not exerting a serious detrimental effect in otherwise normal animals that had a reserve capacity for disposing of it. Suppose a second line carried a dominant gene resulting in inefficient metabolism of the same chemical. Even though the second line was also able to perform more or less normally as long as production of the chemical was not excessive, it can readily be seen that a cross of the two lines could be disastrous. Fortunately, most genes with undesirable effects are recessive. Negative heterosis seems to be rare in animals. There are, however, some apparent examples in laboratory organisms.[1] It should also be pointed out that what is desired may determine whether heterosis is positive or negative in an economic sense. For example, if smallness were desired in a species for some specific purpose and if crosses resulted in size increases, the effect would be economically negative even though we might still think of it as biologically positive.

As a general rule in living organisms, genes with some degree of dominance are more often favorable than are recessives. Thus, crosses would usually be expected to exhibit increased rather than decreased vigor.

The similarities of, and indeed the difficulty in discriminating between, theories that heterosis is dependent upon covering the effects of recessive genes and interallelic interactions between dominants is readily apparent. In both cases maximum heterosis would depend upon the presence of at least one dominant of a maximum number of allelic pairs. If either explanation, or a combination of both, were wholly responsible for heterosis, it should theoretically be possible to select pure lines or breeds which would carry all the desirable dominants in homozygous form and thus have maximum performance. If we concede the theoretical possibility of this, we must also recognize that the development of such lines or breeds would be an extremely slow and difficult process. The principle reason for this is the large number of gene pairs which affect most productive characters. Two animals heterozygous for n pairs of genes can theoretically produce 3^n types of offspring. If only seven pairs of genes were heterozygous, 3^7, or 2,187, different combinations could be expected in the offspring. It ten pairs were heterozygous, 59,049 different combinations would be possible. Actually, gene-pair numbers are probably much larger than this. Animals usually have only a relatively small number of offspring. Thus, probabilities of the perfect combination being produced are small even with many generations of selection. Effects of environment resulting in imperfect heritability would likely make it impossible to identify the perfect combinations with certainty even if they occurred. Further, there is an excellent chance that some undesirable recessives would be fixed in each line or breed purely

[1] See Mason, R. W., et al., "Biometrical Genetics," O. Kempthorne, ed., Pergamon Press, New York, 1960; and Stern, Curt, *Genetics*, 33:215–219, 1948.

as a result of chance genetic drift. Another fact which could slow progress in fixing desired combinations is that desirable and undesirable genes could be linked in unfavorable combinations in the foundation stock with which selection began. Many generations might be required to reach a linkage equilibrium.

So far as is known, no one has as yet succeeded in producing lines, breeds, or varieties of plants or animals which in themselves have all desirable traits and which show no heterosis in crosses. For the reasons given above, this is not surprising, even if the above theories should be largely or wholly responsible for heterosis.

Another possible explanation of heterosis is that for some gene pairs the heterozygote may be more vigorous than either homozygote. That is, for a given gene pair such as *A* and *a*, *Aa* might be superior to either *AA* or *aa*. This situation is called *overdominance*. Much indirect evidence in both plants and animals suggests that overdominance may be an important factor in heterosis in at least some species and for some characters. If this is the case, it would be impossible to produce lines, breeds, or varieties which could produce at maximum rates in pure form. Maximum productivity could be attained only in crosses.

The best hypothesis at present is that all of the foregoing possibilities may be factors in heterosis. Different genetic mechanisms may be of predominant importance in heterosis in different species of plants and animals and for different traits in the same species.

A full understanding of the genetical basis of heterosis will probably have to await extensions of knowledge of the physiology of gene action and the physiological reactions resulting in heterosis. To date, too little is known of the physiology of heterosis, even for simple characters of laboratory organisms, to permit the exposition of theories or hypotheses with likely complete applicability to domestic animals.

The fact that the genetic and physiological bases for heterosis are imperfectly understood stands as a challenge to scientists and probably to optimum use of heterosis in practical situations. It does not, however, hinder plant and animal breeders from taking advantage of heterosis through use of crosses of breeds, varieties, lines, or even species, which have proved through either experimentation or experience to produce at higher levels than either parental type.

CROSSBREEDING

Crossbreeding is the mating of animals from different established breeds. The term technically applies only to the first crosses of purebreds, but it is generally applied also to the more widely used systems involving crisscrossing of two breeds or the rotational crossing of three or more breeds and to crosses of purebred sires of one breed to high-grade females of another.

Crossbreeding for Commercial Production. Crossbreeding for commercial production may be practiced for either or both of two reasons. The first is to take advantage of heterosis, or hybrid vigor, which may make the crossbreds better and more productive than either of the parental breeds, even though the latter are of similar type. This is what a commercial hog producer is doing when he uses two or more American breeds of hogs in crosses. Although differences exist among the breeds which originated in the United States during the nineteenth century, they can be considered as being of the same basic type. The second reason for crossbreeding is to take advantage, as far as possible, of the good qualities of two or more breeds of distinctly different types. This is what the Western sheepman is doing when he crosses his vigorous, hardy, gregarious Rambouillet ewes to Hampshire or Suffolk rams. He hopes to get improved carcass quality and increased growth rate in the lambs from their sires while continuing to take advantage of the good range qualities and mothering ability of the Rambouillet ewes.

In considering crossbreeding, it should be emphasized that maximum productivity in a commercial livestock enterprise usually depends upon maximizing *both* heterosis and the frequency of desirable genes with additive effects. Except for reproductive characters and vigor, most traits of economic importance are influenced more by additive gene action than by heterosis. As an example, growth rate in beef cattle is highly hereditary. Some breeds have average growth rates 25 to 50 per cent, or even more, greater than those of other breeds. Differences between lines of breeding within breeds are usually less. However, variations of 10 to 25 per cent or more have been observed.

As contrasted to these wide differences between breeds and between strains within breeds, increases in growth rate due to heterosis in most crossbreeding experiments have been 5 per cent or less. As a more extreme example, many carcass traits such as back-fat thickness in swine are highly hereditary but exhibit little or no heterosis. For highly hereditary characters such as this, breed and strain differences are transmitted to crossbred as well as purebred progeny.

The foregoing facts have sometimes been advanced as arguments against crossbreeding. This is not justified, since the additional total performance, as compared to that of the best available purebred, which can be attained in crosses as a result of heterosis, is often great enough to be of economic importance. The facts do, however, emphasize the necessity of basing crossbreeding programs only on purebreds of high individual merit for desired highly hereditary characters.

There are production situations in which one breed is so greatly superior to any other for one or more highly desired traits that crosses with other breeds, while above the average of the two parental breeds in value, are still inferior to the best parent. If these highly desired traits were exclusively or very highly related to profitability, there would be no justification for crossbreeding. Profitability from most enterprises depends upon several production factors. As will be discussed later, one of the characteristics of crossbreds is often that they exceed the average of the purebred parents only a little for any one trait, but that by virtue of the small advantages in each trait being cumulative, they rather greatly exceed the parental averages in total production. Therefore, a producer should carefully examine the total performance situation rather than just a single trait before arriving at the conclusion that a crossbreeding program would not be advantageous for him.

Probable genetic bases for heterosis from crossbreeding. As discussed in more detail earlier, inbreeding has not been intense within the pure breeds of farm animals. Cumulatively, however, over the period of perhaps 25 to 50 or more generations since the herdbooks were closed for most popular breeds, the increase in inbreeding levels has been appreciable. Although much genetic variability remains, the increase in inbreeding, together with chance or random fluctuation in gene frequency, has undoubtedly resulted in the fixation in homozygous form of some undesirable recessive genes in most, if not all, individuals of each breed. Similarly, one or the other allele of some gene pairs exhibiting overdominance or heterozygote superiority may have become fixed in most breeds.

The foregoing leads to the supposition that in a very mild way breeds may exhibit some of the same characteristics of inbred lines. Since the outstanding outward effect of inbreeding is to lower performance, especially for certain traits, we could well expect animals of the various pure breeds to have somewhat less than maximum performance levels.

Because of the random nature of gene fixation with inbreeding, it would be very unlikely that the different breeds would have the same undesirable alleles fixed in homozygous form. This would be most true of long-separated breeds with diverse origins and less true of breeds developed from similar foundations within comparatively recent times. Crossing breeds would both cover undesirable recessive genes and increase heterozygosity. Since both these things are probably important as bases for heterosis, the performance of crossbreds would be expected to exceed that of the purebred parents to some degree. The increased performance should be most marked in those traits most depressed by inbreeding, i.e., (1) reproductive qualities including fertility and fecundity of both sexes and maternal qualities of females, and (2) vigor as measured by mortality rates. Greatest increases in performance would be expected from crosses of breeds having the greatest diversity of origin.

Inbreeding levels are still relatively low in most breeds. Also, most popular breeds have reasonably good performance characteristics for most important traits, else they would probably not be widely used. Therefore the theoretically expected increases in performance with crossbreeding are relatively small for any one trait. However, heterotic effects are often additive or multiplicative. For example, if survival were 5 per cent greater and those that survived to a given age were each 5 per cent heavier at that age, then total production per breeding unit would be 10 per cent greater for crossbreds.

Experimental results of crossbreeding. Results of crossbreeding studies are in general accord with theoretical expectations. Crossbreds have usually exceeded the average performance levels of the parental purebreds by margins which are relatively small on a percentage basis for any one character but are often cumulative and large enough to be economically important in terms of total production efficiency. Increases have been most important for characters most depressed by inbreeding.

The increases from crossbreeding often represent large potential increases in net income from a commercial livestock enterprise; i.e., they can be obtained with little or no additional production costs. Thus, the small increases take on added significance when it is realized that an increase of perhaps 3 per cent in total production might well represent an increase of 50 or even 100 per cent in net income.

Heterosis may be apparent in single characters or it may be a function of crossbreds having a more nearly optimum combination of desired characters than either of the parental breeds. There is no one breed of any species that excels in all desired traits. Most traits of productive importance exhibit a degree of dominance. This leads to the supposition that judicious crossing of breeds which complement each other might result in performance somewhat above the parental averages for each trait and considerably above either parental average in total productivity. Much more experimental work is needed in all species before it will be possible to delineate with any degree of precision the usefulness of various breeds in specific crosses.

Advantages from crossing of breeds or inbred lines of farm animals appear to be of sufficient importance that commercial producers should give serious consideration to the development of systematic crossing programs if maximum performance is to be attained. This is particularly true of the species with higher reproductive rates. This is not to say that crossing is necessarily best for all situations. Crossbreeding has been a controversial subject among breeders, commercial producers, and scientists. As with most controversial subjects, it is probable that no one general recommendation will cover all cases.

In the chapters on each class of livestock we have attempted to review the experimental evidence on crossbreeding and also to point out gaps in current knowledge. Each producer must weigh the facts and himself decide whether crossing will have advantages over other breeding systems for his particular situation.

Crossbred animals usually exhibit a mixture of the breed trademarks of the parental breeds. For some people this reduces the satisfaction they get from raising livestock to the point that they abhor crossbreeding regardless of its economic merits. In today's competitive and specialized agriculture, however, most commercial producers will make their decisions on crossbreeding largely on the basis of whether or not they think it will pay.

Crossbreeding for Formation of New Breeds. Early in this century it seems to have been generally accepted by many that breed formation was over, i.e., no more breeds would be formed. However, during the twentieth century numerous new breeds have been formed—all on crossbred foundations. Sometimes this has been thought to be a new system. However, the extent to which crossbreeding entered into the formation of older breeds is often not fully realized. Of the 58 breeds of farm animals discussed by Briggs[1] as being used in America, and for which breed registry societies existed prior to 1920, at least 34 are said to have been based on crosses, "mingling of blood," or "infusions of blood," of two or more older breeds or strains. The formation of new breeds on crossbred foundations is thus not new.

The question of whether or not we need new breeds is controversial. Breeders of established breeds usually hold to the view that we already have enough breeds and that selection and the application of the proper breeding systems can mold them to meet all demands. Certainly, there is much logic to their argument that the same amount of effort expended on the improvement of an existing breed might be more profitable in many cases than the development of a new one.

Attempts to form new breeds are justified, however, if no existing breed satisfactorily meets the needs of a given geographic area or economic situation. Even though selection within an existing breed might

[1] Briggs, H. M., "Modern Breeds of Livestock," 2d ed., The Macmillan Company, New York, 1958.

eventually lead to the development of a new strain with the desired characteristics, the process can be greatly expedited by using a crossbred foundation, provided there are existing breeds whose crosses approximate the desired type. Other matters to be considered are the probable productivity of a new breed in relation to the productivity of crosses between the two existing breeds and the difficulty of maintaining systematic long-time crossbreeding programs.[1]

The necessity for building new breeds would be considerably lessened if existing breed associations would revise their rules to permit the introduction of outside genes where they obviously are needed.

In general, the creation of a new breed is something not to be taken lightly and should be undertaken only when a definite need is apparent. The number of new breeds that could be developed is virtually limitless. The drawbacks to attempting the development of new breeds are:

1. The time and expense involved.
2. The large number of animals that should be raised to give ample range for selection.
3. The difficulty that might be encountered in selling the new breed to producers even if it had real merit.

In spite of the successes some breeders have had in establishing new breeds, many failures have been recorded and doubtless many more have been forgotten.

The whole matter of the creation of new breeds is fascinating. Most breeders lack the time, capital, and equipment to start new breeds successfully, and perhaps these fields should best be left to state experiment stations or to the U.S. Department of Agriculture. However, breeders with the financial resources necessary for success, a sound knowledge of the factors involved, and the capacity for experimentation should not be discouraged if they visualize a real necessity for a new breed.

Work of the past half century in the formation of new breeds under controlled experiment-station conditions has done much to refute ideas formerly prevalent (1) that use of crossbred animals for breeding purposes is followed by wide segregation in economically important traits and a general loss in vigor, and (2) that it would be impossible to attain the uniformity of an established breed among the progeny of a crossbred foundation within the span of a human lifetime.

Most people working with intermated populations of crossbred animals have been impressed by the evidence of segregation. However, from

[1] See Warwick, E. J., in A. O. Rhoad, ed., "Breeding Beef Cattle for Unfavorable Environments," pp. 133–149, University of Texas Press, Austin, Tex., 1955.

published reports it is difficult to find consistent evidence of greater variability among the progeny of crossbred matings in such quantitative characters as growth rates, milk and butterfat production, and fleece characters than is found in long-established breeds.

On theoretical grounds we would not expect greatly increased variability since:

1. No farm animals are homozygous for genes influencing quantitative traits.
2. Such characters are known to be influenced by many pairs of hereditary factors.
3. They are influenced by environment, usually being only 20 to 50 per cent heritable.

To emphasize these points, let us assume that the two foundation breeds crossed were each completely homozygous—one for all the plus genes affecting a character such as size, and the other for all the minus genes—and that the character is not affected by environmental differences. The F_1 would be completely heterozygous but would exhibit no variation. Different genotypes would be formed in the F_2 according to the expansion of the binomial $(a + b)^n$, where a represents genes from one parental breed, b genes from the other parental breed, and n the number of pairs of genes involved. If only two pairs of genes were involved, $\frac{1}{4}$ of the F_2 individuals would be as extreme as each parental breed, and variability would be readily apparent. However, if as many as ten pairs were involved, only 1 in 2,048 of the F_2 individuals would be this extreme, and a high proportion would be near the F_1. With the usual small populations found in farm animals, the probability of the extreme types being produced is low. If as many as 30 or 40 gene pairs were involved, variation would be even less apparent. Now if we superimpose on this the facts that none of the parental types are homozygous and that environment does exert an influence on variability of both parental stocks and segregating generations, it is apparent that variability would be little increased.

This leads to the conclusion that many of the characteristics on which ideas of variability are based, such as color, horns, and details of conformation, must be based on the action of relatively few gene pairs and must be highly hereditary. They would thus be relatively easy to fix in a new breed.

Available data indicate that new breeds can be made as uniform or as nearly uniform in most characters as the established parental breeds in a relatively few generations. Thus, this is not the problem it was once assumed to be in breed formation.

CROSSING INBRED LINES FOR COMMERCIAL PRODUCTION

The hybrid corn used commercially is produced by crossing highly inbred lines. Where adapted hybrids are available, they have usually considerably outyielded the best varieties produced by older breeding methods. This superiority is usually thought of as being due to heterosis. Much of it is due also to successful choices of inbred lines which put together favorable combinations in the hybrids. These favorable combinations occur uniformly in every hybrid plant owing to the homozygosity of the parent lines. The higher yields of hybrids are due largely to uniform production from nearly all plants rather than to extraordinary yields by some plants.

As we have seen, most breeds of farm animals are slightly inbred, they differ somewhat from each other genetically, and crosses between them usually show a small-to-moderate amount of hybrid vigor. This leads to the supposition that crosses between highly inbred lines might exceed the performance of presently existing types.

Before such an hypothesis could be tried it was necessary to develop inbred lines. This is a much more difficult task with farm animals than with corn or other plants for the following reasons:

1. Self-fertilization is impossible with animals, thus more than twice as many generations are required to develop the same degree of inbreeding in animals as in plants.
2. Because of higher costs of maintaining and raising each individual (or genetic unit), the cost of developing inbred animal lines is great.
3. Because of the expense of formation and development, fewer lines can be developed, thus lessening the amount of culling of lines which is possible.

In spite of these difficulties, inbred lines of all species of farm animals have been developed, usually to inbreeding levels of not over 30 to 60 per cent, although a few lines have been carried to much higher levels. More detailed results are given in the chapters on each species, but as generalizations it can be stated that:

1. Productivity is reduced as inbreeding levels increase with some lines being affected more than others. The traits of low heritability related to fertility and viability are usually affected most drastically. Some lines have been lost due to infertility or high death rates, but it has been possible to maintain most of those started to inbreeding levels of at least 40 to 50 per cent.
2. Linecrosses have on the average restored the productivity lost with inbreeding, but the average performance of all linecrosses is apparently

not above that of non-inbred stocks. This leads to the conclusion that inbreeding in itself does not create genetic situations which, in crosses, will uniformly result in enough heterosis to increase productivity above that of outbred animals.

3. Crosses of specific lines have often performed at levels apparently above those of outbred animals. This suggests the potential usefulness of the technique, but as yet it is still in doubt whether results are better from the selected crosses than could have been attained with the same expenditure of time, effort, and money by selection within outbred stocks and the crossing of non-inbred strains within breeds or by breed crosses.

SELECTION FOR COMBINING ABILITY IN CROSSES

From the foregoing, the discerning student will have gathered that, at the present state of knowledge, performance of two or more breeds or lines in crosses is somewhat unpredictable. This is true to a degree for all traits but much more so for those of low heritability. Some lines or breeds appear to "combine well" whereas others do not. This can in most cases be determined only by test crosses. Thus, a breeder attempting to produce lines which will combine well with each other presently has no alternative but to form a large number of lines. He can then test them in crosses and find those which give best results. This is expensive, time consuming, and uncertain.

For highly heritable traits, the straightbred performance of breeds and lines is a fairly good indicator of performance of cross progeny. For the less highly heritable traits, which usually show greatest heterosis, no predictors studied to date have been very reliable in estimating cross-progeny performance. As a general rule, lines or breeds most diverse in origin (i.e., the most generations away from ancestors they have in common) give greatest heterosis in crosses, but this is not universally true.

The discussion thus far has stated in effect that heterosis is useful and that advantage should be taken of it when crosses are found which result in superior performance. By implication it could have been assumed that nothing was known about how to improve cross performance. From a practical standpoint this is currently almost true with farm animals. However, a brief consideration of research aimed at improving heterosis or cross-progeny performance may be of interest. Obviously, critical knowledge on the subject must await a better understanding of the nature of heterosis.

On the assumption that overdominance is of importance in heterosis, a breeding system of *recurrent selection* was proposed.[1] In this system, a highly inbred line, presumably homozygous at most loci, is selected as a tester. A large number of individuals are tested in crosses with this line. Those giving the best results are intermated and a large number of their progeny again tested on the inbred tester. The cycle is repeated over and over.

If heterosis is largely dependent upon overdominance, this procedure should result in the line selected on cross performance becoming homozygous for different alleles than the inbred used as the tester. In other words, where the tester is *aa*, the selected line would become *AA*; where the tester is *BB*, the selected line would become *bb*, etc.

Subsequently, a system of *reciprocal recurrent selection* was proposed[2]. In this system, randomly selected representatives of each of two non-inbred strains are progeny tested in crosses with the other. Those individuals of each strain having the best cross progeny are then intermated to propagate their respective strains. Offspring from these within-strain matings are again progeny tested in crosses with the other and the cycle repeated. The system should lead to improved cross performance whether it is the result of overdominance, dominance, epistasis, or only additive effects.

Both these systems involve progeny testing. Due to the increased generation intervals, this would be expected to result in slower progress than other breeding systems for characters moderate to high in heritability. They would be expected to be more useful than other breeding systems only if overdominance or other nonadditive types of inter- or intra-allelic gene action are important in heterosis.

Experimental tests of these techniques have shown promise in improving egg-laying capacity in two species of the fruit fly *Drosophila*.[3] This trait is low in heritability and shows a high degree of heterosis in crosses. Even with this trait, both groups of workers found that some crosses of highly inbred lines, developed without selection from the same foundations, equaled or exceeded the productivity of strains developed after 16 to 39 generations of recurrent selection by the new techniques. It was also found that for the highly heritable trait of egg size, which exhibits some heterosis in crosses, conventional breeding and selection techniques gave greater improvement than either type of recurrent selection.

Several experiments to test the potential usefulness of these techniques have been started with farm animals. In the only known report to date[4]

[1] Hull, Fred H., *Jour. Amer. Soc. Agron.*, **37**:134–145, 1945.
[2] Comstock, R. E., et al., *Agron. Jour.*, **41**:360–367, 1949.
[3] Bell, A. E., et al., *Cold Spring Harbor Symposium on Quant. Biol.*, **20**:197–212, 1955; and Kojima, Ken-Ich, and Kelleher, Therese M., *Genetics*, **48**:57–72, 1963.
[4] Flower, A. E., et al., *Jour. Anim. Sci.*, **23**:189–195, 1964.

on these trials, most of the progress made in growth of beef cattle could be accounted for by selection of the bulls which were progeny tested on the tester line. In this experiment, bulls were selected on the basis of growth and conformation for progeny test on an inbred tester line. Due to limited numbers, about half the bulls progeny tested had to be used later in the selection line. Thus, selection on the basis of progeny produced in crosses with the tester line was itself not intense. For this reason, as well as the fact that the characters considered in selection were moderate to high in heritability, the results to date should not be interpreted as indicating the method may not have usefulness with farm animals. Rather the results emphasize the desirability of large numbers of animals in such studies.

In some cases[1] sire progenies have been shown to rank in different orders for economically important traits in progeny tests on females of another strain or breed as compared to progeny tests on females of their own strain or breed. This would suggest the potential usefulness of selection on the basis of cross-progeny performance if the strain or breed is going to be used commercially in crosses.

However, at present it must be considered as uncertain whether the techniques of recurrent selection have a potential usefulness in farm animals. If they do, it will likely be in breeds or strains in which performance is already high for highly hereditary traits and in which it is desired to improve the potential performance of their cross progeny for the lowly heritable traits related to fertility and liveability.

SPECIES HYBRIDIZATION

The widest possible kind of outbreeding is the cross of two species. A good example of this is the cross between the jack of the ass species and the mare of the horse species to produce the mule. It seems quite probable that these two species may have descended from common parent stocks far back in the course of evolution. In the dim past, natural selection, working on variation (recombinations, mutations, and chromosomal aberrations), began to set them off into distinct species. They still have many genes in common, but they also have enough different ones so that they are recognized as distinct species. Up to now they have enough similarity so that the sperm of one will fertilize the ova of the other and produce vigorous offspring. The mule has been valued for centuries for its ability to work under difficult conditions and to withstand abuse.

[1] Robison, O. W., et al., *Jour. Anim. Sci.*, **23**:853, 1964 (abstract).

Fig. 9-3. *Top,* U. T. Logan Again, a registered American Jack. *Center,* a draft mare. *Below,* a mule of the type expected from mating animals such as those above. (*Courtesy of Professor H. R. Duncan, Tennessee Agricultural Experiment Station.*)

Genetic differences between the horse and the ass are apparently so great, however, that the male mules are always sterile and the females usually so. The separation of the chromosomes at the time of germ-cell formation is usually abnormal, and gametes are either not formed or are incapable of fertilization. At least three fertile mare mules have been reported, however. In 1920, L. T. Branham of Montalla, Texas, reported that a mare mule in his possession had dropped a living foal to the service of a jack. The contention was supported by affidavits of the owner and some neighbors. The mare was loaned to Texas A. & M. College and, after failing to conceive when bred to a jack in 1921, dropped a living stallion foal in 1923 as the result of being bred to a Saddle stallion the previous year. Other fertile mare mules have been reported in Indiana and Arizona.

In all cases, these apparently fertile mare mules have bred as if their fertile ova contained only horse chromosomes. Thus their breeding behavior has been the same as that of mares. They produced foals that resembled mules when bred to jacks and foals that were indistinguishable from horses when bred to stallions.

An examination[1] of the chromosomes of another allegedly fertile mare mule showed them to be like those of an ass, thus raising the question of whether she was really a mule. A critical review of other reported fertile mare mules led these workers to question whether any cases had been irrefutably proven.

Several other species hybrids have been produced. Among them are:

1. The *hinny* is the reciprocal of the mule, being the result of the stallion-jennet cross. It is said to resemble the horse more than the mule does. It has never become popular in the United States.

2. The *zebroid*, a zebra-horse hybrid, is fairly popular in the tropics because of its docility and its resistance to disease and the effects of heat.

3. The *pien niu* is produced by crossing cattle and the yak in Tibet.

4. The *cattalo* is a term applied to various combinations of domestic cattle and the bison (American buffalo). Male calves from the cross of bison bulls on domestic cows usually die at birth and if they survive are sterile. Females live, are fertile, and produce offspring when bred to domestic bulls. Produce of this and other cross combinations include some males with a degree of fertility so that it is possible to maintain closed populations descending from the hybrids. Much exploratory and experimental work has been done, mostly in Canada, on the potential usefulness of these animals for meat production in cold climates. To date, the trials have not been successful in developing animals which have achieved commercial acceptance.

[1] Benirschke, Kurt, et al., *Jour. Hered.*, **55**:31–38, 1964.

Fig. 9-4. Example of an American Bison-cattle hybrid. This female is ½ American Bison, ¼ Angus, and ¼ Holstein. (*Courtesy of Pierce Rosander, Belle Fourche, South Dakota.*)

5. The humped cattle of India (*Zebu* type) are classed as a different species from European-type cattle. As will be discussed in some detail in the chapter on Improving Beef Cattle, crosses between these two species are quite commonly used in the Southern United States and other subtropical and tropical areas of the world. Their hybrids are fertile in both sexes. There is some doubt that the two groups should be called different species, and their hybrids are usually called crossbreds.

The sheep and goat are related groups, but apparently not so closely related as the above groups. When crosses are made between them, fertilization sometimes takes place and the embryos develop for a time. However, they always die before normal parturition and are resorbed or aborted. When crosses are made between other less closely related species, fertilization does not even occur.

SUMMARY

Outbreeding is a general term applied to any breeding system in which animals mated are less closely related than the average of the population from which they come. Outcrossing combined with selection is a highly useful technique for within-breed improvement for moderately to highly hereditary characters. Heterosis is the extra performance above the averages of parental types often observed in crosses between breeds, inbred lines, or species. Its genetic and physiological bases are not clearly understood, but in within-species crosses it is most apparent for the lowly heritable traits related to fertility and viability. Although not well understood, heterosis can be used to advantage in commercial production through crossing breeds and perhaps inbred lines within breeds. Breeding techniques designed to increase heterosis or improve combining ability of lines may have applicability in domestic animals. Species crosses, of which the horse-ass cross resulting in the mule is the best known, result in much heterosis when species are closely enough related to be cross fertile. Fertility is very low or nil among the progeny of most species crosses.

REFERENCES

Bowman, J. C. 1959. Selection for Heterosis, *Anim. Breed. Abstracts,* **27**:261–272.

Comstock, R. E. 1960. Dominance, Genotype-Environmental Interaction, and Homeostasis, in O. Kempthorne, ed., "Biometrical Genetics," pp. 3–9, Pergamon Press, New York.

de Boer, H. 1960. "Cattle Herd Books in Europe," European Association for Animal Production, Rome, Italy.

Gowen, John W., ed. 1952. "Heterosis," The Iowa State University Press, Ames, Iowa.

Hayes, H. K. 1963. "A Professor's Story of Hybrid Corn," Burgess Publishing Company, Minneapolis.

King, James C. 1961. Inbreeding, Heterosis and Information Theory, *Amer. Nat.,* **95**:345–364.

Lerner, I. Michael. 1954. "Genetic Homeostasis," John Wiley & Sons, Inc., New York.

Lerner, I. Michael, and Donald, H. P. 1966. "Modern Developments in Animal Breeding," Academic Press Inc., London and New York.

Smith, Charles. 1964. The Use of Specialized Sire and Dam Lines in Selection for Meat Production, *Anim. Prod.* **6**:337–344.

Sprague, G. F., ed. 1955. "Corn and Corn Improvement," Academic Press Inc., New York.

CHAPTER TEN

Principles of Selection

Selection has been aptly termed the *keystone of the arch* of animal improvement. Each generation the breeder must select from his herd or other herds those animals which are to be the parents of the next generation. The influence which each of them has on the genetic make-up of the herd will be determined by the number of offspring (samples of genes) which each contributes to the next generation. Selection does not create new genes, but by permitting the possessors of certain genes to leave more offspring, it increases the frequencies of the more desirable alleles.

Selection is not an invention of modern man. It has been going on in nature since life first existed on the earth. In nature the animals best adapted to their environment survive and produce the largest number of offspring. This natural selection, or survival of the fittest, acting upon the variations produced by mutation and recombinations of genetic factors, eliminates the unsuccessful genetic combinations and allows the most successful to multiply. Selection by man represents the addition of new standards—related to an animal's ability to serve human needs—to the natural standard of ability to survive and reproduce.

Selection may be practiced at many stages in the life cycle of an organism. Some individuals may not be born because their potential parents were culled. Others may experience birth but be culled before being allowed to reproduce. Certain individuals may be selected on the basis of their pedigree or own performance, but they may be culled when the information on their progeny test becomes available. Still other individuals may be retained to produce as many offspring as practicable. In farm animals, the selection process also may be influenced by economic considerations. What might be most desirable for genetic improvement may have to be tempered somewhat by economic considerations.

CHARACTERISTICS OF QUANTITATIVE TRAITS

The change in traits due to selection stems directly from changes in the frequency of genes influencing the traits. Selection in practice can seldom be for genes at a single locus. Actually selection cannot be based solely on the effects of the genes which may influence the expression of a single desired trait. Rather the entire individual or zygote must be chosen.

Most of the traits of economic importance for farm animals are quantitative in nature and possess the following characteristics:

1. They are influenced by many genes, most of which probably individually have small effects. Seldom, if ever, is it possible to identify individual gene effects on a quantitative or metric trait.
2. Although we are far from having complete knowledge of the nature of gene action, it appears that additive, dominance, and epistatic effects are all expressed, their relative importance varying from trait to trait.
3. The expression of the genes is greatly affected by environmental influences. Consequently the traits have a continuous distribution with no sharp demarcation between "good" and "bad."

EFFECTIVENESS OF SELECTION

The initial task in selection is to obtain an accurate estimate of the average or additive breeding values of the animals available for selection. The breeding values of the individuals are predicted from available information, such as their own performance and their relatives' performance. Following the approach of Lush, much of our consideration of selection alternatives in this chapter will center around methods for enhancing the accuracy of selection, i.e., increasing the correlation between breeding value of the individual G and the information I used for selection (r_{GI}). Most often I will be the individual's phenotype P.

The simplest quantitative expression for the expected genetic change due to selection is that genetic change equals the heritability of the trait times the selection differential. In mathematical form this can be expressed:

$$\Delta G = h^2(\bar{P}s - \bar{P})$$

where ΔG = expected genetic change

$\quad h^2$ = heritability of trait

$(\bar{P}s - \bar{P})$ = selection differential or difference between mean of selected parents $\bar{P}s$ and average of population $\bar{P}$ for their generation

We shall want to express the above equation in a different form to enumerate more clearly the factors which influence the efficiency of the selection process. The prediction aspect of selection has been emphasized and the above equation for ΔG can be expressed as a prediction equation following the form given on page 139. The predicted average breeding value $\hat{G}$ for the next generation is equal to the mean breeding value $\bar{G}$ of the herd or population plus the average genetic superiority of the animals selected as parents.

$$\hat{G} = \bar{G} + h^2(\bar{P}s - \bar{P})$$

The average genetic superiority can also be expressed as

$$\Delta G = (G_s - \bar{G}) = h^2(\bar{P}s - \bar{P})$$

In each of these expressions, the value for heritability can be considered a regression coefficient. For individual selection it represents the regression of breeding value on phenotypic value (b_{GP}) or the change in average breeding value for a unit of phenotypic selection. Thus we may write

$$\Delta G = b_{GP}(\bar{P}s - \bar{P})$$
$$= r_{GP}\frac{\sigma G}{\sigma P}(\bar{P}s - \bar{P})$$

and the second expression follows from the relation between the regression coefficient and the correlation coefficient, page 139.

Most biological measurements are distributed normally. If the percentage of the population which is saved from a normally distributed population is specified, functional relations are available to express the selection differential in terms of the number of standard deviations the mean of the selected parents exceeds the population mean.[1] Thus the selection differential can be written where i is the selection differential in standard deviation units and σP is the phenotypic standard deviation of the trait.

$$(\bar{P}s - \bar{P}) = i\sigma P$$

The above expressions can now be put together to express ΔG in an alternate manner:

$$\Delta G = r_{GP}\frac{\sigma G}{\sigma P}i\sigma P$$

---

[1] A more thorough coverage of this point will be given in the subsequent section on Intensity of Selection.

which can be reduced to provide the following expression for ΔG:

$$\Delta G = r_{GP}i\sigma G$$

where r_{GP} = correlation between breeding value and phenotype, or
        measurements on which selection is based
    i = intensity of selection in equivalent standard deviation units
   σG = standard deviation of breeding values

The significance of each of these factors will be considered in subsequent discussions.

One further point should be presented at this time. Man is impatient in his selection and seeks to accomplish in a few years what nature may be content to wrestle with for centuries. Hence, a measure of genetic improvement per unit of time is often a useful practical concept.

$$\Delta G_t = \frac{1}{L}\,(r_{GP}i\sigma G)$$

where ΔG_t is the genetic change per unit of time and L is the average generation interval.

Accuracy of Selection. The correlation coefficient measures the association between two variables. If they are closely related, the value for one variable can be accurately predicted from the second variable. The accuracy of the prediction varies directly with the magnitude of the correlation coefficient. For maximum accuracy in selection, the correlation between the average breeding value and the variable(s) on which the selections are based should be as high as practicable.

Increased accuracy in selection can be obtained by comparing breeding stock under as controlled environmental circumstances as can be provided. Corrections also may be made for age of the individual, age of dam, and sex to remove much nongenetic variation. These techniques also tend to increase the heritability of the trait by reducing the environmental component of variance in the denominator of the heritability fraction. Actually the heritability of a trait and the correlation between breeding and phenotypic values are directly related, since $h^2 = (r_{GP})^2$.[1] The accuracy of selection for a trait is tied directly to its heritability. If heritability is high, selection on phenotype will permit an accurate

---

[1] This relationship can be demonstrated by several approaches. Perhaps the most direct at this point is to recall (Chap. 6) that $(r_{GP})^2$ would represent the fraction of the phenotypic variance associated with variation in average breeding values. Thus $(r_{GP})^2 = \sigma^2 G/\sigma^2 P = h^2$.

sorting of breeding values. If heritability is low, many errors will be made when trying to select the best breeding animals based on their own performance.

When the accuracy of selection on individuality is too low to make reasonable genetic improvement, accuracy can be enhanced by:

1. Using additional measurements for the trait from the same individual
2. Using measurements of correlated traits
3. Using measurements on relatives (family selection)

to predict the breeding values. These measurements represent aids to individual selection which will be discussed later.

Intensity of Selection. The intensity of selection is measured by the magnitude of the selection differential. The proportion of the population which must be saved places an upper limit on the size of the selection differential. It is largely determined by the reproductive rate of each species. Nevertheless, some additional opportunity to select can be gained through good husbandry practices which reduce losses from disease and other nongenetic factors. The use of artificial insemination and frozen semen materially increases the selection intensity which can be exercised in the choice of males. Superovulation and ova transfer have been accomplished experimentally in some species, and they are intriguing possibilities for enhancing the selection intensity among females in the future.

The percentage of the population which must be saved is dependent on whether or not a population is expanding, remaining constant, or decreasing in size. In addition, the proportion that must be saved for populations which remain constant in size varies markedly from herd to herd due to managemental difference suggested in the previous paragraph. Table 10-1 gives what are believed to be reasonable estimates for the percentage of the males and females which must be retained in the breeding herd to maintain constant herd size. These values can only be taken as guides since vital statistics on livestock populations are not very adequate.

Selection differentials in standard deviation units for different selection intensities are shown in Table 10-2. They are equal to z/p, or i (intensity of selection), where p is the proportion saved and z is the height of the normal curve at the point where the selected individual with the lowest record falls. Selection of this type is often referred to as truncation selection. All individuals included in the normal distribution above the specific culling level would be selected, and those below this level are culled (see Fig. 10-1).

When the value for z/p, or i, is multiplied by the standard deviation

Table 10-1. *Approximate Percentages of Animals Required for Breeding to Maintain Constant Numbers*

Class of livestock	Percentages needed	
	Males*	Females
Beef cattle	4–5	40–50
Dairy cattle	4–5	50–60
Sheep	2–4	40–50
Swine	1–2	10–15

* Could be reduced to ½ per cent or less with artificial insemination.

of the trait being selected, this gives the maximum selection differential attainable when a given percentage of a herd or population must be retained for breeding. Selection differentials of this magnitude can be attained only if all the animals above a certain point in the frequency distribution are selected and all those below it are culled. Failure to attain this intensity of selection for important characters or for an index can result from carelessness, inadequate evaluation, or from being influenced in selection by some unimportant traits. Values of the selection differential when various percentages of the population are saved are given in Table 10-2.

Genetic Variability. The genetic standard deviation σG is characteristic of the trait itself and it is somewhat difficult to change. Crosses among divergent stocks may be used in certain instances to increase

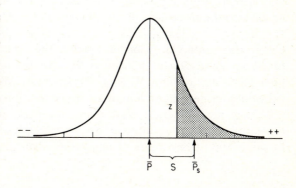

Fig. 10-1. The figure represents truncation selection in which only individuals whose phenotypes are one or more standard deviations (represented by shaded area) are selected. This represents saving approximately 16 per cent of the population (p). The selection differential (z/p) is approximately 1.53 standard deviations above the mean for the original population.

Table 10-2. Selection Differentials Attainable with Truncation Selection When Specific Fractions of a Population Are Retained for Breeding

Fraction saved (p)	Selection differential*	Fraction saved (p)	Selection differential*
0.90	0.20	0.09	1.80
0.80	0.35	0.08	1.86
0.70	0.50	0.07	1.92
0.60	0.64	0.06	1.99
0.50	0.80	0.05	2.06
0.40	0.97	0.04	2.15
0.30	1.16	0.03	2.27
0.25	1.27	0.02	2.42
0.20	1.40	0.01	2.67
0.15	1.55	0.005	2.89
0.10	1.76	0.001	3.37

* Selection differentials in standard deviation units.

the genetic standard deviation and to provide a foundation population with a broader genetic base for selection.

An idea of the importance of the magnitude of the genetic standard deviation on the absolute response to selection can be gained from an example with milk composition in dairy cattle. The heritability of both fat percentage and protein percentage is 0.50 to 0.60. Yet the genetic standard deviation for fat percentage is about 0.30 and for protein about 0.20. Hence, 50 per cent more absolute change in fat percentage would be anticipated with the same intensity of selection for both traits.

Concern is often expressed over the loss of initial genetic variability when selection is continued for many generations. Some increase in homozygosity eventually may result from selection. Also some unavoidable inbreeding will occur when particular individuals are selected and their influence is concentrated. In a program of selection and outbreeding, loss of original genetic variability should not be a major deterrent to selective improvement during the span of most breeding operations.

Generation Interval. Conventionally the generation interval for a breeding population is defined as the average age of the parents when their offspring are born. This of course varies with species. Some practical values for generation length are given in Table 10-3.

Management practices that allow for early breeding of individuals can reduce generation interval. For example, when beef heifers are bred to calve first at two years of age rather than at three years, the female

Table 10-3. Approximate Generation Interval for Various Classes of Livestock

Class of livestock	Average generation length in years		
	Male		
	Mass selection*	Progeny testing†	Female
Beef cattle	3.0	8.0	4.5
Dairy cattle	3.0	8.0	4.5
Sheep	2.0	4.0	4.0
Swine	1.5	3.0	1.5

* Males used early and replaced after two breeding seasons.
† Males to be used three breeding seasons after progeny test is complete.

generation interval can be reduced approximately one year. On the other hand, certain breeding practices such as progeny testing extend the generation interval, and this should be reckoned with in assessing the total merit of such a selection system.

INDIVIDUAL SELECTIONS

Selection on the basis of individual merit is strictly phenotypic. It is the most commonly used basis for selective improvement in livestock. Undoubtedly most of the progress in livestock improvement to date can be credited to individual selection.

Such traits as body type, growth rate, fleece production, and others of a similar nature can be evaluated directly from the performance of the individual animal, if suitable performance records are being kept. Such evaluations (or at least fairly accurate preliminary estimates of them) are usually available by the time initial selection of breeding stock has to be made. Furthermore, an evaluation of the individuality of all animals can be made. In contrast, only a few can be progeny tested.

Individual selection also has shortcomings which can be summarized as follows:

1. Several important traits, including milk production in dairy cattle, maternal abilities in brood cows, ewes, and sows, and egg production

in poultry, are expressed only by females. Thus, selection of breeding males cannot be based on their own performance.

2. Performance records for milk and egg production and other maternal qualities are available only after sexual maturity is reached.
3. In cases in which heritability is low, individual merit is a poor indicator of breeding value.
4. The easy appraisal of appearance (or "type") often tempts the breeder to overemphasize this evaluation in selection. This is true even in meat animals, in which there is a relationship between appearance and carcass value.

In spite of these shortcomings, individual merit certainly must be considered in selection. In general, for traits expressed by both sexes, only animals which are themselves above average should be used for breeding, regardless of the merit of close relatives.

Repeated Observations. The concept of repeatability was introduced in Chap. 7. For some characters, repeated observations are possible. Use of all available records can increase the accuracy of selection for characters affected by temporary environmental conditions by minimizing the effects of these conditions, thus reducing the number of mistakes made in selection. The best estimate of producing ability expressed as a deviation from the herd average is

$$\frac{nr}{1 + (n-1)r} \text{ (individual average } - \text{ herd average)}$$

where n is the number of records or observations per individual and r is the repeatability. Several important traits in farm animals, including fleece weight in sheep, lactation records in dairy cows, and weaning weights of successive calves from the same beef cow, have been shown to have repeatabilities of around 0.40 to 0.60. For a repeatability of 0.40, the accuracy of evaluating an animal on the basis of two records should be increased by 43 per cent as compared with evaluation on the basis of one record, and 67 per cent if three records are available.

The breeder is interested in properly evaluating animals with varying numbers of records, if some are to be culled and some retained. An example of a problem of this kind in beef cattle is given in Table 10-4. The cows listed had records (adjusted to a male sex and mature cow equivalent) as indicated after their calves were weaned. What is the best estimate of the producing ability of each cow for this trait?

Other studies indicate that repeatability of the weaning weight of calves of beef cattle is approximately 0.4. Using this value for repeatability and applying the formula given earlier, we arrive at the figures in the right-hand column as the best estimates of the individual cow's real producing ability. Note that the estimate of real producing ability for cow 18, with one record 47 lb above the herd average, is only slightly higher than that of cow 1, with three records averaging 25 lb above the herd average. This does not say that the real producing ability of cow 18 could not be 47 lb above the herd average. It merely says that, in the absence of further records, the best estimate is that her high record was due in part to a particularly favorable set of environmental circumstances. Future calves by this cow probably will be nearer the herd average. Similar reasoning will permit making comparisons among other animals in Table 10-4.

Repeatability of litter size and weight for successive litters of the same sow is about 0.15 to 0.20. If we assume it to be 0.20, the increased accuracy of evaluations based on two and three records is 67 and 114 per cent, respectively, over that of one record. Comparison of these figures with those given previously for repeatabilities of 0.40 illustrates the fact that increasing the number of records increases the accuracy most if repeatability is low, i.e., when the effects of temporary environmental factors are large.

Additional records for traits with high repeatability are not very useful simply because the first record has already accurately evaluated the animal. Thus one record for a characteristic of high repeatability can give a better indication of real producing ability than will several records for a character with low repeatability.

In view of the finality and permanence with which we commonly consider something like the winning of an important grand championship, it may come as a shock to some students to find that the repeatability of type ratings made at intervals of six months to a year is of approximately the same magnitude as the repeatability of milk production or

Table 10-4. Calf Production Records of Some Beef Cows

Cow no.	No. of records	Average deviation of calf weights from herd average	Probable producing ability as a deviation from herd average
18	1	+47	+18.8
1	3	+25	+16.7
9	2	+22	+12.6
12	1	+20	+8.0
2	1	−14	−5.6
6	3	−16	−10.7
25	1	−30	−12.0

calf-weaning weights. This means that repeated observations on type would be useful in evaluating this character.

A knowledge of repeatability provides a guide to the amount of culling that can safely be done on the basis of one record. With repeatabilities of the order of 0.40, for instance, the lower 10 to 20 per cent of a group can be culled on the basis of one record with little danger of culling an animal that would later be in the upper 25 to 50 per cent of the herd. Doubtful cases can be retained for a second and possibly a third record. Each additional record does add a little to the accuracy of appraisal of the individual. Nevertheless, with repeatabilities of this size most of the value of repeated records will have been obtained from the first two or three, and there is little to be gained in waiting for more.

If each additional record is equivalent to any of the previous records individually, the average of n records should be $\sqrt{n/[1 + (n-1)r]}$ times more accurate for selection than a single record. The correlation between the breeding value and one phenotypic measurement is r_{GP} and this would be increased to $\sqrt{n/[1 + (n-1)r]}(r_{GP})$ using averages. Values of the ratio of the expected accuracy of using averages compared to the accuracy of basing selections on individual records $\sqrt{n/[1 + (n-1)r]}$ are given in Table 10-5.

Waiting for additional records before deciding whether to keep or cull an animal can be costly if inferior animals are kept for extended periods. It can lengthen the generation interval and possibly reduce total annual progress in selection for some traits. This could happen if a dairy-cattle breeder insisted on having five or six records from a cow before saving one of her sons. Thus, the advantages of getting additional

Table 10-5. Relative Accuracy of Selection Expected When Using Averages of Records As Compared to Single Records As Compared to Single Records for Individual Selection

Number of records (n)	Repeatability (r)				
	0.10	0.30	0.50	0.70	0.90
2	1.35	1.24	1.15	1.08	1.03
3	1.58	1.37	1.22	1.12	1.04
4	1.75	1.45	1.26	1.14	1.04
5	1.89	1.51	1.29	1.15	1.04
10	2.29	1.64	1.35	1.17	1.05

records must be balanced against the costs and disadvantages of getting and using them. In addition there are some suggestions in the literature that all lactations of dairy cattle may not provide equal information. The first lactations have had higher heritabilities than other records in a number of studies. Permanent changes in the animal due to such things as udder damage may prevent all records from being of equivalent value in predicting future performance. Additional records would be necessary and desirable to evaluate lifetime productivity, but the average of the two or three first expressions of the trait probably include most of the information that can be obtained for predicting genetic merit for individual expressions of the trait.

PEDIGREE SELECTION

The pedigree of an animal is a record of the animals which are related to him. If only the genealogy of the individual is given, the pedigree is of extremely limited value. From a practical standpoint, knowledge of the productivity of the ancestors is necessary if the pedigree is to be useful. Such information formerly was available only in dairy-cattle pedigrees, but now it is becoming available for beef cattle and swine.

We pay attention to the pedigree because we do not have adequate information on the merit of the individual. In some instances some selections may need to be made before the individual expresses the traits(s), such as feed-lot gain and feed efficiency. In other instances the trait may be sex-limited in its expression, such as milk yield and sow productivity, and selection of males prior to the progeny test must rely on pedigree information. If the individual's phenotype is known with precision, little is gained by attention to the pedigree. Of course, nothing would be gained from pedigree information if the exact breeding value of the individual were known. In fact, attention to pedigree information could introduce errors in selection, when the evaluation of the individual was already complete.

Pedigree information is valuable because each individual receives half of its genes from each parent.[1] It must be remembered, however, that each parent transmits only a sample half of its genes to each offspring. At loci where heterozygosity exists, chance at segregation will determine which gene is transmitted from that locus. Even if heritability were 1.0, permitting us to know the exact breeding values of each parent, the correlation between a parent's breeding value and that of the progeny would only be 0.50. The same factor prevents our approaching

[1] With the exception of sex-linked genes.

Fig. 10-2. Pictorial representation in which the relative size of the ancestors is proportional to their contribution to the inheritance of an individual. The folly of putting much emphasis on one or two remote ancestors in evaluating a pedigree is evident.

a correlation of 1.0 between the average of the parents and the offspring. Its upper limit in a random-mating population is 0.71, and this is one of the limitations of pedigree selection. Of course, with lower heritability values the expected correlations between pedigree information and the individual's breeding value would be much less than the 0.50 or the 0.71 given above.

Probably the two greatest dangers of pedigree selection are (1) undue emphasis on relatives, particularly remote relatives, with the result that intensity of individual selection is reduced, and (2) unwarranted favoritism toward the progeny of favored individuals.

The history of purebred breeding has been marred by numerous instances of unwise pedigree selection on the basis of remote relatives or meaningless family names. The Duchess family of Shorthorns is one of the most notorious examples. The noted pioneer Shorthorn breeder, Thomas Bates, founded the Duchess family. In their time, the members of the Duchess family were the best of the breed. Unfortunately, years after the death of Bates, American breeders began placing undue emphasis on the Duchess strain and propagated them irrespective of their individual merit. This craze culminated in the famous Mill, New York, sale in 1873, in which 109 head of Duchess cattle averaged $3,504 and one cow sold for $40,600. Indiscriminate inbreeding and lack of selection made the strain virtually extinct a few years later.

Guides to Pedigree Evaluation. The appropriate attention to the performance of each ancestor depends on the correlation of the ancestor's phenotypic merit and the breeding value of the pedigreed individual. The emphasis which should be given individuals in the pedigree also varies with the heritability of the trait. Table 10-6 shows the relative weights that should be given to the performance of each parent and each grandparent when the phenotypes of each of these is based on a single record. Study of Table 10-6 makes clear that:

1. Ancestors more closely related to the individual (parental $R = 0.50$, grandparental $R = 0.25$) should receive most emphasis in pedigree appraisal.
2. When the heritability of the trait is low, the more remote ancestors should receive relatively more emphasis, but when heritability is high they provide almost no new information.
3. Pedigree selection is much more accurate when the heritability of the trait is high. The correlation between pedigree information and the individual's breeding value approaches the theoretical 0.71 as heritability approaches 1.0.

Although there are distinct limitations to pedigree selection, it definitely has a place in breeding practices. Pedigree selection is particularly advantageous for initial selection for traits which are expressed by only one sex. Such selection can also be done early and inexpensively. Such pedigree evaluation is now most important in the selection of dairy sires for progeny testing. With progeny-testing programs and artificial insemination, only a most limited number of male calves can be progeny tested. It is extremely important that the pedigree selection be as accurate as it can be reasonably made to insure that the best potential sires have an opportunity to be progeny tested.

Table 10-6. Relative Emphasis on Parents and Grandparents in Pedigree Evaluation

Ancestor	Heritability of trait				
	0.10	*0.20*	*0.40*	*0.67*	*1.00*
Parent	0.151	0.205	0.275	0.350	0.500
Grandparent	0.071	0.091	0.103	0.088	0
Correlation*	0.265	0.364	0.489	0.599	0.707

* Correlations between the appropriately weighted phenotypes in the pedigree and the individual's breeding value for outbred pedigree.

PROGENY AND SELECTION

It has been said: "Individuality tells us what an animal seems to be, his pedigree tells us what he ought to be, but the performance of his progeny tells us what he is." The idea of progeny testing is not new, having been advocated 2,000 years ago by the Roman, Varro. Robert Bakewell is reported to have used it in the eighteenth century by letting out bulls and rams on an annual basis. Then he could later use those which proved to be outstanding transmitters.

Progeny testing has been much heralded in both the popular and the scientific press. Oftentimes exaggerated claims as to its accuracy and usefulness have been made. Progeny testing attempts to evaluate the genotype of an animal on the basis of its progeny's performance. Each parent contributes a sample half of its genes to each offspring. Thus an effort to evaluate an individual (usually a male) on the basis of one or a few offspring can be misleading. Chance at segregation may result in any one or a few offspring receiving a better- or poorer-than-average sample of genes from the parent.

The biometric relations between individuals involved in a progeny test of a sire in arrow form are given in Fig. 10-3. It can be seen here how the progeny test differs from pedigree appraisal in that repeated samples of genes of the individual being progeny tested are figuratively observed in the progeny. Hence, under ideal conditions the repeated

Fig. 10-3. Diagram showing the genetic relations in a progeny test. The genotypes of the progeny (G_1, G_2, . . . , G_n) represent independent samples of genes from the sire (G_s) and the females to which he is mated (G_{D_1}, G_{D_2}, . . . , G_{D_n}). The square root of heritability (h) represents the impact of the individual progeny's genotypes on their phenotypic values (O_1, O_2, . . . , O_n).

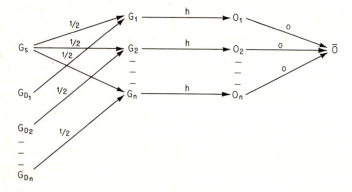

sampling overcomes the limitations which chance at segregation imposes on pedigree selection.

It can be noted from Fig. 10-3 that the progeny average $\bar{O}$ might also be influenced by the contribution of the other parent to the offspring. In practical progeny-testing programs this feature is usually taken into account by mating a male, for example, to a random sample of available mates. In other situations an adjustment can be made for the probable contribution of the other parent, such as was formerly attempted in the daughter-dam comparison with dairy-cattle progeny tests.

It is important that all of the progeny and not just a selected sample be included in the progeny-test appraisal. Omitting the poor progeny is unfair and misleading. Similar poor progeny are just as likely to be produced among the next group of progeny.

The theoretical accuracy of the progeny test as compared with individual merit has been examined in detail by Lush.[1] When there is no correlation between the phenotypes of the progeny due to nongenetic effects, the correlation between the sire's breeding value and the average of n offspring is

$$r_{G\bar{O}} = \frac{h}{2} \sqrt{\frac{n}{1 + (n - 1)t}}$$

where h is the square root of heritability and $t(h^2/4)$ is the observed correlation between the progeny. In contrast the correlation between an individual's own performance and its breeding value r_{GP} is the square root of heritability h. Thus the relative accuracy of the progeny test and individual merit can be examined by comparing h with the value of the above correlation.

Ratios of the accuracy of the progeny test to individual selection are given in Table 10-7. It can be noted that when heritability is low, fewer progeny are required to make the progeny test equivalent in accuracy to individual selection. If there is an environmental correlation among the progeny due to nongenetic factors, the accuracy of the progeny test is much reduced. Such a situation would exist when several progeny-tested sires are being compared, and their progeny had been tested at separate locations. Feeding and management influences, as well as other environmental factors, can contribute to progeny-group differences or to similarities among progeny. These reduce the accuracy of the progeny test. Such a situation is illustrated in the right portion of Table 10-7. A contribution to the correlation among progeny members of 0.10 due to nongenetic factors has been assumed.

[1] Lush, Jay L., *Jour. Dairy Sci.*, **18**:1–19, 1935.

*Table 10-7. Comparative Accuracy of Progeny Test and Individual Performance**

$n\dagger$	h^2 values, $t = h^2/4$			h^2 values, $t = h^2/4 + 0.10$		
	0.10	0.25	0.50	0.10	0.25	0.50
1	0.50	0.50	0.50	0.50	0.50	0.50
2	0.70	0.69	0.66	0.66	0.65	0.63
3	0.85	0.82	0.77	0.77	0.75	0.72
4	0.96	0.92	0.85	0.85	0.82	0.77
5	1.07	1.00	0.91	0.91	0.87	0.81
10	1.43	1.26	1.08	1.08	1.01	0.91
20	1.84	1.51	1.22	1.22	1.10	0.97
50	2.37	1.75	1.32	1.32	1.18	1.02
100	2.68	1.87	1.37	1.37	1.21	1.06

* The tabular values are the ratio of the accuracy of selection on the progeny test to selection on individual merit or

$$\sqrt{\frac{n}{\{4[1 + (n - 1)t]\}}}$$

† n represents the number of progeny.

Although the progeny test can provide an extremely accurate appraisal of an individual's breeding value, its major limitation is in the time and expense required to obtain the information. Waiting for progeny-test information on a sire before he is used extensively prolongs the generation interval. Thus the increased cost and generation interval must be balanced against the additional accuracy of the progeny test.

For an effective large-scale progeny-testing program, several requirements must be met. First, many more animals must be progeny tested than are required for eventual use. Unless such an initial provision is made, there will be little opportunity for selection after the progeny test is available. Effective progeny-testing programs for males usually will require that four or five individuals be progeny tested for each progeny-tested sire that is chosen for extensive later use.

Secondly a procedure must be available to accurately assess the progeny-test information. Unless the progeny test is designed to provide a critical evaluation of the progeny-tested individuals, there is little justification for undertaking the progeny testing. Testing of progeny at several locations using artificial insemination and the adoption of comparison of performance with contemporary animals can measurably increase the accuracy of progeny-test evaluations.

Another extremely important requirement for effective progeny testing is that there must be a means of utilizing the superior animals extensively once they have been located. Artificial insemination and frozen semen have been most important in this respect for cattle. If a progeny-tested male can be used widely, then the industry can afford an extensive investment in progeny-testing programs designed to locate such individuals.

The inability to obtain many progeny from females is one reason why planned progeny testing of females is not of major importance in livestock. In addition, the progeny-test information accumulates so slowly that by the time cows and ewes have three or four tested progeny, many useful years of the dam's life has passed. This does not mean that progeny performance should not be used in selection, but its value is definitely limited. Selection of females must rely quite heavily on individual merit.

Several studies have made it quite clear that effective progeny-testing programs require large populations. In dairy cattle the most efficiently designed progeny-testing program that is practical for herds of 100 to 150 cows would only about equal a system in which sire selection is based only on records of their dams and other close relatives. The same conclusion was arrived at by Warwick[1] for several highly heritable traits in beef cattle under natural breeding. Yet planned progeny-testing programs with artificial insemination in a population of as many as 10,000 breeding females should lead to an average annual genetic improvement in milk yield equivalent to 1.5 to 2.0 per cent of the mean in dairy cattle. In contrast, annual improvement without progeny testing would be expected to be only 0.7 to 1.0 per cent of the mean.

FAMILY SELECTION

Family names are used in at least two senses in animal breeding. The family name has been traced through the dam (in a few breeds through the sires) in an unbroken line of females to a foundation female who was outstanding enough to have a family named for her. Obviously, if the foundation female is very many generations back in the pedigree, her genetic superiority has been halved so many times that she cannot now be source of many genes that would be common to current members of such a family. Such family names have about as much significance as human family names. Captain John Smith was probably a great man, but does that automatically make all the Smiths great?

Family names lend themselves readily to speculation. However, most breeders will do well to pay attention to individual excellence and the

[1] Warwick, E. J., *Natl. Acad. Sci., Natl. Res. Coun. Publ.* 751, 1960, pp. 82–92.

excellence of near relatives, rather than to remote relatives and genetically insignificant family names.

Some linebred families are in existence in which there has been linebreeding to the foundation male or female with the result that the family is kept highly related to it. Such families may have genetic significance. The Anxiety 4th Herefords constituted probably the best-known family of this kind in meat animals. There is danger of paying too much attention to linebred families in certain pedigrees, rather than the excellence of the individuals in the pedigree.

The other use of the family is to represent a group of animals having a common genetic relationship. Generally half-sib and full-sib families are most common in animal populations. With random mating these have *inter se* relationships R of 0.25 and 0.50 respectively. Such family members are collaterally and not directly related. An individual receives no genes from collateral relatives, but their average performance is an indication of the genes transmitted to each family member by one or more common ancestors. Thus family selection can be represented as a part of pedigree selection. The information on the family may provide information to predict the breeding value of an ancestor to indicate the genes probably transmitted to the individual in which we are interested.

In most situations the problem is to select replacement animals from a population composed of several groups of half sibs, each the progeny of one sire. In swine, litters of full sibs are also usually available. Ordinarily the different sires used will have transmitted differently, some groups of half sibs averaging higher in merit than others. The same will be true of the various litters. The question is should selection be (1) entirely on a family basis, saving only the animals from the families with the best averages, (2) entirely on individuality, saving only the best individuals regardless of the average performance of others in the family, or (3) on some basis which combines emphasis on the individual and the family? In the last case some plan of saving the better individuals from the better families would be followed.

Lush[1] made a careful analysis of this problem and pointed out that, if either family or individual selection is absolutely ineffective, the other will be also. The relative effectiveness of the three methods depends upon (1) R, the genetic correlation, or relationship, among family members, (2) t, the phenotypic correlation among family members, and (3) n, the number per family. A comparison of the accuracy of the above three alternatives was made by evaluating the correlation between the individual's breeding value, its own performance, its family average, and the combination of its own performance and family average.

[1] Lush, Jay L., *Amer. Nat.*, 81:241–261, 1947.

When selection is based on the family average alone, its accuracy is $[1 + (n-1)R]/\sqrt{n[1+(n-1)t]}$ times that of individual selection. For circumstances in which this factor is less than 1.0, selection on individual merit is more accurate; when the expression is larger than 1.0, family selection would be expected to be more accurate. Figure 10-4 shows the values of the above expression over the range of values for R and t for selected values of family size. It is evident that individual selection is more accurate when t is large and R is small. The small values of R reflect that there is little family distinctness. On the other hand, family selection is more effective when t is small but R is large. If family selection is to be effective, a first requirement is to have distinct families among which to select. This sounds trite but it continually needs emphasizing in view of the traditional loose definition of the family by livestock breeders. From a genetic viewpoint, when individuals have relationships of less than 0.25, the so-called *family average* contributes little. Family selection will seldom, if ever, be superior to purely individual

Fig. 10-4. In comparing the efficiency of individual selection versus family selection, individual selection is more accurate than family selection for combinations of R and t to the left of the diagonal lines; but family selection surpasses individual selection for situations to the right of these lines. [*Lush, Jay, L., Am. Nat., 81:* 249 (*1947*).]

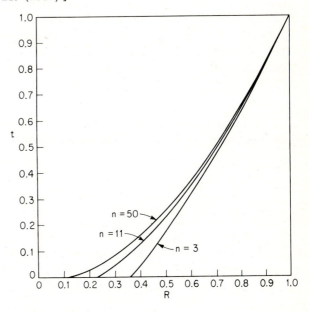

selection among families of half sibs, and will be superior among families of full sibs only when t is low.

We should reemphasize the point made in Chap. 8. One of the most important uses of inbreeding is to produce distinct families. An *inter se* relationship of 0.50 is the maximum attainable for farm animals without inbreeding. With cattle and sheep, full sibs are so infrequent that inbreeding must be utilized even to develop families with relationships which are much above 0.31. In poultry large families of full sibs are available and selection for most traits is based on family performance.

If the information on the individual and its family average is combined properly, the combination will always be at least as accurate as individual selection alone. Yet one may wonder whether sufficient accuracy is gained to justify the combination. Some of the comparative situations are shown in Table 10-8.

An optimum combination of emphasis on family averages and individual merit should be

$$\sqrt{\frac{1 + (R - t)^2}{1 - t} \frac{n - 1}{1 + (n - 1)t}}$$

times as effective as purely individual selection. This expression is equal to 1, and the two types of selection are equally effective if $R = t$. If R and t are unequal, a combination of family and individual selection will be more effective than individual selection alone. If t is less than R, the family average should be given positive emphasis in selection. If t is larger than R, we know that some environmental factor is producing at least part of the resemblance. This leads to the conclusion that a good individual in a better-than-average family may be good partly

Table 10-8. *Effectiveness of a Combination of Family and Individual Selection As Compared with Purely Individual Selection under Certain Assumed Conditions*

t	$R = 0.25$			$R = 0.50$			$R = 0.60$		
	$n = 8$	$n = 12$	$n = 20$	$n = 8$	$n = 12$	$n = 20$	$n = 8$	$n = 12$	$n = 20$
0.1	1.050	1.063	1.079	1.316	1.390	1.472	1.465	1.567	1.680
0.2	1.004	1.005	1.006	1.153	1.178	1.203	1.258	1.299	1.339
0.4	1.034	1.038	1.041	1.016	1.017	1.019	1.060	1.066	1.071
0.6	1.188	1.201	1.212	1.017	1.018	1.019	1.000	1.000	1.000
0.8	1.614	1.641	1.665	1.216	1.227	1.236	1.101	1.107	1.111

because of favorable environment and that we should discount the record. Although it may seem paradoxical, family averages should be given a negative emphasis in selection under such conditions. Table 10–8 gives the relative effectiveness of combination and individual selection at several levels of R, t, and n. It can readily be seen that combination selection is of greatest value when R is relatively high and t is low or when t greatly exceeds R.

SELECTION FOR SEVERAL TRAITS

Our previous considerations have centered about the improvement of one primary trait in the selection program. Such a situation will be experienced only rarely in practice. Even if one trait is of paramount importance, other qualities, such as fertility, need to be emphasized. Usually several qualities are of direct economic importance; in sheep, wool clip and lamb weights; in swine, growth rate, litter size, feed efficiency, and carcass merit; in dairy cattle, milk yield and composition; and in beef cattle, growth rate and carcass merit.

Selection should be directed only toward the traits of real importance, as selection for more than one trait reduces the selection pressure on any single trait. The reason why the selection differential for a specific trait is reduced by selecting simultaneously for several traits can be understood more clearly from the following illustration. If 10 per cent of a population must be saved for breeding purposes, the selection differential would be 1.76 standard deviations, when selection emphasized only one trait. However, if selection is for two independent traits, the probability of an individual being in the upper 10 per cent for both traits is the product of the two proportions 0.10 × 0.10, or 0.01. To get 10 per cent of the individuals which rank as high as possible in each of the two traits, it is necessary that the product equal 0.10, or $\sqrt{0.10} \times \sqrt{0.10} = (0.316)(0.316)$. Hence, individuals representing the upper 31.6 per cent for the two traits must be selected. The expected selection differential for the best 31.6 per cent from a normal curve is about 1.13 standard deviations rather than 1.76, or only about 64 per cent as large as could have been attained for each trait separately.

The reduction in intensity of selection is not quite so severe when an index or total score is used. Deficiencies in one character then can be overcome by superiority in another. When selection is directed toward improving n traits which are equally important and independent, the intensity of selection will be $1/\sqrt{n}$ as intense as if selection was for only one trait.

The foregoing is not to argue that selection should be for only one

Table 10-9. Relative Intensity of Selection for Individual Traits When n Equally Important Independent Traits Are Selected

n	Relative intensity
1	1.00
2	0.71
3	0.58
4	0.50
n	$1/\sqrt{n}$

trait. It is essential to emphasize more than one trait in most breeding operations, as was pointed out earlier. Unimportant traits should not be emphasized, but if all traits being selected are really desirable, index or total score selection on all of them will give a larger increase in total merit than emphasizing only one or two.

The most important consideration is that of being certain that the traits fully merit consideration. Selection for traits that are economically unimportant, extremely low in heritability, or for details of color marking or conformation is often excused on the basis that it does no harm since they have no relation to economic value. This is fallacious reasoning, since such selection is indirectly harmful through reducing the selection intensity for more important traits which will respond to selection.

Methods for Selection.   Hazel and Lush[1] have examined the theoretical accuracy of three methods of selection to improve several traits. These were the tandem method, the method of independent culling levels, and the total score or index method. In the tandem method, selection is practiced for only one trait at a time until a satisfactory level is reached, then a second trait is considered, etc. The efficiency of this method is dependent on the genetic relationships among the traits. Should there be negative relationships, the subsequent selection for a second trait could undo the progress previously made in earlier selections. Of course, strong positive genetic correlations among the traits would mean that selection for one trait would improve the correlated traits.

With independent culling levels, selection may be practiced for two or more traits simultaneously. A minimum standard is set for each of the n traits, and all individuals below that level for any one trait are culled without regard to their merit for other traits. This is one of the

[1] Hazel, L. N., and Lush, Jay L., *Jour. Hered.*, 33:393–399, 1942.

major disadvantages of this method; the superiority for one trait does not have an opportunity to offset lack of merit in another trait.

Using an index or total score for selection allows one to add up the merit for each trait and arrive at a total score. Such an approach was found to be more efficient than either of the other two methods. In practice both the latter methods will probably have to be used, since in most herds at least some culling must be done before all the information needed for the optimum use of the total-score method is available. For example, in beef cattle, weaning weight and rate of gain after weaning are both economically important traits. In using the total-score method of selection, we would want to give points for performance in each trait. However, if our facilities for testing postweaning rate of gain would not permit testing all animals, we would have to cull at weaning those below a certain weaning weight.

Selection Index. The approach to selection using an index or total score stems from the work of Smith[1] and Hazel.[2] Such a procedure utilizes scores or index values for each individual, so that the index values are as closely correlated to the individual's composite breeding value as is possible to obtain with a linear combination of traits.

Why is an index essential for efficient selection for several traits? First, the traits considered usually are not all of equal economic importance. Some differential weighting in accord with the net economic return expected from a unit of improvement of each trait is needed. Secondly, all traits do not have the same heritability, and the same intensity of selection will not be expected to give the same proportionate improvement for each trait. Thirdly, there may be phenotypic and genetic interrelationships among the traits. Emphasis on one trait may affect change in another, and these interrelationships must be appropriately considered.

In the development of a selection index for an animal enterprise, certain information is needed:

1. The relative net economic importance of a change in each of the traits. This information effectively defines the goal of the selection program. A composite of the several traits weighted by their relative net economic importance is the goal of improvement rather than a single trait.
2. The heritability or the magnitude of the genetic variance for each of the traits and the phenotypic variances for each trait.
3. The phenotypic and genetic covariances among each trait in the index. For alternate computational approaches, the phenotypic and genetic correlations may be used.

[1] Smith, H. F., *Ann. Eugenics,* 7:240–250, 1936.
[2] Hazel, L. N., *Genetics,* 28:476–490, 1943.

The computation of selection indexes involves the technique of multiple regression (several independent or X variables), whereas we have only considered the simple linear regression (Chap. 6). The index computed is of the form:

$$I = b_1 X_1 + b_2 X_2 + \cdots + b_n X_n$$

where the X's represent the phenotypic values for the different traits, and the b's are the weights given to each of the traits. In statistical terms the b's are multiple regression coefficients which maximize the correlation between index values and composite breeding values.

The selection index approach has permitted a clearer focus on the parameters that are important when selection involves many traits, each possessing considerable additive genetic variance. Indexes have been used more widely with sheep and swine than with beef and dairy cattle. Perhaps this has resulted because several traits are important to the economic success of these enterprises, and it has been more necessary to use some index or total-score method for selection. Yet with all livestock breeding enterprises the failure to use the index arises primarily because the information required to calculate an effective index, including all traits, is not available or accurate enough.

The relative economic importance of traits may vary for different enterprises. Heritability values and the genetic and phenotypic interrelationships also may differ for various populations, making it difficult to recommend generalized selection indexes. The net economic weights and the heritabilities of the several traits are usually the most important parameters. Lush has reported an early index for swine based on market score, litter weight, and sow productivity computed on the basis of the economic weights and the heritabilities of the traits. This index was recomputed considering the genetic and phenotypic interrelationships among the traits. The expected accuracy of the first index was 90 per cent as accurate as the second. Indexes which have been developed for various classes of livestock will be discussed in the later chapters devoted to these.

Computation of a Selection Index. This section may be passed over without losing any essential continuity in the chapter on Selection. Although the development of a selection index is somewhat complicated, the inquiring student may desire to gain an acquaintance with the computational techniques.

Our example will be for a sheep enterprise in which the economic value of the enterprise is primarily determined by the weight of lamb and the weight of fleece produced by the ewe. Hence the worth W of a ewe is defined in terms of these two measures.

$$W = a_1 X_1 + a_2 X_2$$

where X_1 represents lamb weight, X_2 represents fleece weight, and the a_1 and a_2 represent the relative net economic importance of these two traits.

Long-time market trends and a study of the production aspects of the enterprise would be required to ascertain the net return (return after deduction of production costs) from a unit increase in a particular trait. We shall assume that the net return for each additional pound of lamb is \$0.05 and the net return for an additional pound of fleece is \$0.20. Then a_1 is 1.0 (0.05/0.05) and a_2 is 4.0 (0.20/0.05). The goal of our selection program is then defined in terms of W.

$$W = X_1 + 4X_2$$

Instead of merely desiring to improve one of these traits, we desire to improve both traits with the combination of weights, 1.0 and 4.0. The genotype we seek to improve by selection is also a combination of breeding values for X_1 and X_2. Actually we shall express it as the genotype for worth g_w.[1]

$$g_w = g_1 + 4g_2$$

Thus we seek to improve g_w, a linear combination or composite of the two traits.

In order to compute our index we now need the other information previously itemized. Our selection index will be of the form

$$I = b_1 X_1 + b_2 X_2$$

The b's are obtained by solving the following simultaneous equations:

$$b_1 \text{ var } x_1 + b_2 \text{ cov } x_1 x_2 = \text{cov } g_w x_1$$
$$b_2 \text{ cov } x_1 x_2 + b_2 \text{ var } x_2 = \text{cov } g_w x_2$$

where var x_1 and var x_2 are the phenotypic variances of the two traits, cov $x_1 x_2$ is the phenotypic covariance between the two traits, and cov $g_w x_1$

---

[1] As was conventional in Chap. 6, Variation, the lower-case letters refer to deviations from the mean. Since the variance involved deviations from the mean, actually var X = var x.

and cov $g_w x_2$ are the covariances of the individual traits with the composite breeding value g_w.

The right-hand members of these equations deserve some further explanation. From suitable analyses of variance and covariance, the genetic variances of the two traits and their genetic covariance can be computed. For our current example:

$$\text{var } g_1 = 32.0 \qquad \text{var } g_2 = 0.60 \qquad \text{cov } g_1 g_2 = 1.50$$

then the two right-hand members of the equations can be computed

$$\text{cov } g_w x_1 = a_1 \text{ var } g_1 + a_2 \text{ cov } g_1 g_2 = 32.0 + (4)(1.50) = 38.0$$
$$\text{cov } g_w x_2 = a_1 \text{ cov } g_1 g_2 + a_2 \text{ var } g_2 = 1.50 + (4)(0.60) = 3.9$$

The equations to be solved for the b's are:

$$b_1 95.00 + b_2 3.00 = 38.0$$
$$b_1 3.00 + b_2 1.50 = 3.9$$
$$b_1 = 0.3505 \qquad b_2 = 1.8990$$

For simplicity in using the index, the b's can be coded by dividing each by 0.3505 so that the index used would be

$$I = x_1 + 5.4 x_2$$

Index values would be computed for each ewe and selections then would be based on the values of I.

It should be apparent that inadequate relative net economic values, or variances and covariances required to compute an index, could introduce errors into the index. Large volumes of accurate data representative of the population to which the index is applied are necessary to provide sound information to compute the selection index. Generally the genetic covariances (cov $g_1 g_2$, etc.) are the least adequate items of information needed in the computations. Studies have shown that indexes computed from erroneous or inadequate information can be ineffective in selection. When the phenotypic and genetic parameters are poorly estimated, weighting the traits on the basis of the values for their relative net economic worth may be more desirable.

ENVIRONMENT AND SELECTION

The question of what environment a breeder should provide for his animal is a broad one, and no definitive answer is available. It has sometimes been assumed that animals in breeder herds should be maintained in the best practical environment, and the commercial producer using the descendents of these animals should strive to improve his environment toward the level of the breeder herds. From a practical economic point of view, excellent management and care will probably pay well in improving the level of performance and appearance of breeding stock that will be available for sale.

The concept is also expressed by some that a poor environment will put a virtual ceiling on the expression of inherent differences in ability to gain, type, or production, making selection ineffective by reducing genetic variation. Hammond[1] recommended providing the animals an environment which would permit maximum expression of the traits. Selections would be made from among the animals with maximum performance, even though later descendents were to perform commercially under poorer conditions.

Hammond's proposition would be satisfactory if genotype-environmental interactions were not important. This would mean that a specific change in the environment would result in a proportionate change in the phenotype for the entire array of genotypes. If such is not the situation, genotype-environmental interactions would be expressed. A practical approach to genotype-environmental interaction is to examine if the productivity of individuals or strains rank them in the same order in different environments. If such were the case, selection, though not necessarily being equally effective, would be for the same goals and changes in the environment and would not determine which animals would be selected. Conversely, if individuals or strains changed their rank in different environments, a genotype-environmental interaction would be suggested. Selection for improvement in one environment would not necessarily result in improved performance in another. Hammond's proposal would be preferred if there were no genotype-environmental interactions and if the heritability were higher in the optimal environment.

Correlated Selection Response. The concept that a trait measured in two different environments can be considered as two different traits has been developed by Falconer.[2] As an example, the growth rate of animals reared on a high plane of nutrition would be considered one trait, and the growth rate of these animals reared on a low plane of nutrition a second trait. If the genetic correlation between growth rate in the two environments was high, very little interaction of genotypes and environment would be suggested. The change in growth rate after

[1] Hammond, J., *Biol. Rev.*, **22**:195–213, 1947.
[2] Falconer, D. S., *Amer. Nat.*, **86**:293–298, 1952.

selection in either environment can be examined in terms of correlated responses to selection. A correlated response is defined as one in which selection is primarily for one trait, but due to a strong genetic correlation a change occurs in a second trait.

The above idea can be represented by defining one trait as the primary trait and the other as the secondary trait. It is assumed for illustration that genetic improvement in the primary trait is of fundamental importance. The direct response to selection can be represented as

$$\Delta G_D = h_1{}^2 i\sigma P_1 = h_1 i\sigma G_1$$

and the correlated response can be expressed as

$$\Delta G_C = r_{g_1 g_1} h_2 \frac{\sigma G_1}{\sigma P_2} i\sigma P_2 = r_{g_1 g_2} h_2 i\sigma G_1$$

where D and C respectively represent the direct and correlated responses to selection, and subscripts 1 and 2 represent the primary and secondary traits respectively. Then the ratio of the correlated to the direct response is

$$\frac{\Delta G_C}{\Delta G_D} = \frac{r_{g_1 g_2} h_2 i\sigma G_1}{h_1 i\sigma G_1} = \frac{r_{g_1 g_2} h_2}{h_1}$$

From the above expression it is apparent that if $r_{g_1 g_2}$ is 1.0 and $h_1 = h_2$, the direct and correlated responses would be equal. The correlated response would be less than the direct response if $r_{g_1 g_2}$ is less than 1.0 or if $h_1 > h_2$. The product of $r_{g_1 g_2} h_2$ must be larger than h_1 for the correlated response to exceed the direct response.

When the alternatives of direct or correlated response to selection are considered for a specific population, the intensity of selection i will not be affected by the trait selected. For example, whether we select for increasing growth rate using the gains to one year of age or the conformation of the animals at one year as a correlated trait, we would still have to save the same percentage of the herd. However, when the response to selection for a trait in two different environments is considered, the intensities of selection must be taken into consideration. Then the ratio of the correlated to direct response to selection is

$$\frac{\Delta G_C}{\Delta G_D} = \frac{r_{g_1 g_2} h_2 i_2}{h_1 i_1}$$

In addition to its implication regarding genotype-environmental interactions, the above development clarifies what characteristics a trait must have to be a reliable selection criterion for another trait. It must have a high genetic correlation with the primary trait and also have a high heritability. Much of the practical methodology in animal breeding has been that of seeking such measurements when the desired trait is difficult to measure with precision. Few such traits have emerged and most selection is based on the primary trait or measurement. Of course, a combination of the direct and indirect measure in an index should provide the most effective selection criterion when both measurements can be obtained.

Experimental Evidence. Fowler and Ensminger[1] selected swine using an index which gave major emphasis to rate of gain to determine if selection for growth rate was as effective when swine were limit-fed as when they were liberally fed. Foundation stock of a Danish Landrace $\times$ Chester White cross was divided into two equivalent lines. One line was full-fed from weaning to 150 days, limit-fed from 150 days to parturition, and full-fed during lactation. The second line was given 70 per cent of the feed consumed by the liberally fed line during each of the above periods. Progress from six generations of selection for rate of gain was very close to that expected from the selection differentials and the heritability in each line.

Samples of the pigs from the two lines were tested on both feeding regimes after six generations of selection. The previously limit-fed line gained faster on the liberal ration than animals of the line selected for fast gains on full feed. Likewise when animals of both lines were tested on the limited ration, pigs from the limit-fed line achieved the faster gains. Although the differences between the lines were not great, a genotype-environmental interaction was indicated. The results did not support Hammond's contention. Actually the line selected under the more rigorous condition performed better. The authors interpret the differential response as being the result of selection for appetite alone in the full-fed line but for both appetite and efficient feed utilization in the limit-fed line.

Selection experiments with mice for growth or unrestricted and restricted diets have been reported by Falconer.[2] In these experiments selection on the restricted plane of nutrition appeared to improve the genetic potential for growth under either feeding regime. Selection on the unrestricted diet was not effective for improving growth rate on the restricted diet. Some other studies have suggested that genotype-environmental interactions may be present, but their magnitude is such

---

[1] Fowler, S. H., and Ensminger, M. E., *Jour. Anim. Sci.*, 19:434–449, 1960.
[2] Falconer, D. S., *Genet. Res.*, 1:91–113, 1960.

that few are of practical significance. Our interpretation of available evidence on this important question is that adaptation tends to be fairly general for most characters. Only in exceptional circumstances would selection in one environment be totally ineffective for performance in another. Evidence that is available does indicate the desirability of selecting under conditions reasonably like those of commercial animals of the area to fully justify the practice.

In spite of available experimental data, however, the most common procedures for rearing animals in the best-known sheep and beef-cattle breeder herds to date in the United States have been very nearly in line with Hammond's viewpoint. Seedstock herds and flocks have for the most part been maintained in areas where feed conditions are good and other environmental conditions are not extreme. Furthermore, many animals are often forced artificially with nurse cows or highly concentrated feeds, with the result that there can be little direct selection for milking ability of dams and other important performance traits. Animals from these herds have then provided foundation stock for commercial production in all areas of the country. This situation, together with the nationwide promotional activities of many of our breed associations, has sometimes tended to result in animals of the same general type being raised in virtually all parts of the country regardless of their adaptation. A critical look at this situation by the breeders would appear long overdue.

With dairy cattle, the breeding operation is very rapidly centering about the programs developed by artificial insemination studs. The selections are based on progeny performance in the tested herds throughout the area. Thus selections are based on information from a broad cross section of environments in practical, producing herds.

RESULTS OF SELECTION

Selection ·experiments with livestock are expensive to conduct and require several years to obtain definitive results. Yet evidence is accruing to attest to the effectiveness of selection in changing animal populations. The interest in selection even in theoretical studies has appeared to gain a new impetus. Considerable selection results with *Drosophila* have been reported during recent years. In addition, research with laboratory mammals has increased in tempo.

Undoubtedly the classical demonstration of the potential for selection has come from the Illinois selection experiment for high and low oil content. In the last published report[1] the oil content of the high line

[1] Leng, E. R., *Ztschr. Pflanzenzücht.*, 47:67–91, 1962.

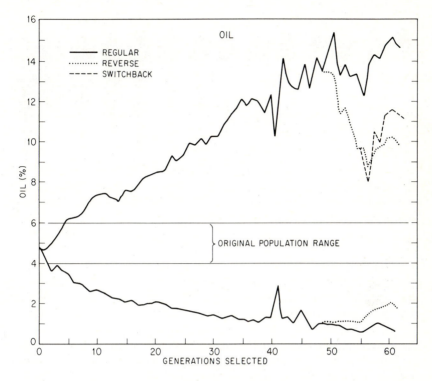

Fig. 10-5. Response to 61 generations of selection for oil content of corn. Actual data are indicated by the solid lines. Reverse selections are shown as broken lines. Note that the selection response for oil is a symmetrical with a plateau in the low line at about one per cent. [*Leng, E. R., Ztschr. Pflanzenzücht.,* **47**:67–91 (1962).]

had increased to 15.56 per cent, and the content in the low line was only 1.01 per cent. Further decline in the low line has been hampered by poor germination as the oil content became too low. From an initial oil content of 4.68 per cent, the above changes are most striking. Even in more recent years there has been a continuing response to selection.

Selection with Laboratory Mammals. The more dramatic results with laboratory mammals serve to remind us that directed selection can bring about change. Castle[1] pioneered work with quantitative traits in his selection for variation in the hooded (spotting) pattern in rats. Races with the extremes in hooding as shown in Fig. 10-6 were produced

[1] Castle, W. E., "Genetics and Eugenics," pp. 237–240, Harvard University Press, Cambridge, Mass., 1930.

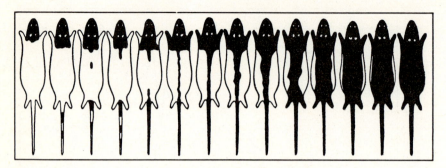

Fig. 10-6. Series of grades for classifying the plus and minus variations of the white spotting pattern of hooded rats in Castle's selection studies. (*From Castle, Genetics and Eugenics, Harvard University Press.*)

by selection. At the time Castle conducted this research, there was a scientific controversy over the fundamental effect of selection. Some persons maintained that the genes themselves were changed by the selection process. Castle's comment on this point is illuminating:

The question now arose whether the observed changes had occurred as a result of change in the single unit-character or gene clearly concerned in the case, or whether this was due to other agencies. To test the matter the selected races, now modified genetically in opposite directions, were crossed repeatedly with a non-hooded (wild) race. The recessive hooded character disappeared in F_1 but was recovered again in F_2 in the expected 25 per cent of this generation. These extracted hooded individuals, following each cross, were less divergent than their hooded grandparents from the ordinary hooded pattern. After three successive crosses (six generations) the whitest individuals extracted from the dark hooded race were no darker than the darkest individuals extracted from the white hooded race. In other words, repeated crossing with the non-hooded (wild) race had caused the changes in the hooded character, which had been secured by selection, largely to disappear. The conclusion was drawn that the hooded allelomorph itself had remained unchanged throughout the selection experiment, and that the phenotype had been altered by associating with the hooded gene a different assortment of other genes in each of the selected races, these serving as genetic modifiers. In the course of the selection experiment a mutation was observed to occur in the plus series to practically the Irish stage. This is not included in the summary but is mentioned to show how it is possible for selection to be aided in its progress by the occurrence of contemporaneous genetic changes, no less than by the sorting out of variations originally present in the foundation stock. Apart from the mutation mentioned the results of selection in this case show conclusively that the changes obtained had not occurred in the gene for the hooded pattern, but in the residual heredity. Other cases

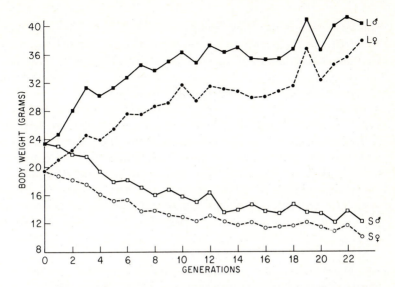

Fig. 10-7. Response to individual selection for 60-day body weight in mice. [*McArthur, J. W., Genetics, 34:196 (1949)*.]

of apparent gradual change in unit-characters under the action of selection may be explained in a similar way. Accordingly, we are led to conclude that unit-characters or genes are remarkably constant and that when they seem to change as a result of hybridization or of selection unattended by hybridization, the changes are rather in the total comples of factors concerned in heredity than in single genes.

Several published reports are available on the long-term response to selection for body size in mice. Goodale[1] selected for large body size at 60 days in albino mice. Although no control or low line was carried, the change in body weight from 23.6 to 32.2 grams leaves little doubt that selection was responsible for genetic change. McArthur[2] conducted two-way selection from a heterogeneous base population. After 21 generations of selection, mice from the large line at 60 days were 3.3 times as heavy as those from the small line. More recent experiments from selection studies confirm the general response to selection for body size presented by McArthur.

Selection with Farm Mammals. Designed selection experiments with livestock are not very numerous, although an increasing number of selec-

[1] Goodale, H. E., *Jour. Hered.*, 24:101–112, 1938.
[2] McArthur, J. W., *Genetics*, 34:194–209, 1949.

tion experiments are now underway. The selection for rate of growth under two feeding regimes by Fowler and Ensminger[1] has already been mentioned. Selection for rate of gain was effective in both lines, and the progress was close to what would have been expected from the selection differentials and the heritability estimates.

Two other experiments with swine have been reported from Alabama[2] and Illinois.[3] At Alabama, selection was for high and low efficiency of feed utilization in two lines arising from the same foundation stock in the Duroc breed. All pigs were individually full-fed a mixed ration from 72 days of age until they attained a live weight of 225 lb. After five generations of selection, the efficient line required 25 lb less feed per 100 lb of gain. The Illinois experiment consisted of two-way selection for gain in Hampshire swine. After eight generations of selection, the rapid-gaining line was 61.8 lb heavier than the low line at 180 days.

More recently the effect of selection for back-fat thickness in Yorkshire and Duroc swine has been undertaken at Beltsville, Maryland. A summary of the results of this experiment is given in Chap. 14.

Computing Selection Differentials and Estimating Genetic Progress. Often a breeder may feel that he is making little or no progress in improving his herd. Trends in a population, and particularly in individual herds, are difficult to disentangle. Changes with time are due to both genetic and environmental factors. Some perspective into whether or not one might expect any genetic improvement can be gained by computing the selection differentials and multiplying them by appropriate heritability values. Such an analysis will not give the answer as to what genetic change took place, but when the selection differentials are accurately determined, one can see if they are large enough to reasonably expect any genetic change.

The selection differential is the difference between the performance of those individuals which are selected to be parents and the average of all individuals in their generation. Because of the varying percentages of the offspring which are needed by different species to maintain the population, their selection differentials can vary in accordance with their reproductive rates. Furthermore, there will be a difference in the selection differentials that can be attained for males and females, as was suggested by the information in Table 10-1.

Computing the selection differentials and the expected or probable

[1] Fowler, S. H., and Ensminger, M. E., *Jour. Anim. Sci.*, **19**:434–449, 1960.
[2] Dickerson, G. E., and Grimes, J. C., *Jour. Anim. Sci.*, **6**:265–287, 1947.
[3] Krider, J. L., et al., *Jour. Anim. Sci.*, **5**:3–15, 1946.

genetic improvement is relatively easy when individual selection for a trait is practiced. Progress per generation ΔG is expressed:

$$\Delta G = h^2 \frac{S_m + S_f}{2}$$

The selection differentials for males and females are represented by S_m and S_f respectively, and h^2 is the heritability of the trait. Such extrapolations cannot be made over an indefinite period, but they can extend for two or three generations with reasonable validity.

The expected genetic change per year ΔG_t can be computed if the generation intervals for the two sexes are known.

$$\Delta G_t = h^2 \frac{S_m + S_f}{2} \frac{1}{(L_m + L_f)/2}$$
$$= h^2 \frac{S_m + S_f}{L_m + L_f}$$

It is clear from the above expression that if the selection program is one that lengthens the generation interval, such as with progeny testing of males, L_m would increase. The expected genetic change per year would be reduced, unless S_m was increased sufficiently by progeny testing to offset the increase in the generation length for the sires.

The selection differential computed in the conventional manner, the practiced selection differential, may not be equal to the realized selection differential. This distinction between the practiced and realized selection differentials must be made because each parent does not contribute the same number of progeny to the next generation. The realized selection differential is obtained by weighting the deviations of the individual parents from the mean of their generation by the number of progeny which are measured. Such a weighting takes into account differences in fertility, natural selection, and accidental deaths.

Computing selection differentials and expected genetic progress is much more complex than indicated above in most situations. Reproductive rates are too low in cattle, horses, and sheep to permit replacing all breeding animals each generation, and the generations overlap. Dickerson and Hazel[1] have developed expressions for computing annual selection differentials in these complex situations. Rendel and Robertson[2] have extended the thinking to account for the fact that each selection

---

[1] Dickerson, G. E., and Hazel, L. N., *Jour. Agr., Res.*, **69**:459–476, 1944.
[2] Rendel, J. M., and Robertson, A., *Jour. Genet.*, **1**:1–8, 1950.

differential may have a different effective heritability, depending upon whether it is based on individual performance, repeated observations, or progeny-test results. Presentation of the details of these approaches is beyond the intent of this text, but the inquiring reader is directed to the two papers cited above for additional information.

SUMMARY

Selection is the process of deciding which animals in a generation will be allowed to become parents of the next generation and how many progeny they will be permitted to leave. Selective improvement in quantitative traits is dependent on the heritability of the trait and the amount of selection practiced, the selection differential. Individual merit is the most important basis for selection, although information on ancestors, collateral relatives, and the progeny test are valuable aids to individual selection for specific traits. Pedigree, family, and progeny information are essential for selection of traits expressed in only one sex. Simultaneous selection for several traits reduces the intensity of selection for any one trait. Nevertheless if the several traits are important to the economy of the livestock enterprise, maximum total improvement can be made by using an index which optimally weights each trait. Even though current evidence does not indicate that interactions of genotypes and environments are of widespread practical importance, the soundest recommendation is that selection should be practiced in the same type of environment in which the commercial stocks will be raised. Although there are few results of critical selection experiments with livestock, many experiments with laboratory mammals attest to the changes that can be made by selection. Significant population trends in many traits over many years also make it difficult to avoid the conclusion that selection may have contributed much to these trends of improvement.

REFERENCES

Books

Dunn, L. C., ed. 1951. "Genetics in the 20th Century," The Macmillan Company, New York, (see especially Chap. 22, pp. 493–525).

Hammond, J. 1952. "Farm Animals: Their Breeding, Growth and Inheritance," Edward Arnold (Publishers) Ltd., London.

Lush, Jay L. 1954. "Animal Breeding Plans," 3d ed., The Iowa State University Press, Ames, Iowa.

Lerner, I. M. 1951. "Population Genetics and Animal Improvement," Cambridge University Press, London.

Lerner, I. M. 1958. "The Genetic Basis of Selection," John Wiley & Sons, Inc., New York

Articles

Dickerson, G. E., and Hazel, L. N. 1944. Effectiveness of Selection on Progeny Performance as a Supplement to Earlier Culling in Livestock, *Jour. Agr. Res.*, **69**:459–476.

Hazel, L. N. 1943. The Genetic Basis for Constructing Selection Indexes, *Genetics*, **28**:476–490.

Morely, F. H. W. 1951. Selection for Economic Characters in Australian Merino Sheep, (1) Estimates of Phenotypic and Genotypic Parameters, *Sci. Bul. Dept. Agr.*, NSW 73, 45 pp.

Lush, Jay L. 1935. Progeny Test and Individual Performance as Indicators of an Animal's Breeding Value, *Jour. Dairy Sci.*, **18**:1–19.

Lush, Jay L. 1947. Family Merit and Individual Merit as Bases for Selection, *Amer. Nat.*, **81**:241–261 and 362–379.

Rendel, J. M. and Robertson, A. 1950. Estimation of Genetic Gain in a Closed Herd of Dairy Cattle, *Jour. Genet.*, **50**:1–8.

CHAPTER ELEVEN

Improving Dairy Cattle

Dairy cattle are found on about one-half of the farms in the United States, and their milk and meat production account for about 15 per cent of the total cash farm income. The family cow has almost disappeared, and larger herds with specialized operations provide most of the nation's milk supply. During the past 10 years, the average production of all cows has increased 2,038 lb of milk and 67 lb of fat. Now cows on Dairy Herd Improvement Association (D.H.I.A.) tests average an unprecedented 11,976 lb of milk and 457 lb of fat per year. The number of dairy cattle two years or over in milk has declined from over 21 million to just over 16 million during the same period. Only about 5 to 10 per cent of these animals are registered, but they represent the base for further genetic improvement.

In one respect, the improvement of dairy cattle is somewhat easier than with other classes of farm mammals, for a measure of a cow's producing ability may be secured over her whole productive life. We can thus have tangible evidence of her average yearly lactation yield as well as that of her female offspring. In another respect, dairy-cattle improvement is more difficult than in other classes of livestock, since producing ability is poorly predicted from the exterior form. The male cannot express the quality for which the species is kept; i.e., the bull yields no milk, so his genetic merit can be measured only through the performance or transmitting ability of his offspring.

Milk production is the end result of a long chain of events caused by manifold and complex physiologic functionings, which probably means that many genes and many kinds of gene interactions are involved. What we are trying to do is to direct the genetic change in very complicated physiologic processes in a species which has many chromosomes and few offspring. The environment also can play tricks and lead us astray. Milk production is a highly elastic trait, easily influenced by men, feeding, fussing, fertilizing—and health as well as heredity. Finally, the cow is forced to live, produce, and reproduce under conditions very far removed from what might be called natural ones.

In spite of the difficulties, enormous progress has been made in improving the productiveness and type of our dairy animals. We have no pictures of the conformation of the cows that our savage, cave-dwelling, or nomadic ancestors tended, but it seems certain that they would win nothing but derision or a laugh if exhibited in a modern show ring. The earliest cows probably produced only enough milk to nourish their calves, whereas today many thousands of D.H.I.A. cows average over 12,000 lb of milk yearly, and the recorded production by a single cow stands close to 43,000 lb of milk in a year.

Although we must look to improved breeding practices for sustained lasting improvement in our cattle, it must be pointed out that the man just beginning to specialize in dairying must depend on good feeding, good management, and judicious culling for immediate improvement. The long generation interval in cattle does not permit a rapid realization of the benefits of a well-planned breeding program. Most breeders will

Fig. 11-1. Ayrshire cow, Morrow Jim's Florence that produced 25,975 lb milk and 944 lb of fat in 305 days, 2×, at 9 years 10 months. (*Courtesy Ayrshire Breeders' Association.*)

find individual selection useful, but it can be enhanced appreciably by considering records of performance of collateral relatives as well as ancestors and progeny.

GOALS IN DAIRY-CATTLE IMPROVEMENT

In developing our breeding goal, we first must examine the source of receipts from the dairy enterprise. Just what does contribute to returns? Gross income arises from the sale of:

1. Milk
2. Beef and veal
3. Breeding stock

These are direct sources of economic return. Other items contribute to economic returns as they influence the yield and quality of milk or reduce the costs of production for the items which can be marketed.

Returns from milk sales are most heavily influenced by volume, although the variation in percentage of fat and potentially protein or PLM (protein-lactose-mineral) above or below the market standard may affect returns. The beef and veal receipts can be considered as serving to offset the cost of herd replacements. Probably the returns from this source may approximate one-half the cost of developing herd replacements.

What about the returns from the sale of breeding stock? Such returns assume major importance only in specialized breeding establishments. In commercial herds the value of a cow must be measured in terms of her beef value plus the anticipated returns from milk. It is often inferred that good type can be marketed through breeding stock, yet Bayley and co-workers[1] suggest that for as many as 95 per cent of the dairymen, milk yield is at least 20 times as important to their income as type. The commercial dairyman has a most limited market for exceptional type in his grade cows. Even in specialized breeding herds the marketing channels are developed only after considerable effort and advertising, and at least acceptable production must accompany the type rating.

Excessive attention to qualities which dairymen have no reasonable chance to market must be avoided. Our expected genetic change per year for most traits is so meager that we must use the potential for genetic change wisely. It would appear that in most situations commercial dairymen must emphasize more milk of acceptable composition.

[1] Bayley, N. D., et al., *U.S. Dept. Agr. Tech. Bul.* 1240, 1961.

Fig. 11-2. Brown Swiss cow, Mable's Tamarind Violet. Grand Champion National Show 1962. She also produced 20,942 lb of milk and 1,041 lb of fat in 305 days, 2×, at 8 years 4 months of age. (*Courtesy Brown Swiss Breeders Association.*)

Conformation and other traits merit emphasis as they reduce production costs by enhancing productive life, reducing feed costs, and reducing labor requirements.

Heritabilities and Interrelationships. In addition to the net profit expected from each trait, knowledge of their heritabilities and interrelationships are essential to determine how the traits should fit into the plan for improvement. Numerous studies have been reported on the heritabilities of most of the important traits in dairy cattle, and what appear to be average values are summarized in Table 11-1. Information is missing in some instances on the relationships with milk yield, and certain other values are based on rather meager evidence.

MILK YIELD

Presently high milk yield of satisfactory composition is the most important single factor in insuring high economic returns. Except in unusual situations, the return over feed costs moves steadily upward with increasing production. Milk yield is also highly correlated with the total production of the individual milk components. One of the major deterrents

Table 11-1. Summary of Heritability of Individual Traits and Their Correlation with Milk Yield

Trait	Range of heritability	Correlation with milk yield	
		Phenotypic	*Genetic*
Milk yield	0.20 to 0.30		
Fat (%)	0.50 to 0.60	−0.15 to −0.35	−0.20 to −0.50
PLM* (%)	0.45 to 0.55	−0.10 to −0.30	−0.20 to −0.45
Protein (5)	0.45 to 0.55	−0.10 to −0.30	−0.20 to −0.45
Fat yield	0.20 to 0.30	0.90 to 0.95	0.85 to 0.95
PLM yield	0.20 to 0.30	0.90 to 0.95	0.85 to 0.95
Protein yield	0.20 to 0.30	0.90 to 0.95	0.85 to 0.95
Total solids yield	0.20 to 0.30	0.90 to 0.95	0.85 to 0.95
Feed efficiency	0.30 to 0.40	0.50 to 0.60	0.50 to 0.60
Mastitis	0.10 to 0.30	?	?
Mature size	0.30 to 0.50	0.15 to 0.30	−0.20 to +0.10
Milking qualities	0.30 to 0.40	0.05 to 0.20	?
Productive life (length)	0 to 0.10	0.15 to 0.20	?
Reproductive efficiency	0 to 0.10	0.10 to 0.25	?
Type score	0.15 to 0.30	0 to 0.15	0 to 0.20

* Protein-lactose-mineral.

to further increases in average milk yields is that only about 20 per cent of the cows now have production records.

Many dairymen appreciate the need for testing, but they too often let the press of other details lead to putting the testing aside. Nothing constructive can be accomplished in this manner, and herd improvement is at a standstill. Some dairymen may weight their milk, but not on a regular basis. The high daily yield of a particular cow at the start of lactation will leave a vivid impression in the dairyman's mind, but the persistency of the cow during the entire lactation can be evaluated only when complete lactation records are kept. For more rapid progress in dairy breeding, we must have more records and factual information on which to base our judgment in selection. These seem to come very slowly, and too often the records we get are made under conditions which are more likely to serve advertising needs than genetic ones. Fortunately, the emphasis on testing the entire herd rather than a few special cows is providing useful information on those herds which are being tested. The heritability of milk yield is in the medium range (0.20 to 0.30), and selection should be effective.

Standardization of Production Records.   The lactation production of a cow results from the interplay of heredity and environment. For accuracy in selection, it is important that the record reflect as precisely as practicable the cow's genetic potential for milk yield. The actual records themselves may be poor indicators of breeding value, since so many environmental influences have a marked effect on a cow's performance during a particular lactation. Some of the more important items which deserve consideration are length of the lactation period, number of times milked daily, age of the cow at freshening, length of the preceding dry period, the season of freshening, and days open.

Lactation Length. The D.H.I.A. Sire-proving Program and the sire programs of the Purebred Associations have adopted 305-day lactation records. This seems a logical base for measuring production, since a cow produces most profitably when she freshens each year. Breeders should strive for a 12- to 13-month calving interval. By using the production for the first 305 days, cows can be bred to calve each year and still provide for a 6- to 8-week dry period. Though the actual production for the first 305 days is desired, often only a completed 365-day record is available, with no opportunity to compute the actual 305-day record. In such instances the 365-day record can be converted to a 305-day basis by using the factor given in Table 11-2. Use of production for only the first 305 days also reduces to a considerable extent the variation resulting from the influence of gestation.

Although the D.H.I.A. Sire-proving Program and the sire programs of the dairy-breed associations are used on 305-day lactation records, there is often a real need to use records of less than 10 months' duration. Some workers have advocated using records of less than 305 days to reduce the influence of variation in time cows become pregnant after calving. Others have suggested using short-time records in sire proving to make the proof available sooner and shorten the generation interval. Furthermore, a preliminary proof based on incomplete daughter records is often desired to provide information for a tentative decision on a bull. In the routine summary of a bull's proof there may be one or

Table 11-2. Factors for Computing Lactation Records of over 305 Days to a 305-day Basis

305–308	1.00	337–340	0.92
309–312	0.99	341–344	0.91
313–316	0.98	345–348	0.90
317–320	0.97	349–352	0.89
321–324	0.96	353–356	0.88
325–328	0.95	357–360	0.87
329–332	0.94	361–364	0.86
333–336	0.93	365	0.85

Table 11-3. Factors for Converting Incomplete Milk Records to a 305-day Basis*

Days	<36 Mo.	≥36 Mo.	Days	<36 Mo.	≥36 Mo.
30	8.32	7.42	170	1.58	1.48
40	6.24	5.57	180	1.51	1.41
50	4.99	4.47	190	1.44	1.35
60	4.16	3.74	200	1.38	1.30
70	3.58	3.23	210	1.32	1.26
80	3.15	2.85	220	1.27	1.22
90	2.82	2.56	230	1.23	1.18
100	2.55	2.32	240	1.19	1.14
110	2.34	2.13	250	1.15	1.11
120	2.16	1.98	260	1.12	1.09
130	2.01	1.85	270	1.08	1.06
140	1.88	1.73	280	1.06	1.04
150	1.77	1.64	290	1.03	1.03
160	1.67	1.55	300	1.01	1.01

* From *Dairy-Herd-Improvement Association Letter*, vol. 41, No. 6, for Holsteins, 1965.

more daughters or dams lacking complete records. In many cases this is due to culling for low production during the early months of lactation, and the omission of these records fails to provide an unselected summary. Factors for selected days in milk as taken from values compiled by the Dairy Herd Improvement Section, U.S. Department of Agriculture, are presented in Table 11-3. Separate factors for first lactation animals, those freshening under 36 months of age, and later lactation animals were developed because of the higher persistency of the first lactation animals. Only selected values for milk yield for the Holstein breed are presented here. Values for each of the dairy breeds for both milk and fat are given in the U.S. Department of Agriculture publication.

Frequency of Milking. Although some breeders find it desirable to milk their cows three times daily, well over 95 per cent milk their cows twice daily. Records made with three milkings per day need to be converted to a twice-daily basis, as found on the average dairy farm. Numerous studies have been made to evaluate the effect of frequency of milking on the total milk production during a lactation. Although the results of these studies are not in complete agreement, there is sufficient evidence to indicate that on the average a cow milked three times a day will produce 15 to 20 per cent more milk than if she is milked twice a day. The younger cows seem to show a greater increase in production from more frequent milking than the older cows. Factors

Table 11-4. Factors for Reducing Records of 305 Days or Less to a Twice-a-day Milking Basis

Number of days milked, 3✕	2–3 years of age	3–4 years of age	4 years of age and over
5–15	0.99	0.99	0.99
16–25	0.98	0.99	0.99
26–35	0.98	0.98	0.98
36–45	0.97	0.98	0.98
46–55	0.97	0.97	0.97
56–65	0.96	0.97	0.97
66–75	0.95	0.96	0.96
76–85	0.95	0.95	0.96
86–95	0.94	0.95	0.96
96–105	0.94	0.94	0.95
106–115	0.93	0.94	0.95
116–125	0.92	0.93	0.94
126–135	0.92	0.93	0.94
136–145	0.91	0.93	0.93
146–155	0.91	0.92	0.93
156–165	0.90	0.92	0.93
166–175	0.90	0.91	0.92
176–185	0.89	0.91	0.92
186–195	0.89	0.90	0.91
196–205	0.88	0.90	0.91
206–215	0.88	0.89	0.90
216–225	0.87	0.89	0.90
226–235	0.87	0.88	0.90
236–245	0.86	0.88	0.89
246–255	0.86	0.88	0.89
256–265	0.85	0.87	0.88
266–275	0.85	0.87	0.88
276–285	0.84	0.86	0.88
286–295	0.84	0.86	0.87
296–305	0.83	0.85	0.87

for reducing records to a twice-a-day milking basis, as derived from studies of D.H.I.A. and herd-test records, are given in Table 11-4.

Age at Freshening. Cows gradually increase their lactation milk yields from the time they first freshen as two-year-olds until they are six to eight years of age. Thereafter senility begins and the amount of fat and milk produced during each lactation gradually declines.

Age-conversion factors are used to bring the 305-day production

Table 11-5. Age-conversion Factors for 305-day Production Records

Age	Ayrshire	Brown Swiss	Guernsey	Holstein and Red Dane	Jersey
1–9	1.36	1.48	1.31	1.37	1.32
1–10	1.34	1.47	1.28	1.35	1.30
1–11	1.32	1.46	1.26	1.33	1.28
2–0	1.30	1.45	1.24	1.31	1.27
2–2	1.28	1.43	1.22	1.29	1.25
2–4	1.26	1.39	1.20	1.26	1.23
2–6	1.24	1.35	1.18	1.24	1.21
2–8	1.22	1.31	1.16	1.22	1.19
2–10	1.20	1.27	1.14	1.20	1.17
3–0	1.18	1.23	1.12	1.10	1.15
3–2	1.16	1.20	1.10	1.16	1.13
3–4	1.14	1.18	1.09	1.14	1.11
3–6	1.13	1.16	1.08	1.12	1.09
3–8	1.12	1.14	1.08	1.11	1.08
3–10	1.11	1.12	1.07	1.10	1.07
4–0	1.11	1.10	1.06	1.08	1.06
4–3	1.08	1.08	1.05	1.05	1.05
4–6	1.06	1.07	1 04	1.04	1.03
4–9	1.05	1.05	1.03	1.03	1.03
5–0	1.03	1.04	1.02	1.02	1.02
5–6	1.02	1.02	1.01	1.02	1.01
6–0	1.00	1.00	1.00	1.00	1.00
7–6	1.00	1.00	1.01	1.00	1.01
8–0	1.01	1.01	1.02	1.01	1.02
9–0	1.02	1.01	1.02	1.02	1.02
10–0	1.03	1.02	1.04	1.04	1.04
11–0	1.04	1.04	1.06	1.06	1.06
12–0	1.06	1.06	1.08	1.09	1.08
13–0	1.07	1.08	1.10	1.12	1.10
14–0	1.09	1.10	1.12	1.15	1.12

records to the basis of what a cow would be expected to produce if she were mature. The age-conversion factors given in Table 11-5 were taken from those developed by Kendrick for use in the D.H.I.A. Sire-proving Program. These factors as well as those for frequency of milking are based on averages of a large group of records. Although individual lactation records on an age-corrected basis will undoubtedly continue to be used, it should be remembered that correction factors can be used with much greater reliability when several individual age-corrected records are averaged. This point should be emphasized for all the factors used to bring records to a standardized basis.

Since the factors which have been presented are multiplicative factors, it makes no difference which factor is applied first when two or more factors, such as age and frequency of milking, are required for the same record.

What accounts for the increase in production up to six to eight years of age? Without doubt the major reason is the increase in size that parallels advancing age. Along with the increase in body weight, further development and growth of the udder take place with successive freshenings. The age-conversion factors partially correct for each of these influ-

Fig. 11-3. Differences in the rate of increase in milk and fat production with age for the five major dairy breeds (based D.H.I.A. age-conversion factors).

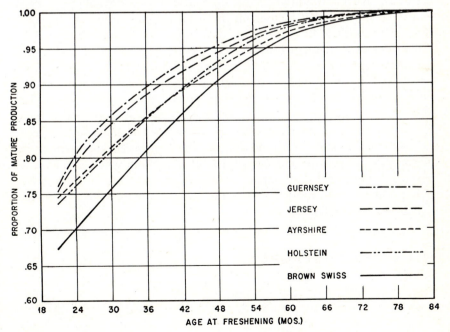

Table 11-6. Factors for Adjusting 305-day Milk Yields for Variation in Days Open*

Days open	Factor	Days open	Factor
24 or less	1.09	86–95	1.01
25–30	1.08	96–105	1.00
31–35	1.07	106–115	0.99
36–45	1.06	116–135	0.98
56–65	1.05	156–175	0.97
66–75	1.03	176–205	0.96
76–85	1.02	206 or more	0.94

* 100 days open taken as standard.

ences, but, as has been pointed out, they may inadequately adjust for age or size in individual cases.

Days Dry and Days Open. Days dry prior to calving do not greatly influence lactation yields unless the dry period is quite short, 30 days or less. However, pregnancy does depress yield, particularly after about 150 days of gestation. This variable may be measured practically either as days carried calf or days open from parturition to conception. Both of these measures are superior to calving interval in that they permit the inclusion of information on the final lactations of cows leaving the herd in a barren condition. Days open also has an advantage in herd management, in that it focuses on the need for conception at a specific time after calving. In two studies, days open accounted for from 4 to 7 per cent of the variance in 305-day records. Even in extremely well-managed herds days open could have an important effect on milk yield.

Table 11-6 gives factors for correcting 305-day, $2\times$, mature-equivalent milk yields to a standard of 100 days open.[1] With an average of 100 days open, a 12- to 13-month calving interval would be maintained. In sire-proving and other studies, days open are not now routinely considered when records are standardized. This can be an important cause of variation, especially among records of elite cows where an unusual effort is made to get these animals bred. When there are herd reproductive problems, the production average of the progeny of a sire also may be measurably affected.

MILK COMPOSITION

Interest in milk composition is growing, and certain markets are beginning to give some attention to the PLM (protein-lactose-mineral) con-

[1] Smith, J. W., and Legates, J. E., *Jour. Dairy Sci.*, **45**:1192–1198, 1962.

tent of milk as well as fat. Some few markets are emphasizing the protein content, but most of the marketing differentials for milk are still largely determined by fat composition. While the market changes are extremely slow, all evidence suggests that the nonfat constituents will become economically more important in the future.

Current evidence suggests that the heritability of fat protein and PLM percentages are about 0.50. With the high heritabilities of the milk constituents, a rather accurate assessment of the merit of a cow can be obtained from a single lactation. Testing for fat percentage is now the common practice. Since fat percentage is positively correlated with the other constituents, maintaining a minimum fat level which satisfies the market requirement should help maintain protein and PLM at reasonably satisfactory levels.

Low variability for PLM and protein percentages will limit change in these constituents. Actually larger increases in total pounds of these constituents can be gained by emphasizing milk yield than by selecting for percentage composition.

Each of the major constituents are positively correlated both genetically and phenotypically as shown in Table 11-1. The genetic relationship between protein and PLM appears to be high enough to permit selection

Fig. 11-4. Guernsey cow, Yellow Creek Handy Sue (Excellent five times). She produced 17,864 lb of milk and 1,122 lb of fat in 365 days, 2×, at 4 years. (*Courtesy of the American Guernsey Cattle Club.*)

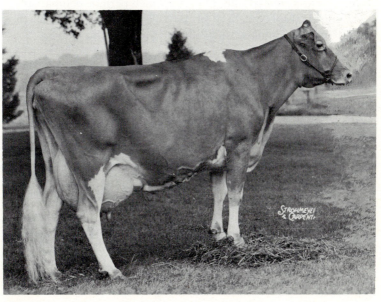

Table 11-7. Factors to Convert Production to
Various Breed Standard Fat Composition

Breed	Average test*	Conversion factors	
		Milk	Total fat
Ayrshire	4.03	0.398	14.9
Brown Swiss	4.08	0.395	14.8
Guernsey	4.75	0.360	13.5
Holstein	3.64	0.422	15.8
Jersey	5.11	0.343	12.9

* D.H.I.A. averages from U.S. Department of Agriculture for cows freshening 1962–1963.

for PLM to be reasonably effective in selecting for protein content. Perhaps the ideal would be to have the energy secreted as fat converted to protein. Nonetheless the strong positive relationships among the constituents suggest that selection for either protein or PLM would be expected to increase fat percentage also.

The energy value of milk is extremely closely correlated with the fat content of milk, and in many instances the yields of cows differ in composition within and between breeds. Gaines and Davidson[1] have suggested a method for converting milk of any test to a given standard fat-corrected milk (FCM), which permits comparisons on a common energy basis per unit of milk. From the results of their studies the following formula was developed:

$$4\% \text{ FCM} = 0.4 \text{ total milk} + 15 \text{ total fat}$$

In addition to expressing milk on a 4 per cent FCM basis, milk could be converted to other desired standards such as the average test for the breed using the above relationship. Taking the Holstein average of 3.64 per cent in Table 11-7, 1 lb of 3.64 per cent milk is equivalent 0.95 lb of 4 per cent milk ($1 \times 0.40 = 0.40$, and $0.0364 \times 15 = 0.55$, then $0.40 + 0.55 = 0.95$).

By dividing the factors for 4 per cent FCM by 0.95, the factors for 3.64 milk in Table 11-7 are obtained. The factors to convert the production of the other breeds to the standard for the breed were derived in a similar manner.

[1] Gaines, W. L., *Ill. Agr. Expt. Sta. Bul.* 308, 1928.

TYPE

Animals have long been evaluated on the basis of their conformation. In Roman animal husbandry nearly 2,000 years ago, the bovine was used primarily as a draft animal. The ideal animal was to be well made: sound, deep-bodied; long, thick neck; broad, high shoulders; wide, deep body; good rump; short, straight legs; good hoofs; and a smooth, soft hide. Included with the above points, which obviously have some bearing on power, were stipulations that the animal have dark horns, hairy ears, flat nose, black muzzle, tail reaching to the heels, and preferably be black in color—points which might be described as fancy rather than utilitarian. With this as a beginning, it is not surprising that dairy-cattle scorecards continue to carry aesthetic as well as utilitarian requirements.

Prior to 1929 the show ring and scorecard were the only yardsticks for measuring type or conformation in dairy cattle. In 1929 the Holstein-Friesian Association of America inaugurated a type-classification program, and the other breed associations adopted programs shortly thereafter. These programs represent the most systematic approach yet made toward measuring and recording the type of large numbers of dairy cattle. During the early years, only the over-all rating of a cow was given, viz., Excellent, Very Good, Good Plus, Good, Fair, and Poor. An over-all rating such as Good did not point out whether the cow was off in general appearance, body capacity, dairy character, or mammary system. More recently, ratings have been made on the individual scorecard breakdown. This has represented a sound advance, but even yet, when a bull's daughter's udders average only Good, we do not know whether they scored low because they were small, unbalanced, poorly attached, or showed poor teat placement.

Still another drawback to using type to evaluate animals is that animals look different on different days and in various stages of lactation. Classification ratings change with age and, besides, different classifiers vary in their emphasis on and evaluation of borderline defects. The repeatability of type ratings is only about 0.40, yet most breed type-classification programs do not lower a cow's over-all rating even though ratings can be raised.

During each calving interval, a cow usually undergoes a cyclic change in physical apperance. In the majority of cases the cow puts on flesh during the dry period and has a certain bloom, which is lost during lactation. The size and appearance of the udder also changes. Data collected at the West Virginia Station[1] indicated that classification ratings

---

[1] Hyatt, George, Jr., and Tyler, W. J., *Jour. Dairy Sci.*, 31:71, 1948.

were higher on cows classified during the first three months of a lactation or just before freshening as compared with the middle segment of the lactation. Work at the same station indicates that different classifiers varied most in their appraisal of defective feet and legs and udders.

An objective consideration of conformation should emphasize those qualities related to serviceability and productivity. Feet and legs are often referred to, and recent reports indicate that about 1 per cent of the cows in herds leave because of reported feet and leg troubles. Such an evaluation may not fully reflect the practical importance of good feet and legs, since foot trimming can be an important chore, especially with loose-housing arrangements.

Udders are needed that will be useful for a long time, but they do not necessarily need to be fancy. Although lower udders are more prone to mastitis,[1] there is some evidence suggesting that animals with deeper udders are higher producers. Swedish workers[2] found wide variations in yield between the fore and rear quarters of the udder, but very little difference between the yields of the left and right sides of the udder. The heritability of differences between the front and rear halves of the udder were about 0.75, whereas the differences between the left and right halves of the udder were largely developmental and environmentally caused. Selection for more fully developed forequarters might be expected to yield good results when decisions are based on actual measurements of yield for the front and rear quarters. External judging of the udder permitted identification of the most extreme types, but special investigations have shown that there is a rather poor agreement between udder scores and actual measurements of yields.

Type and Production.   The question as to whether or not there is a relationship between type and production is raised continually. Long before herd-classification programs were initiated, breeders sought a physical measurement which could be used to predict a cow's production. All the measures tried were less reliable for predicting future production than a cow's previous lactation record. The relationship between classification ratings and production records has been studied by several workers. There is general agreement from these studies that type and production go together when applied to a large number of animals. However, such general relationships must be subjected to closer scrutiny when the individual cow is considered.

The following summary and discussion largely involve the association

[1] Young, C. W., Legates, J. E., and Lecce, J. G., *Jour. Dairy Sci.*, 43:54–62, 1960.
[2] Johansson, Ivar, "Genetic Aspects of Dairy Cattle Breeding," pp. 116–122, The University of Illinois Press, Urbana, Ill., 1961.

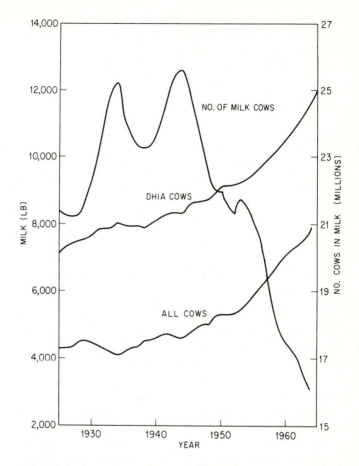

Fig. 11-5. Increase in average milk production per cow from 1926 to 1965 for all cows kept for milk in the United States and for all cows on D.H.I.A. test. Note the sharp increase in average production per cow and the decline in the number of cows during the last decade.

between type and fat production. The relationship of type to fat yield is presented here since it was possible to draw upon the results from many more studies than were available for milk yield alone. Conclusions based on the relationship between type and fat yields are practically equally applicable to the relationship between type and milk yield.

The correlation between lactation production and type score has been studied by several workers. The summary in Table 11-8 shows this average relationship when comparing type and production of cows in the same herd. In each case the correlation is positive but indeed small

Table 11-8. Summary of General Relationship between Type and Production

Classification rating	305-day, 2×, M.E. fat production		
	Ayrshires*	Holsteins†	Jerseys‡
Excellent	415	508	483
Very good	394	478	460
Good plus	376	456	448
Good	361	435	434
Fair	363	414	420

* First records of 5,177 cows.
† Average of all available H.I.R. records for 59,642 cows.
‡ Highest records for 14,147 cows.

when we realize that positive correlations can range from 0.0 to 1.0. Many studies suggest an average correlation of about 0.15. For purposes of comparison, one production record shows a correlation of about 0.40 with a subsequent production record in the same herd, making it two or three times as valuable in predicting future productivity as a single type rating.

With this low relationship between type and production, there is much opportunity for cows of poor type to be high producers and for cows of good type to be poor producers. The average difference in production for cows differing by one type grade is given in the study of Tyler et al.,[1] in which the correlation between type and fat production was 0.18. With this correlation, each change of one type grade on the average was accompanied by a 15.3-lb change in lactation fat production. Other things being equal, suppose the correlation were perfect, 1.00 instead of 0.18. Each change of one type grade on the average then would be associated with a change of 86 lb of fat per lactation.

One might suspect that inasmuch as the final rating has woven into it emphasis on breed character and other fine points, a judge looking solely for those qualities believed to be associated with production could predict production more accurately than the over-all rating indicates. During the early years of classification programs, only the over-all rating was given a cow. More recently, the components of type for each animal have been scored.

From an analysis of Ayrshire type and production records, general

[1] Tyler, W. J., et al., *Jour. Anim. Sci.*, 14:1189, 1955.

quality or dairy character showed a correlation of 0.14 with production in contrast to 0.08 for over-all rating. From a similar analysis with Holsteins, dairy character showed a correlation of 0.25 with production as compared with 0.18 for over-all rating. All the other components of types showed smaller correlations with production than did dairy character. Legs and feet and rump scores showed the least association with production.

It has been suggested that some components of type are more closely related to lifetime production than to lactation production. Feet and legs, which were indicated as having a low correlation with lactation production, and udder attachment are commonly cited as important in lifetime production. The statement that an unusually high proportion of those cows producing 100,000 lb of milk are classified Excellent or Very Good, as well as tabulations that Excellent cows average more records per cow, are often cited to clinch the argument that type and high lifetime production are closely associated. However, the facts to substantiate these relationships are yet to be gathered and presented systematically. It is well to ask whether many of the cows with exceptional type may have remained in the herd despite mediocre production. Thereby, they may have been permitted to build up a creditable lifetime production, whereas cows with equally good production and lasting qualities, but of poor type, may have been culled.

Will selection for superior type bring about improvement in production? Whether or not selection for superior type can be expected to influence production is dependent on the genetic association between the two traits. Four studies which have been made using Holstein, Ayrshire, and Jersey classification and production records show that type and production are essentially independent genetically (Table 11-1). Although there is no genetic antagonism between good type and high production, selection for type alone will have little direct influence on production, and, conversely, selection for production alone will have little direct influence on over-all type rating. If both are desired, a breeder must emphasize both type and production in his selections. The price which is exacted for attention to both traits is that the intensity of selection for each trait is reduced. Only rarely will animals giving promise of highest transmitting ability for production be animals which also have superior type.

BODY SIZE

Varying viewpoints are held regarding the importance of body size in cattle improvement. The general practice has been to give preference to the larger cow when all other things are equal. Much confusion has resulted in the study of body size and its relationship with production since the effects of age usually have not been simultaneously considered. When age differences are ignored, phenomenal changes in production appear to be associated with size changes. On the other hand, the increase in milk yield per 100 lb of body weight for cows of the same age is only about 200 lb.[1] The genotype for body size and the genotype for milk yield do not appear to be closely associated.

Our present system of selecting cows and sires on total lactation yield takes no account of difference in maintenance requirements for the large and small cows. In fact, continued emphasis upon total yield per cow regardless of size could tend to develop low inherent efficiency by increasing the size of the cow. Studies to date have suggested that from a physiologic point of view small cows are more efficient in converting feed energy into milk energy. It has been postulated that this could have possibly resulted from our emphasis on total production, requiring the small cow to produce total fat or milk equivalent to that of her larger herd mates in order to remain in the herd.

When costs for maintaining the additional 100 lb of body weight are taken into account, the added 200 lb of milk usually will fail to offset these costs. From a practical viewpoint, increased managemental efficiency favors the large cow. Since it takes little extra time to milk, feed, clean, and manage a large cow producing more milk than a small cow, it is obvious that labor, management, and other overhead costs should be less per unit of milk produced from large animals of comparable efficiency. Some have suggested that larger cows possess more stamina and should have longer productive lives, but there is little evidence to support this view. Current evidence indicates that dairymen should not emphasize increased body size unless it is accompanied by increased milk yields in excess of the average of 200 lb per 100 lb increase in body weight.

UDDER AND MILKING QUALITIES

As dairying continues to move toward larger mechanized operations, the importance of milking rate is being forcefully brought into focus. If 7 as compared to 5 minutes are required per milking, a cow would be involved in the milking operation an additional 20 hours during a 305-day, $2\times$, lactation. Peak flow, the maximum milk yield for any one minute during the milking, has been widely used as a measure of milking

[1] McDaniel, B. T., and Legates, J. E., *Jour. Dairy Sci.*, 48:947–956, 1965.

rate. Most workers agree that it is less influenced by the skill of the milker than is average rate of flow. Machine time, which represents the milking time up to the start of machine stripping, also has been used to measure milking rate. Some consideration also must be given the total milk yield for the milking in making comparisons among cows. While little systematic selection has been exercised on milking qualities, both peak flow and milking time have heritabilities of about 0.35.

Increasing requirements for high lactation yields place added stress on udders and udder attachments. Low and pendulous udders are more subject to injury and prone to attacks of clinical mastitis. The importance of the teat sphincter as a barrier to udder infection also has been stressed. A higher incidence of mastitis has been found among fast-milking cows and among cows with higher peak milk flow. However, this association is not strong enough to suggest breeding for slower milking to reduce the incidence of mastitis.

REPRODUCTIVE EFFICIENCY

Cows must calve regularly; i.e., they must be efficient reproducers if they are to maintain a high average daily milk yield. The increased use of artificial insemination during the past ten years has stimulated maintaining better breeding records and focused attention on reproductive troubles in dairy cattle. Several measures of reproductive efficiency, such as services per conception, calving interval, days from first breeding to conception, and indexes based on a calf each twelve months, have been studied to determine whether or not selection for improved reproductive performance can be effective. Table 11-9 gives a summary of the repeatability and heritability values obtained for several of these measures of reproductive efficiency.

The low heritability values point out that most of the variation in the commonly used measures of reproductive efficiency is nongenetic. Genetic influences have been shown to be more important for several special conditions. Work at Wisconsin[1] has shown that the cystic-ovary condition is influenced by the genetic make-up of the cow. The heritability of the occurrence of the cystic condition sometime during the cow's life was 0.43. A number of lethal and semilethal genes also exert some influence on the measures of reproductive efficiency mentioned above, since they influence embryonic mortality. Other lethal or semilethal genes are responsible for a portion of the calves which are dead at birth. Certain other genes may prevent fertilization, such as in the white

[1] Casida, L. E., and Chapman, A. B., *Jour. Dairy Sci.*, 34:1200–1204, 1951.

Table 11-9. Repeatability and Heritability of Measures of Breeding Efficiency in Dairy Cattle

Measure used	Repeatability	Heritability	Source
Services per conception	0.22		Berge
Calving interval	0.13		Ostergaard
Services per conception	0.08		Olds and Seath
Services per conception	0.07		Leonard
Calving interval	0.07		Asdell
Nonreturns to first service	0.05	0.004	Dunbar and Henderson
Calving interval		0	Dunbar and Henderson
Services per conception	0.12	0.07	Pou et al.
Days first service to conception	0.11	0.07	Pou et al.
Services per conception	0	0.03	Legates
Calving interval	0.13	0	Legates
Days open		0.09	Smith and Legates

heifer disease that is often observed in White Shorthorns. Ovarian and testicular hypoplasia (underdevelopment) and sperm abnormalities of genetic origin have also been reported.

The use of the measures of reproductive efficiency shown in Table 11-9 is restricted because they are concerned with variations among fertile cows. In practice, however, nature has selected against and continues to select against poor reproductive efficiency. Those cows which fail to settle are eliminated from the population naturally; hence our real opportunity to improve fertility by selection lies in being able to distinguish the differential fertility among those cows which are capable of dropping offspring. The precision with which we can make these distinctions is restricted, as the generally low values for repeatablity and heritability indicate. Application of genetic principles in selecting for fewer services per conception or for a shortening of the calving interval offer little promise to the dairy-cattle breeder.

DISEASE RESISTANCE

The use of preventive and therapeutic measures has been relied upon to control most of our cattle diseases. Selection for resistance to specific diseases has been a much more attractive approach to disease control in plants than in animals. When inexpensive control and preventive measures are available, it is difficult to justify, practically, breeding for disease resistance in animals of so high individual economic value and of so slow a reproductive rate as cattle. In addition, with so many desired qualities to select for, only those diseases which seem to defy effective prevention and control can be justifiably included in the selection goal.

Mastitis. Mastitis in dairy cattle represents a disease problem which is difficult to prevent or control effectively, since so many different pathogens can produce the disease. Many workers have pointed out special instances which indicate that the susceptibility to mastitis manifested by related animals might have a genetic basis. The results of Murphy's study given in Table 11-10 is an example of this type of evidence. In addition, two studies have yielded heritability estimates for mastitis resistance of 0.27 and 0.38. Reports and observations indicate that pendulous udders predispose to mastitis by permitting the teats and udder to be subject to injury and infection. Udder size, strength of attachment, and the strength of the teat sphincter may be genetically influenced and related to mastitis susceptibility.

Should future studies justify major emphasis on mastitis resistance in dairy-cattle selection, the progeny testing of bulls for their transmission of genes influencing resistance offers a possible breeding approach to reducing the incidence of mastitis. Such a program has its limitations, however, since a satisfactory phenotypic evaluation of the daughters of a sire by present methods must be delayed until they have completed their second or third lactations. Without an early individual test for mastitis, selection must be based on pedigree or family information, with the bull's dam or his sisters receiving major emphasis.

*Table 11-10. Incidence of Mastitis in Two Cow Families, Showing the Percentage of the Milking Time Occupied by Infection**

Dams		Daughters		Granddaughters	
Cow no.	*Infection per cent of milking time*	*Cow no.*	*Infection per cent of milking time*	*Cow no.*	*Infection per cent of milking time*
283	100.0	469	50.0	558	66.7
		505	70.4	614	20.8
		593	83.3		
247	0.0	480	12.5	638	4.2
		515	1.8	646	50.0
		559	12.5		
		604	16.7		

* Murphy et al., *Cornell Vet.*, **34**:186, 1944.

Other Diseases. A condition known as *crampy* has been recognized in cattle for many years. The condition is characterized by intermittent spastic contractions of the muscles of the back and rear legs of adult animals. Definite evidence of a hereditary influence on this trait has been established and in some respects it is recessive in its expression.[1]

Susceptibility to actinomycosis, commonly called *lumpy jaw*, also is genetically conditioned. The incidence in the Guernsey breed was seven times as great as in other breeds. Some limited evidence is also available to suggest hereditary differences in susceptibility to milk fever. Jerseys have a higher incidence than do other breeds.

OTHER TRAITS

From a practical standpoint, a dairyman may desire many other qualities in his cattle. He will want cows that are not too high-strung and nervous. Animals that are generally vigorous, maintain high production for a long time during each lactation, and remain productive for several years will be desired. Longevity is difficult to select for, since family selection must be used, and generally the selection pressure cannot be intense.

In some regions strains of cattle which can tolerate high environmental temperatures are desired. In beef cattle the introduction of Zebu cattle not only improved heat tolerance but also developed a resistance to ticks which cause Texas fever. Emerging from this effort was the development of the Santa Gertrudis breed, which on the average had ⅜ blood of Indian cattle and ⅝ Shorthorn blood.

Attempts to duplicate this with dairy cattle have been made by crossing Red Sindhis with our European dairy breeds. The problem of incorporating heat tolerance in dairy cattle, however, appears to be a task of greater proportion than it was with beef cattle. Studies to date have suggested that the ability of animals to maintain lower body temperatures, and hence more heat tolerance, is more closely related to lower heat production than to improved heat dissipation. Much heat is produced in milk secretion, and the demand for higher milk production makes the problem more acute. Although the problem cannot be stated with certainty at present, it appears likely that efficiency of feed utilization, with a greater proportion of the ration going to milk energy rather than to heat energy, may enter into the development of heat tolerance in our dairy animals.

Dairy-cattle breeders generally have not directly considered the prob-

[1] Becker, R. B., Wilcox, C. J., and Pritchard, W. R., *Fl. Agr. Expt. Sta. Bul.* 639, 1961.

lem of efficiency of feed utilization and conversion to milk. Work with monozygotic twins and controlled-feeding trials to point up the possibilities in this field are almost certain to be intensified.

FEMALE SELECTION

Improvement depends, first, on being able to recognize those animals which are genetically superior and, second, upon how effective we are in permitting these superior animals to reproduce. Rating and selecting of females involves, for the most part, selection among cattle within one herd. If the animals in the herd are handled as a unit, so that one is not deliberately given special attention and another neglected, the average production for the herd will be a wise base upon which to compare the cows in the herd. Even within the herd, cows will not have the same opportunity to repeat their performance with respect to the herd average each year. Certain uncontrollable environmental conditions for individual cows within the herd will be better during one year than during other years.

Selection for Future Production. High records within a herd result because the cows making the high records have high inherent abilities, and also because they had a better-than-average opportunity, insofar as the herd environment was concerned. The superiority due to the cow's own ability will tend to be repeated in subsequent lactations. However, a new set of the uncontrolled environmental conditions will be experienced, which in succeeding lactations should be more like the average for the herd. These principles will be keenly appreciated if a breeder will divide his herd into a high and a low half on the basis of production in one year and note the average production of these two groups during a second year. This was illustrated in Fig. 7-6 of Chap. 7.

The regression of the future records on the present records or the degree to which a record is repeated should be known in order to select effectively those cows which are expected to perform best during the next year. This actually involves predicting what Lush[1] has termed the cow's "most probable producing ability" for the next lactation. When the cow has only one record, her most probable producing ability expressed as a deviation from the herd average is equal to

$$\text{Herd average} + r(\text{cow's average} - \text{herd average})$$

[1] Lush, Jay L., "Animal Breeding Plans," 3d ed., p. 173, 1945.

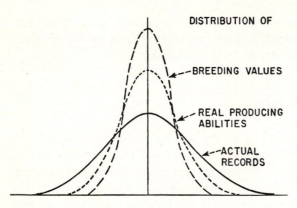

DISTRIBUTION OF

BREEDING VALUES

REAL PRODUCING
ABILITIES

ACTUAL
RECORDS

Fig. 11-6. Normal curves with standard deviations in proportion to 100, 66, and 53, representative of the probable distribution of actual records, real producing abilities, and breeding values. Cows with the best records generally are not as good as their records, and cows with the worst records generally are not as poor as their records. [*From Lush et al., J. Dairy Sci.,* **29:** 719 (1943).]

where r is the value for repeatability. This value shows how much reliance can logically be placed on the cow's average to indicate how much of her superiority or inferiority, as compared with the herd, we can expect to be exhibited in her next lactation. Following the above expression, if a cow produced 16,000 lb of milk in a herd averaging 14,000 lb, assuming a repeatability of 0.50, we would predict that she would produce 15,000 lb of milk in her next lactation. Of the 2,000 lb the cow was above the herd average, about 1,000 lb would be attributed to the cow's inherent ability, and the other 1,000 lb would represent superiority resulting, on the average, from fortuitous temporary environmental conditions.

When a cow has several records, more dependence should be placed on her average as an indicator of what she would produce in her next lactation. This added reliance is demonstrated by the change in the formula for computing the cow's most probable producing ability. When the cow has several records, her most probable producing ability is

$$\text{Herd average} + \frac{nr}{1 + (n - 1)r} \text{ (cow's average } - \text{ herd average)}$$

If we assume the cow used in our previous example had three records rather than one, we would anticipate that 0.75 rather than 0.50 of her

Fig. 11-7. Holstein-Friesian cow, Harborcrest Rose Mellie (Excellent), Grand Champion female, National Holstein Show and All-American aged cow. She has produced 24,921 lb of milk and 1,242 lb of fat in 365 days, 2×, at 8 years. (*Courtesy of Holstein-Friesian Association of America.*)

superiority would be exhibited in the next lactation. Using the above formula, cows having different numbers of lactation records can be compared equitably.

The herd average varies from year to year; consequently a more realistic and accurate approach is to compare each cow's record with the herd average for that particular year. Each cow is evaluated in terms of how much above or below the herd average she will be expected to produce. The differences between the cow's n records and the herd average would be averaged and multiplied by $nr/[1 + (n - 1)r]$ to predict her expected production above or below the herd average:[1]

$$\frac{nr}{1 + (n - 1)r} \qquad \frac{\Sigma(\text{cow's record} - \text{yearly herd average})}{n}$$

Ranking cows in the herd on the basis of their expected producing abilities is an effective way to assemble information for culling. The

[1] When computations are made with electronic computers, deviations from seasonal means such as suggested in the later section of this chapter in "Methods of Expressing the Progeny Test" would be desirable.

standards for culling will be relative rather than rigid. A herd under selection for many years should be able to meet a higher standard than younger herds. During years when a small number of replacements is available, culling will not be so intense. In any case, when herd size is being increased, only a few of the poorer cows can be culled.

Heifers which did extremely poorly during the first lactation under apparently normal conditions should be culled. The doubtful or border-line cases should be allowed a second chance. However, by the time the second lactation is four or five months along, it should be fairly evident whether or not these cows should be culled.

Opportunity to Cull. In many instances a breeder may not be able to cull as intensely as he would like, since economic considerations enter the picture. When he is setting his marketing base, he may elect to retain certain cows that he otherwise would not want to keep in his herd. During periods when the cost of raising a heifer to milking age is considerably more than her beef value, unpromising heifers would be culled a bit more strictly. Since culling animals without production records is less accurate than culling after a record on the cow is available, genetic improvement may be slowed down some by this practice, but

Fig. 11-8. Jersey cow, Beacon Bas Patience (Excellent), Grand Champion at All-American Jersey Show in 1964. She has produced 21,024 lb of milk and 1,051 lb of fat in 365 days, 3×, at 6 years. (*Courtesy of the American Jersey Cattle Club.*)

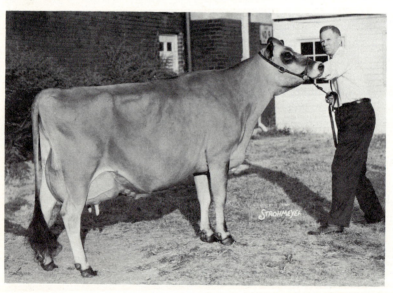

probably not enough to make it worth while to incur the extra cost of raising the questionable heifers.

At other times the opportunity to select will be limited only by the replacements which are available. A summary of over 2¾ million cows in the United States points up the following reasons why cows leave

Percentage of Cows Leaving Herds for Various Reasons°

Reason	Per cent
Dairy purposes	5.1
Low production	7.3
Udder troubles	2.5
Abortion	1.5
Sterility	1.8
Death	1.1
Old age	0.6
Other	1.7
Total	21.6

* From Asdell, S. A., *Jour. Dairy Sci.*, **34**:531, 1951.

the herd. Out of every 100 cows in the herd, approximately 22 are replaced each year. Of these 22 cows, 10 leave the herd for reasons not directly associated with production. If the cows sold for dairy purposes are included among those that could be culled for low production, about one cow in eight presently are culled for low production. An informative breakdown of the opportunity for improvement by selection in an organized progeny-testing plan given by Rendel and Robertson[1] points out the percentage of genetic improvement that can be expected from the following sources:

	Per cent of improvement
Dams of future herd replacements	6
Dams of future young sires	33
Sires of future herd replacements	18
Sires of future young sires	43

Although this particular analysis is based upon a population of 2,000 cows using artificial insemination, it points up vividly that the major

[1] Rendel, J. M., and Robertson, A., *Jour. Genet.*, **50**:1–8, 1950.

opportunities for genetic improvement come from the choice of the sires and dams of our future herd sires.

Estimating Breeding Values for Females. Much has been written about cow families and their importance in determining the breeding value of an animal. So far as the family consists of a cow, her dam, and her daughters, the concept has much utility. However, the usual definition of a cow family as all the female ancestors arising from one foundation female has received more attention than is justified. The use of sires across all of the so-called *families* makes it possible that such a family may include individuals less closely related to other members within the same family than they are to members of another family within the herd.

Analyzing the herd into its foundation lines in order to obtain a clearer picture of the female relationships within the herd and also to note how the herd has developed is of considerable interest. Nevertheless, in using family relationships to appraise critically the probable breeding value of a young heifer or a young bull, animals of less than 25 per cent relationship to the individual in question cannot be justifiably included. The meager amount of information which more remote relatives add is seldom sufficient to counterbalance the risk of deception which environmental changes during the longer time interval may inject.

Particularly when selecting dams of young bulls, it is worth while paying attention to the cow's relatives' performance in addition to the cow's own records. Only close relatives are worthy for consideration in most situations. Generally, these would be the cow's dam, her daughters or full sisters, having a relationship of 0.50, and her maternal and paternal sisters with relations of 0.25 to the cow.

Most of the emphasis is given to the cow's records for highly heritable traits, but the performance of the relatives deserve relatively more attention when the heritability is low. As a general rule for milk and fat yields, the performance of the dam and each daughter of a cow should receive about one-third as much emphasis as the cow's own performance in predicting the cow's breeding value. When it is desirable to combine the information for close relatives and the individual into one figure which is correlated with the animal's breeding value, a selection index may be used.[1,2]

One such index utilizes all available lactations on the cow and her close relatives, and the phenotype of each individual is expressed as its most probable producing ability. The factor $n/[1 + (n-1)r]$ is

[1] Legates, J. E., and Lush, Jay L., *Jour. Dairy Sci.*, **37**:744, 1954.
[2] Deaton, O. W., and McGilliard, L. D., *Jour. Dairy Sci.*, **48**:365, 1965.

the key when using all records rather than only the first record of each individual, since it compensates for the decrease in the variation of averages of *n* records as compared to single records. The index values *I* for all animals in a herd can be determined systematically according to the following formula:

$$I = X_1 + 0.4X_2 + b_3X_3 + b_4X_4 + b_5X_5$$

In this procedure, X_1 and X_2 are the most probable producing abilities of the cow X_1 and her dam X_2, and X_3, X_4, and X_5 are sums of the most probable producing abilities of the cow's daughters, maternal sisters, and paternal sisters, respectively. Sums rather than averages are used here to avoid additional divisions when electronic computers are not available to handle the computations. The values for b_3, b_4, and b_5 vary as the number of daughters and sisters vary. These *b* values given in Table 11-11 are regression coefficients indicating the emphasis to be given each relative's or group of relatives' most probable producing abilities.

The *b* values decrease in size as the numbers increase, because the sums for X_3, X_4, and X_5 are used. Considerable complexity also is avoided in the index by simply including each full sister as a maternal and again as a paternal sister. Such an index procedure usually would give

Table 11-11. Values of b_3, b_4, and b_5 for Different Numbers or Daughters (n_3), Maternal Sisters (n_4), and Paternal Sisters (n_5)

n_3	b_3	n_5*	b_5
1	0.38	1	0.05
2	0.36	2	0.08
3	0.34	3	0.07
4	0.31	4–5	0.06
5	0.29	6–7	0.05
		8–11	0.04
n_4	b_4	12–15	0.03
1	0.10	. . . . .	. . . .
2–3	0.09	20	0.02
4–5	0.08	>20	0.01

* The individual for which the index value *I* is desired is included in n_5; thus when n_5 equals 1, X_1 and X_5 are the same.

a 10 to 20 per cent greater accuracy in selection than that expected when only the cow's records are used.

MALE SELECTION

The importance of sire selection was indicated by the percentages of expected improvement given in the previous section. Choices of sires control the last three items, making up over 90 per cent of the opportunity for improvement by selection. Only a small proportion of the potential males are needed with artificial insemination, and a single sire can leave from 20,000 to 40,000 progeny per year. Extensive programs for planned progeny testing have been developed by artificial insemination organizations to locate genetically superior sires. Since so few sires can be progeny tested, it is extremely important that the procedures used to select young bulls ensure that those with the best potential have an opportunity to be progeny tested.

Selecting a Young Bull. Notwithstanding the often-quoted recommendation to use well-proved sires for most rapid herd improvement, all bulls are at one time unproved. As such, they must first be selected on the basis of whatever sound information is available, which usually must come from the pedigree. In using this information we must remember that the immediate close relatives must be given most attention. Two-generation pedigrees which have complete and unselected information are sufficient, although the names of the animals in the third generation are preferred by many to give the relationship to other animals possibly known by a prospective buyer.

The advice to select the son of a well-proved bull is sound. If the bull's sire is proved, the young bull's paternal sisters can be inspected. On the dam's side of the pedigree, the dam's performance as well as the performances of the maternal and full sisters of the bull should be investigated to be sure they meet the standards of production and type which are desired. There is nothing genetically wrong with saving the son of a young cow, except for the fact that generally she must be evaluated on the basis of one or two of her own records.

Cows for planned matings are now being sought by artificial breeding associations. Close scrutiny is given to young sire prospects. Pedigree indexes are computed for these young bulls to evaluate their probable breeding merit.

Progeny Testing. Several circumstances favor the progeny testing of dairy sires. Milk production is a sex-limited trait, and male's breeding

NAME: MONCONY ARISTOCRAT *ASTRONAUT*   NO: 1401506

BORN: February 21, 1961   SEX: Male

BREEDER: Hill Brothers, Spencerport, New York

BAKER FARM NOTEWORTHY 893803 (G)

USDA Sire Summary (10–60)
48 non-AI daus. W.H.M.
av.   13,826M 3.5%   481F
Unweighted H.M. diff.   +554   −2

26 classified daus. av.   80.8
  2E, 5VG, 9GP, 7G, 3F

COUNCIL ROCK WORTHY ARISTOCRAT

1124780 (E)

USDA Sire Summary (4–63)

584 AI daus. W.H.M. av. 14,036M 3.7%   521F
Adj. diff. from breed av.   +433   +23

641 classified daus. av.   80.8
  92VG, 323GP, 212G, 14F

COUNCIL ROCK MISTRESS BONHEUR

2464060 (VG)

9–1	307d	2x	15,509M	4.1%	639F
10–1	365d	2x	18,240M	4.0%	726F

SIR BESS ORMSBY FOBES DEAN 761977

USDA Sire Summary (7–63)
1736 AI daus. av.   12,602M 3.7%   466F
 515 AI daus. W.H.M. av. 13,497M 3.7%   500F
Adj. diff. from breed av.   +652   +33

459 classified daus. av.   82.1
  1E, 14VG, 84GP, 124G, 16F

MONCONY DEAN JADE 3573017 (VG)

2–7	322d	2x	13,520M	4.2%	579F
3–7	344d	2x	16,730M	4.2%	708F
4–8	305d	2x	17,830M	4.2%	755F
5–9	305d	2x	18,450M	4.2%	767F
6–9	305d	2x	18,390M	4.1%	758F
7–11	305d	2x	17,340M	3.8%	664F
9–0	305d	2x	15,650M	4.1%	645F

7 recs. av. 305d, 2x, M.E. 17,387M 4.1% 723F
  V85, E, V, +, +, V, V, +

MONCONY PRINCESS K.B. NINA ORMSBY

2597621

3–0	346d	2x	13,970M	4.2%	590F
4–7	348d	2x	15,670M	4.3%	682F
5–9	319d	2x	16,970M	4.0%	609F
6–10	365d	2x	18,290M	4.1%	764F

8 lactations total   121,500M 4.20% 5181F

FULL SISTERS TO *ASTRONAUT*
MONCONY ARISTOCRAT JANETTE (GP)
2–2   305d   2x   13,270M   3.9%   522F
MONCONY ARISTOCRAT JEWELETTE
                    (VG)
2–7   305d   2x   13,916M   4.1%   566F
MONCONY ARISTOCRAT JULIANN (G)
2–2   301d   2x   9,910M   3.9%   382F

FOUR MATERNAL SISTERS TO JADE
                 (1VG, 3GP)
4 av. 305d, 2x, M.E.   15,387M 4.0%   627F

Fig. 11-9. Pedigree of an outstanding young sire prospect giving complete production information and avoiding pedigree filler. The pedigree index on this young bull indicates that his daughters should average 825 lb milk and 35 lb of fat above their herdmates before the effect of the genetic trend is discounted.

value must be predicted from the performance of close relatives and progeny. The rapid acceptance of artificial insemination and the advancement of techniques for the freezing and storage of bovine semen have greatly extended the use of outstanding progeny-tested sires.

Proving bulls through natural service in one or two herds was the original scheme for progeny testing. Many young bulls are still selceted for proving in individual breeder's herds. When the sire is to be returned to service in the herd after his progeny test is available, such a progeny test is meaningful. However, most herds are too small to carry out an effective progeny-testing program.

With a 50-cow herd or even a 100-cow herd, what are the possibilities for an effective progeny-testing program? In individual herds, usually records on approximately 10 daughters are desired for the progeny test. When all daughters are kept in the herd, experience has shown that at least three cows must be pregnant to provide one daughter with a satisfactory record for use in the proof. Thus to test one young bull

Table 11-12. Growth of Organized Artificial Breeding in the United States[*]

Year	Cows bred	Sires in service	Cows bred per sire
1939	7,539	33	228
1949	33,977	138	246
1941	70,751	237	299
1942	112,788	412	274
1943	182,524	574	318
1944	218,070	657	332
1945	360,732	729	495
1946	537,376	900	597
1947	1,184,168	1,453	815
1948	1,713,581	1,745	982
1949	2,091,175	1,940	1,078
1950	2,619,555	2,104	1,245
1951	3,509,573	2,187	1,605
1952	4,295,243	2,325	1,848
1953	4,845,222	2,598	1,865
1954	5,155,240	2,661	1,937
1955	5,413,874	2,450	2,210
1956	5,762,656	2,553	2,257
1957	6,055,982	2,651	2,284
1958	6,645,568	2,676	2,483
1959	6,932,294	2,460	2,816
1960	7,144,679	2,544	2,808
1961	7,482,740	2,486	3,010
1962	7,748,687	2,456	3,155
1963	7,673,582	2,559	2,999
1964	7,747,953	2,538	3,053
1965	7,879,982	2,316	3,402

[*] From D.H.I.A. letter, April–May, 1965. Reported on basis of cows enrolled in associations.

Fig. 11-10. Interior of a bull barn for a large artificial breeding organization. Facilities are available for handling over 600 bulls in various stages of progeny testing at this location. (*Courtesy of American Breeders Service.*)

each year, successful matings to 30 females will be required. This would mean that 60 per cent of the 50-cow herd and 30 per cent of the 100-cow herd would be used to test one new bull each year. But when only one bull is tested at a time, we have a poor basis of comparison. Also if the bull does not compare favorably with the previous sire, he must be discarded, and we start all over again. Most individual herds cannot prove enough sires to ensure steady long-range improvement.

Planned progeny-testing programs in conjunction with artificial insemination offer exceptional opportunities to improve the accuracy of sire evaluation. Use of a sire in many herds under varying managemental conditions provides an "acid test" of a bull's transmitting ability. Some of the advantages of such a program are:

1. A proof which will more reliably indicate what future-sired daughters will produce can be obtained at an early age.
2. The risks of sire proving are spread among many herds, and for the

industry as a whole the number of daughters from poorer sires is reduced.

3. Since the daughters of the bull are scattered in several herds, there is less likelihood that herd trends and those special environmental circumstances which cannot be appraised adequately will unduly influence the proof.

Some of the major limitations to such a program are:

1. The cost of holding young bulls may be high.
2. A sound testing program with education and cooperation of the cooperating breeders is essential to its success.

The cost and time involved in any sire-proving program will be great. However, the use of outstanding bulls proved in artificial insemination offers the best opportunity for future improvement, even though no short cuts are now available to obtain such bulls.

Methods of Expressing the Progeny Test.   Since the principle of what is now the basis for the Equal-parent Index was first suggested, numerous proposals have been made for evaluating the progeny test. The daughter average, standardized on a 305-day, $2\times$, mature-equivalent (M.E.) basis, is the simplest measure of the progeny test. If the number of daughters is adequate and all daughters are included without selection, this provides a sound basis of evaluation when the feeding and management are to be the same for future daughters. Since the daughter average does not take into account the merit of the dams to which a bull is mated, many breeders prefer some form of daughter-dam comparison to evaluate a bull. The daughter average and the daughter-dam comparison have been widely used in summarizing the progeny test.

Many have recognized that the conventional daughter-dam difference is markedly influenced by changes in herd management. In many cases, the difference of the daughters compared to their dams directly reflects the managemental trend in the herd. Generally comments on the daughter-dam difference are prefaced by stating that if the conditions were the same for the daughters and their dams, the difference reflects the influence of the sire. But one cannot ensure that the conditions will be the same for the daughters and their dams. Most of the dams make their records two or three years before their daughters. For most herds only about one-third of the dams make records concurrently with their daughters. Herd opportunity is so subject to change with the herdsman, the feed supply, and the weather and economic conditions that it is practically impossible to provide the same opportunity for daughters and dams.

For illustration, let us look at an actual example of the daughter-dam comparison for a bull on a 305-day, 2×, M.E. basis.

41 daus. av.	13,904M	3.66%	504M
41 dams av.	11,732M	3.52%	410M
	+2,172M	+.14%	+94M

This appears to be a strong production proof, but the information must be interpreted properly. The animals represented in this proof completed records in several herds. During the interval when the daughters were dropped and when they completed their records, the average production in these herds increased by about 1,700M and 60F. The 41 daughters by this bull made up less than 5 per cent of the records involved, hence they could not have greatly influenced the over-all production trend.

		305-day, 2×, M.E.		
Year	No. of records	Milk, lb	%	Fat, lb
1	530	11,196	3.51	391
2	522	11,715	3.45	403
3	517	12,286	3.48	427
4	588	13,035	3.52	457

We recognize that managemental changes in these herds could have contributed to the daughter-dam difference. If the trend had been downward, in all probability the difference would have been negative. How are we to gain a more accurate assessment of this sire's transmitting ability? Our most promising practical procedure to provide this assessment of the relative transmitting ability of a sire is the herdmate comparison, which is now in general use in U.S. Department of Agriculture sire summarization.

First let us consider what the herdmate's average is and the purpose the herdmates serve. Herdmates of a cow include all animals by other sires which freshen in the same season of the year as the daughters of the sire being summarized. For U.S. Department of Agriculture procedures, the herdmate average includes animals which freshen two months before the individual daughter freshens, the month the daughter freshens, and two months after the daughter freshens. These animals should be fed the same feed, experience the same climatic stresses, and be milked by the same personnel as the daughters of the sire being summarized. For the herdmate comparison to be effective, the sire's daughters and the herdmates *must* be provided the same opportunity.

Let us now look at this same sire on the basis of the herdmate comparison. In contrast to the daughter-dam comparisons, the daughters have now been compared with animals freshening in the same herd, year, and season. The trend is not added to the difference in this case, and a realistic picture of what the sire can be expected to transmit to future daughters is presented.

41 daus. av.	13,904M	3.66%	504F
Herdmate av.	13,265M	3.58%	469F
Difference	+639M	+.08%	+35F

Due to the heavy reliance on the daughter-dam comparison in the past, a question may arise. Yes, this is fine, but hasn't the contribution of the dams of the daughters (mates of the sire) been overlooked? The dams of the herdmate also should be brought into the picture, since they contribute to the genetic make-up of the herdmates. Our main concern is to determine how the daughters of a particular sire perform in comparison with the daughters of other sires. If the dams of each of these two groups of animals, the sire's daughters and the herdmates, were representative samples from the herd(s) in which the sire was used, no correction needs to be made. The genetic contribution of the dams of these two groups of animals would cancel out.

Studies of data in which the daughters of a sire are in several herds suggest that the adjustment for the dams would not be important.[1] When the herdmate comparison is compiled from a single herd, there may be more need to consider selection of dams. Occasionally sires are introduced into a herd to be mated to the daughters of a specific bull. However, since the heritability of milk yield is only about 0.25, on the average only one-eighth of the average difference in the production of the mates of the sire and the dams of the herdmates would contribute to the herdmate difference.

Herdmate comparisons have now replaced daughter-dam comparisons

[1] Miller, R. H., and Corley, E. L., *Jour. Dairy Sci.*, **48**:580–585, 1965.

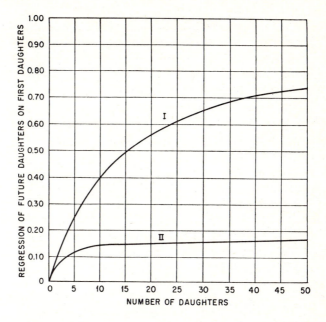

Fig. 11-11. Variation in the regression of milk or fat yields of future daughters in artificial insemination on the first tested daughters (I—where first daughters are tested in many herds, II—where first daughters are in one herd). The value for the regression coefficient represents the fraction of the superiority of the first daughters that will be expected to be manifestly an indefinitely large number of future artificially sired daughters.

in the compilation of the U.S. Department of Agriculture Sire Summaries. The principle is similar to that illustrated in our previous example, although certain helpful refinements have been introduced with the aid of high-speed computers. Some of these refinements are

1. The herdmate average is corrected for variation in the number of herdmates for each daughter. When there is a small number of herdmates, the herdmate average is regressed markedly toward the breed season average.
2. The production level of the daughters is adjusted to account for differences in the production level of the herdmates of different bulls. Research has shown that 0.9 of the difference in production between a group of herdmates and the breed average is reflected in the production level of a sire's daughters.
3. The predicted difference computed for milk and fat takes into account the variation in the number of daughters in a sire's summary.

The following herdmate-comparison procedure is now used by the U.S. Department of Agriculture to compile sire summaries on a national basis from D.H.I.A. records. Herd-year-seasons are based upon a five-month moving average. The appropriate average for the daughter of the sire is obtained by averaging all records of daughters of other sires calving in the herd in the same month as the daughters, in the previous two months, and in the succeeding two months. For example, the herdmate average for a sire's daughter calving in March would be obtained by averaging the production of all cows, except her paternal sisters, calving in that herd from January through May. All comparisons included in these summaries are on a within-breed basis.

Because the usefulness of the herdmate average is dependent in part upon the number of herdmates available, the herdmates' production is adjusted to place the herdmate data of daughters on a more comparable basis. For each individual lactation of a sire's daughter, the herdmates' average is corrected as follows:

$$\text{Adj. herdmate av.} = \text{breed-season av.} + \frac{\text{No. herdmates}}{\text{No. herdmates} + 1}$$
$$(\text{herdmate av.} - \text{breed-season av.})$$

The breed-season averages are calculated from the nationwide 305-day-$2\times$-ME. D.H.I.A. lactation averages for cows of the breed for the previous two years. This is done for both milk and fat. If a daughter has several records, the adjusted herdmate averages for all her lactations are averaged.

When a sire's daughters are distributed in many herds, additional refinements can be added to the daughter herdmate information to increase the accuracy of comparing different sires. The procedures described here, as well as the adjustment for herdmate averages, follow those developed by Henderson and co-workers at Cornell University. The daughter average increases as the production level of herds in which the bull is used increases. This is because about 0.9 of the difference of the herdmates from the breed average is reflected in the production level of the daughters of a sire. The following procedure takes this condition into account:

$$\text{Adj. daughter av.} = \text{daughter av.} - 0.9 (\text{adj. herdmate av.} - \text{breed av.})$$

This step serves to decrease the average of the daughters of a sire used in high-producing herds and to increase the average of the daughters of a sire used in low-producing herds. The breed average production used is compiled from all D.H.I.A. cows of the breed calving for the previous two years.

Fig. 11-12. An outstanding example of the widespread influence which a sire can exert on a breed through artificial insemination. Sears Farm Dean Ada Imperial probably sired over 100,000 calves during his lifetime. His 8,089 artificially sired daughters which were tested in D.H.I.A. averaged 15,079M and 540F on a 305-day, 2×, M.E. basis. His daughters averaged 1,116M and 35F above their herdmates.

In order to compare bulls which have widely varying numbers of offspring, it is necessary to adjust the above value for each sire in accordance with the number of progeny available. The "Predicted Difference" is calculated for the A l data of a sire as follows:

$$\text{Predicted difference} = \frac{\text{No. of daus.}}{\text{No. of daus.} + 20} (\text{adj. daughter av.} - \text{breed av.})$$

This *predicted difference* is an estimate of the expected average superiority or inferiority of many future daughters of the sire when these daughters are unselected and perform at standards of average D.H.I.A. herds of the specific breed.

CROSSBREEDING

Crossbreeding experiments with dairy cattle were undertaken in the early 1900s just after the discovery of Mendelism. Some of the experiments were motivated by a desire to gain an understanding of the Mendelian nature of inheritance in cattle. Of special interest was whether or not dominance was the rule for the desired trait. Four early experiments initiated in Denmark in 1906 and in the United States in 1911, 1912, and 1913 are discussed in a review by Robertson.[1] However, crossbreeding did not become commonplace. Much early stigma against the practice was voiced during the height of the "purebred era." Fear was expressed that mongrel stocks would result from such crossing.

Crossbreeding study was started by the U.S. Department of Agriculture in 1939. This study involved the Holstein, Jersey, Guernsey, and Red Dane breeds. Purebred sires were used in various sequences on two-breed crossbred cows to produce three-breed and eventually four-breed crosses. Unfortunately, production data on the purebred foundation cows and purebred daughters of the sires used were obtained mostly at field stations or on twice-a-day milking, whereas production records on the crossbreds were obtained at Beltsville, Maryland, on three-times-a-day milking. Thus, direct comparisons of records of purebred dams with those of their crossbred daughters and between purebred and crossbred daughters of sires may be misleading. Besides, only a few sires of each breed were used, so the general applicability of the results is uncertain. The crossbreds, including two-, three-, and four-breed crosses, were high producers, and there are suggestions of hybrid vigor in at least some of the crosses.

Crossbreeding at the South Carolina Station involved Holstein, Jersey, Guernsey, and Brown Swiss breeds. Numbers in the experiment were small, but records of the crossbreds were compared with those of their dams, and comparisons of maternal and paternal purebred half sisters were made in the same herd. Results of two-breed crosses indicate that the crossbreds were generally above the average of the two parental breeds, and in a few cases exceeded the better purebred parent in total production of fat and 4 per cent fat-corrected milk. Fat test was generally slightly below the average of the two parental breeds. Production of three- and four-breed crossbreds varied with the transmitting ability of the particular sires used, but it averaged slightly higher than that of their two- and three-breed dams.

The Animal Breeding and Research Organization, at Edinburgh, initiated a crossbreeding study using the Ayrshire, Holstein, and Jersey breeds in 1950. Published reports on milk yield are not yet available. Several other crossbreeding experiments are in progress in the United States. Results from many of these are just becoming available. The location of these experiments and the breeds included are as follows:

[1] Robertson, A., *Anim. Breed. Abstracts*, **17**:201–208, 1949.

Beltsville	Ayrshire, Brown Swiss, Holstein
Georgia	Brown Swiss, Holstein, Jersey
Illinois	Guernsey, Holstein
Purdue	Holstein, Red Dane
South Carolina	Brown Swiss, Holstein

Preliminary results at Beltsville indicate 8 to 18 per cent heterosis for milk, fat, PLM, and protein production for first lactation F_1's. At Georgia, 7 to 11 per cent heterosis for Holstein $\times$ Jersey and Brown Swiss $\times$ Jersey F's were reported for first-lactation milk yields. The F_1's at Illinois exceeded their parental average by 11.6 per cent. First lactation milk yields of the F's were 2 per cent above the parental mean, but later lactation averaged 3 to 4 per cent below the parents.

Almost without exception, crossbred animals have exhibited greater reproductive fitness than the parent breeds. At Illinois, crossbreeds completed 14 per cent more first lactations, 20 per cent more second lactations, and 24 per cent more total lactations than purebreds. Crossbreds also had greater livability, as losses from birth to calving were 17.4 per cent for purebreds as compared to 8.2 per cent for crossbreeds.

Some degree of heterosis has been evident for most traits studied, but superiority over the parental mean may not be a satisfactory practical criterion. Crossbreeding does become more important as increments of heterosis are expressed by several traits, i.e., reproduction, viability, and production. Heterosis of net merit may result even though the cross does not exceed the better parent in any one trait. The average merit of the breeds available for crossing is also a major practical concern This is undoubtedly one reason why crossbreeding of dairy cattle in the United States is not used more extensively. Milk volume is of major economic significance and advanced programs of progeny testing in the Holsteins have provided many bulls with superior transmitting ability for milk yield. Many times a bull of inferior merit would have to be used if a systematic crossbreeding program were adhered to rigidly. Even when the advice to use the best sire available is followed, in commercial herds in which milk yield is of paramount importance, crossbreeding would be used only limitedly.

INBREEDING

Inbreeding experiments with Guernseys and Holsteins were initiated in 1912, and they were continued until 1943 by the U.S. Department of Agriculture. Birth weight of the calves decreased markedly and mortality for the first year after birth was 15 per cent for the inbreds and 8 per cent for the outbreds. With moderate inbreeding, there was little apparent decrease in production, but as inbreeding became more intensive, production declined rather drastically.

An inbreeding experiment was initiated in 1931 at New Jersey. Four Holstein lines and one Jersey line were started, but three of the Holstein lines were discontinued early in the experiment. The Holstein and Jersey lines were moved and have been continued at California. Decreases of 210 lb of milk and 4.9 lb of fat per 1 per cent inbreeding were found in the Holstein herd. Fat test also declined with inbreeding.

Mild inbreeding and selection has been carried out in the Iowa Holstein herd since 1930. The latest published report showed that the average inbreeding coefficient had reached 0.074 with a maximum of 0.34. For each one per cent increase in inbreeding, the milk yield for the first lactation decreased 54 lb. Early growth was depressed and the attainment of mature size was delayed in the Iowa and California herds by inbreeding.

Studies of records in Wisconsin, Nebraska, and Britain have also shown that milk yields are depressed by inbreeding. The results show expected decreases in production ranging from −0.25 to −0.60 per cent of the mean yield. Although formation of inbred lines for eventual crossing does not appear to be practically feasible for the present, knowledge of the magnitude of inbreeding depression is important in projecting progress from selection. In closed or semiclosed herds some inbreeding is unavoidable. The effective population size must be large enough to keep the inbreeding depression from equaling the gain from selection.

SUMMARY

Dairy cattle represent an important segment of the livestock economy, and production per animal has been increasing at a rapid rate. Many traits merit consideration in a program for improvement, but high milk yield of acceptable composition is most important. Because of the high replacement rates, most females are milked for at least one lactation, and individual performance is the primary basis for female selection. Pedigree selection must be used in choosing young males for progeny testing. Extensive progeny-testing programs have been developed by artificial insemination cooperatives. Such programs can provide for an accurate appraisal of sires and permit maximum genetic improvement in large cattle populations. In the evaluation of the progeny test, the herdmate comparison has emerged as a sound and practical procedure. It has almost supplanted the long used daughter-dam comparison. Inbreeding is generally to be avoided, especially in commercial herds. Extensive crossbreeding research is in progress, but critical results are not yet available.

REFERENCES

Books

Espe, D., and Smith, V. R. 1952. "Secretion of Milk," The Iowa State University Press, Ames, Iowa.

Gilmore, L. O. 1952. "Dairy Cattle Breeding," J. B. Lippincott Company, Philadelphia.

Johansson, I. 1961. "Genetic Aspects of Dairy Cattle Breeding," The University of Illinois Press, Urbana, Ill.

Lush, J. L. 1945. "Animal Breeding Plans," 3d ed., The Iowa State University Press, Ames, Iowa.

Articles

Asdell, S. A. 1951. Variations in Amount of Culling from D.H.I.A. Herds, *Jour. Dairy Sci.*, **34**:529–534.

Brody, S., and Ragsdale, A. C. 1935. Evaluating the Efficiency of Dairy Cattle, *Mo. Agr. Expt. Sta. Bul.* 351.

Cannon, C. Y., and Hansen, E. N. 1939. Expectation of Life in Dairy Cows, *Jour. Dairy Sci.*, **22**:1025–1032.

Dickerson, G. E. 1941. Estimates of Producing Ability in Dairy Cattle, *Jour. Agr. Res.*, **61**:561–586.

Gaines, W. L. 1928. The Energy Basis of Measuring Yield in Dairy Cattle, *Ill. Agr. Expt. Sta. Bul.* 308.

Hancock, J. 1954. Monozygotic Twins in Cattle, *Advances in Genet.*, **6**: 141–181.

Harvey, W. R., and Lush, J. L. 1952. Genetic Correlation between Type and Production in Jersey Cattle, *Jour. Dairy Sci.*, **35**:199–213.

Hyatt, George, Jr., and Tyler, W. J. 1948. Variations in Type Ratings of Individual Ayrshire Cows, *Jour. Dairy Sci.*, **31**:71.

Johansson, I. 1950. The Heritability of Milk and Butterfat Yield, *Anim. Breed. Abstracts*, **18**:1–12.

Legates, J. E., and Lush, J. L. 1954. A Selection Index for Fat Production in Dairy Cattle Utilizing the Fat Yields of the Cow and Her Close Relatives, *Jour. Dairy Sci.*, **37**:744–753.

Lush, J. L., and McGillard, L. D. 1955. Proving Dairy Sires and Dams, *Jour. Dairy Sci.*, **38**:163–180.

—— and Shrode, R. R. 1950. Changes in Milk Production with Age and Milking Frequency, *Jour. Dairy Sci.*, **33**:338–357.

Rendel, J. M., and Robertson, Alan. 1950. Estimation of Genetic Gain in Milk Yield by Selection in a Closed Herd of Dairy Cattle, *Jour. Genet.*, **50**:1–8.

———— and ————. 1950. Some Aspects of Longevity in Dairy Cows, *Jour. Expt. Agr.,* **18**:49–56.

Robertson, A., and Rendel, J. M. 1950. The Use of Progeny Testing with Artificial Insemination in Dairy Cattle, *Jour. Genet.,* **50**:21–31.

————. 1954. Artificial Insemination and Livestock Improvement, *Advances in Genet.,* **6**:451–472.

Seath, D. M. 1940. The Intensity and Kind of Selection Actually Practiced in Dairy Herds, *Jour. Sci.,* **23**:931–951.

Shrode, R. R., and Lush, J. L. 1947. The Genetics of Cattle, *Advances in Genet.,* **1**:209–261.

Young, G. B. 1953. Genetic Aspects of Fertility and Infertility in Cattle, *Vet. Rec.,* **65**:271–276.

CHAPTER TWELVE

Improving Beef Cattle

Great changes in physical form have been wrought in meat animals during the two centuries since the man who has been called the father of animal husbandry, Robert Bakewell, began his pioneer labors in their improvement. During most of this period, breeders and improvers of meat animals were forced to proceed principally upon empirical bases. Since the rediscovery of Mendel's laws in 1900, especially during the decades since 1920, results of breeding experiments with meat animals have become available in increasing numbers to serve as additional guides. The purpose of the present chapter is to survey the general field of meat-animal breeding with special emphasis on beef cattle. Subsequent chapters will deal with sheep and swine. In these chapters we will review the present state of knowledge in the field, make such recommendations as seem to be justified on the basis of presently available evidence, and consider modern developments in relation to historical aspects of animal-breeding practices. It must be emphasized that from a scientific point of view this is a relatively young and rapidly developing field. Thus, it can be confidently predicted that new evidence in future years will greatly extend and probably modify present concepts.

OBJECTIVES IN BREEDING MEAT ANIMALS

Basically the task of the breeder of meat animals is to produce animals which will more efficiently convert vegetable products (some edible by man and some not) and inedible animal products into palatable, nutritious human food. In a world with a constantly increasing human population and a consequent narrowing of the margin of safety between world food needs and potential world production, it is essential that this conversion be made as efficiently as possible. From the standpoint of the individual commercial livestock producer, profit depends upon two things: (1) cost of production and (2) quality of product as reflected in selling price. Thus, his interests in efficiency are identical with those of the population at large so far as the kind of animals is concerned.

In many ways meat-animal selection is simpler than that of dairy cattle or egg-producing poultry. The principal favorable factors are:

1. Many characters of economic importance, such as growth rates and body form, can be observed in both males and females.
2. Many such characters can be observed in young animals before sexual maturity is reached.
3. For at least certain characters (carcass quality is the best example), the external apperance of the live animal is a better indicator of the animal's real value than it is in dairy cattle.

Arrayed against these advantages are certain disadvantages which limit progress. Of these the fact that there is no single end point which can be used to measure value is perhaps the most important. As pointed out in an earlier chapter, selection for more than one thing at a time reduces the intensity of selection which can be directed toward a single goal. Since meat animals must be selected simultaneously for economy of production and for quality of product, and since both of these depend upon a number of factors, some of which may be biologically antagonistic, numerous difficulties are apparent. Added to these is the fact that under usual procedures, quality of product can best be evaluated only in the carcass and that an animal studied in the carcass cannot be used for breeding! Research is constantly attempting to find ways to surmount the last factor. Techniques for "looking under the skin" to estimate potential carcass quality show considerable promise in several cases. Present-day techniques for long-time preservation of semen open the door to the possibility of storing large quantities of semen from young males, slaughtering them for carcass evaluation, and then inseminating females with the semen of those proved outstanding. This has been done experimentally, but its practical possibilities are uncertain.

Carcass evaluation is complicated by differences in consumer preference. These preferences vary with respect to weights of cuts, desired proportions of fat, methods of processing and curing, season of the year, relative prices of other meats, and a variety of other factors. In the long run, variations in consumer preference are probably beneficial to the industry. Without them we would have trouble "licking the platter clean" in the manner of the proverbial Spratt family.

FUNCTIONS AND GOALS OF SEEDSTOCK
AND COMMERCIAL HERDS

Seedstock herds exist solely for the purpose of providing the basic stocks by use of which commercial herds can accomplish their purposes of economical production of quality products. Ultimately all genetic improvement in commercial herds depends upon improvement in seedstock herds. Thus, the greater part of the responsibility for genetic improvement rests with the latter type of herd. Seedstock breeders occupy the most challenging spot in the entire livestock production chain. This is true whether the majority of commercial herds follow the practice of producing high-grade animals of a single breed or strain or whether they follow a crossbreeding scheme. In either case sires used are produced in seedstock herds.

The foregoing represents the basic breeding system upon which most meat-animal production in the United States rests. There are, however, many deviations from it. Some large commercial producers produce all or most of their sires by breeding their best females to the best available males or merely by selecting males from the general commercial herd. In either case they are playing the roles of both seedstock and commercial producer. The desirability of these practices depends upon (1) whether the quality of sires produced is as good as (or better than) could be obtained from a breeder specializing in seedstock production, and (2) relative costs as compared to purchase of sires of equal quality.

In popular usage the term *purebred* is often thought of as being synonymous with seedstock. The term *purebred* merely means that an animal is registered or eligible to be registered in a herdbook maintained by a breed association. Such animals may or may not be of seedstock quality. Conversely, animals not eligible for registration may have qualities which would make them capable of contributing needed qualities to commercial herds. Thus, we think the term *seedstock* is more appropriate.

Some purported seedstock herds are in reality not capable of contributing needed characters to commercial herds. In some cases, the herds are merely hobbies of the owners and selection is for traits with little or no relation to economic value. In other cases the objectives of the owners are sound but bad judgment or bad luck has resulted in the production of animals of too low genetic value to contribute to most commercial herds.

Whether breeds or strains to be used in crossbreeding programs should be intentionally selected for specific qualities with the two or more breeds used in a crossing program complementing each other or whether all should be selected for all-round merit is not a settled question. Most broiler production in chickens is based upon crossing specialized strains—the female parent stocks being from strains of reasonably good egg production to keep chick production costs within reasonable limits. Male parent stocks are expected to contribute growth and desired carcass-quality characters to the commercial bird. The egg-production potential of the male stocks can be, and often is, very low.

The feasibility of this general procedure is intertwined with economics, i.e., costs of maintaining breeding stocks in relation to potential genetic advantages.

In the larger meat animals, considerable interest developed among animal-breeding research people in the formation of lines or strains for use in specific crosses after the theoretical possibilities for progress through this breeding approach were pointed out,[1] and it was suggested that maximum performance might come from crossing divergent strains—one with high fertility to serve as the female parent and one with high gaining ability and feed efficiency as the male parent strain.[2] One theoretical study has suggested that progress for total performance might be increased somewhat by use of specialized sire and dam lines in selection for meat production. However, except in rather extreme cases of genetic negative relationships among important characters, it was concluded that advantages were not likely to be great enough to warrant this approach to genetic improvement.[3]

To date, experimental results are not available with the larger farm animals which encourage single-trait selection for strains to be used in crosses.

The system of using strains which are selected for all-round merit (even if used in crossing) is the most prevalent in the United States. Even with strains developed specifically for commercial use in cross-breeding programs, it seems probable that the economical production of breeding stock would necessitate the maintenance of reasonably high reproductive rate, efficiency of growth, and meat quality in each strain. The factor of meat quality is probably of particular importance since, for most carcass characters, hereditary influences appear to be predominantly additive in nature. Thus, a heterotic response could not be expected to compensate for carcass inferiority in one strain being used in a cross.

In view of the above considerations, most of the material in these chapters assumes breeding systems in seedstock herds aimed at the production of animals of all-round merit, whether their descendants will be used straight or in crosses commercially. As a result of past selection

---

[1] Hull, Fred H., *Jour. Amer. Soc. Agron.*, **37**:134–145, 1945.
[2] Dickerson, G. E., *Iowa Agr. Expt. Sta. Res. Bul.* 354, 1947.
[3] Smith, Charles, *Anim. Prod.*, **6**:337–344, 1964.

and differences in foundation stocks and/or genetic differences resulting from random drift, existing breeds, and in some cases strains or lines within breeds, do differ considerably in many characters. Commercial livestock producers should take advantage of these differences and combine breeds and strains in combinations which will give the greatest total productivity.

In Great Britain considerable stress has been placed upon the development of strains or breeds adapted to particular areas. These adapted breeds are used to furnish the basic stocks for commercial production, and, often in crossbred combinations, in other areas as well as the area of origin.

RELATION OF ENVIRONMENT TO SELECTION IN BEEF CATTLE AND OTHER MEAT ANIMALS

Animals of improved meat types tend to resemble unimproved types in appearance when raised under suboptimal nutritional conditions. Hence the origin of the old saying that "half the breeding goes in the mouth." The extent to which differences in plane of nutrition can influence apparent beef conformation is illustrated in Fig. 12-1. The animal breeder is vitally interested in two questions: (1) how properly to evaluate known differences in environment so far as their effects on potential breeding stock are concerned, and (2) what type of environment should be used in breeding foundation stocks for ultimate commercial use.

The old saying that "fat is a good color" shows that the first of these problems has long been recognized. From the standpoint of the sales value of a breeding animal, it often happens that putting on a degree of finish far and above what the animal needs for proper development will improve the appearance (apparent conformation) enough to be a profitable investment for the seller. This is apparently the principal reason for the high and often excessive degrees of finish so often seen on meat animals in shows and sales rings.

How can the practical breeder, interested in true improvement more than in immediate sales value or show-ring glory, logically make allowance for differences in environments to which various animals have been subjected during growth? There is no absolute rule, but the growth rate and apparent conformation of an animal should be discounted if it is known to have been forced unduly. Conversely, an animal reared under poorer-than-average conditions should be given some credit over and above what its record and appearance would entitle it to. In some cases it is possible to determine the average effect of a certain environmental factor and adjust for it.

The question of what environment should be provided for seedstock

Fig. 12-1. Level of feeding greatly affects the appearance of beef cattle. These 18-month-old Angus heifers are by the same sire and were similar in appearance at 4 months of age. From 4 to 18 months of age the heifer on the left had a nurse cow plus a full feed of grain; the center heifer had her mother's milk to a normal weaning age plus a full feed of grain thereafter; the heifer on the right had her mother's milk to weaning and thereafter was raised principally on hay and pasture. (*Courtesy of Dr. H. J. Smith, Tennessee Agricultural Experiment Station.*)

herds has been considered in some detail in Chap. 10. In many respects the practices currently used with beef cattle are directly opposite to the procedures suggested by both experimental results and theoretical considerations. The problem is not a simple one, however, of selecting for adaptation to a single environment.

Breeding herds of commercial beef cattle are usually maintained principally on roughage, which is often of low quality, under range conditions or farm environments. These are not adequate to produce animals with enough finish to meet market demands. Therefore, it is common to use such areas only as breeding grounds. Either immediately after weaning or after an intervening period of growth on pasture or other high-roughage rations, most potential market animals are provided with high-concentrate rations during a finishing period before marketing. In many cases they are moved hundreds or even thousands of miles into areas with surplus concentrates for this finishing period.

For maximum total productivity, genetic potential must be present to (1) survive, reproduce, and grow under suboptimal conditions and (2) finish readily and produce desirable carcasses when given the opportunity.

Therefore, the logical system to use, both in developing and evaluating future sires for the seedstock herd itself, and in producing sires for use in commercial herds, would seem to be to maintain the seedstock breeding herds largely on pasture under farm or range conditions typical of commercial practice in the area and raise the potential sires to weaning there. A preliminary culling could be made at weaning, eliminating those which are undersized or otherwise undesirable and especially those out of dams giving insufficient milk or being deficient in other maternal qualities essential to high productivity under commercial conditions. The remaining animals should be put on feed tests and final selections made on the basis of a combination of records for (1) preweaning performance on range or pasture and (2) rate and efficiency of gain and apparent conformation after a feeding period. Existing data strongly suggest that the feed test need not be, and probably should not be, at the high levels of nutrition usually used in finishing rations for cattle.

IMPROVING BEEF SEEDSTOCK HERDS

Improvement of beef-cattle seedstock herds depends upon long-time adherence to selection criteria emphasizing traits of economic importance in the industry. In the material which follows we will discuss the importance of a number of traits and the possibilities of improvement.

Reproductive Efficiency. High reproductive rates are essential if a beef herd is to operate with maximum income and profit. Many environmental influences are known to affect fertility. If there are hereditary influences resulting in low fertility, selection against them should be automatic since animals of low fertility will normally leave few offspring. In the few studies made on calving interval and other measures of fertility in beef cattle, heritability has been found to be very low. This is consistent with several reports in dairy cattle.

The foregoing statements indicating probable ineffectiveness of selection for fertility within breeds must be taken with some reservations. In two separate studies under range conditions[1] higher calf crops were raised by cows of large or conventional size as contrasted to small and comprest cows. The larger animals were apparently better adapted to the range conditions under which the comparisons were made and as a result were more fertile. Further investigations of heritability of various components of fertility should be made under a variety of environmental conditions.

There are a few reports of specific defects and conditions influencing reproduction that have a hereditary basis in dairy cattle. The condition

[1] Knox, J. H., *Jour. Anim. Sci.*, 16:240–248, 1957; and Stonaker, H. H., *Colo. Agr. Expt. Sta. Bul.* 501-S, 1958.

of gonadal hypoplasia in a Swedish dairy breed is of special interest.[1] Unilaterally affected cows apparently gave milk with a higher fat test and had a selective advantage. This resulted in increasing the gene frequency for the defect and in an increased proportion of bilaterally affected animals which were sterile. Although no such case is known in beef cattle, it would seem to be partial insurance against future trouble for beef cattlemen to cull animals from parents with subnormal fertility.

Birth Weight. Calves heavier at birth tend to grow faster later, but the value of birth weight as a predictive factor for later performance is relatively low, accounting for less than 25 per cent of the variance in any performance item measured later. Since birth weight is itself of no economic value, its usefulness in a performance-testing program is limited to its value as an indication of other performance factors. If later records could not be obtained, birth weights would be of some usefulness in selection. When later records can be obtained, birth weight is of limited usefulness.

As shown in Table 12-2, birth weight is moderately heritable, and increases should be possible with selection. Within limits in some breeds, this probably would be desirable, but because of the possible increase in difficult births with extremely heavy calves, major emphasis in selection should not be put on this character.

Weaning Weight. Weaning weight is important to all beef cattlemen because a beef cow has to be fed the year round whether she weans a light or a heavy calf, and the cost of keeping her will vary but little with the weight of the calf raised. If, for example, it costs $75 annually to keep a cow and the herd is producing a 90 per cent calf crop, a 500-lb calf at weaning will have cost only 16.7 cents per pound to raise, while a 350-lb calf will have cost 23.8 cents per pound. Obviously, in many cases this can mean the difference between profit and loss.

Cows tend to repeat their performance for weaning weight of calves from year to year[2] with 40 to 50 per cent of the variance in calf weights being accounted for by differences between cows in most herds. This repeatability is high enough that cows raising a first calf or first and second calves distinctly below average in weight (and where accident or sickness is not known to have influenced calf weights) can be safely

[1] Eriksson, Karl, "Hereditary Forms of Sterility in Cattle," Hakan Ohlossons Boktyckeri, Lund, Sweden, 1943.
[2] See Taylor, J. C., et al., *Jour. Anim. Sci.*, 19:700–708, 1960, for one report and references to earlier studies.

Fig. 12-2. A forceful illustration of the differences often found in maternal qualities of beef cows. *Above,* a grade Hereford cow whose 8 calves raised in 8 years gained an average of 1.82 lb daily from birth to weaning. *Below,* a cow of the same age in the same herd who raised only 3 calves in 5 years and whose calves gained an average of only 1.23 lb daily from birth to weaning. (*Courtesy of Tennessee Agricultural Experiment Station.*)

culled with little chance of culling a cow that would be a high producer with later calves.

Weaning weight is a composite character in that it is affected both by the direct influence of the mother (amount of milk provided, intrauterine environment, and possibly care taken of the calf) and by the inherited growth capacity of the calf. Since the dam gives a calf half its genes, she exerts both a direct and a transmitted influence. It has often been assumed that growth rate from birth to weaning is largely a function of milk production of the dam. Marked differences have been found between cows in milk production. There are fairly high correlations between milk production and gain in weight during the preweaning period. The over-all effect of milk production on calf weight in one study[1] is illustrated by data from Hereford cows and their calves:

Table 12-1. Relation of Milk Production to 8-month Weights of Calves from Hereford Cows

Maximum daily milk production of dams, lb	Number of cows	Milk produced, lb		Calf weights at 8 mo., lb	
		Average	Range	Average	Range
Less than 6.5	12	5.0	2.3–6.4	354	259–446
6.5–12.9	37	8.8	6.6–12.9	405	294–514
13.0 and over	6	15.6	14.4–17.7	475	435–528

The importance of maternal ability of the dam, whether it be due to milk production or other factors, is futher emphasized by some experimental results from the Lufkin, Texas, substation:

Breed of sire	Breed of dam	Av. calf weaning wt., lb
Hereford	Brahman × Hereford	465
Brahman × Hereford	Hereford	345

In this test the calves were similar genetically, all having ¾ Hereford and ¼ Brahmam breeding. The differences in weaning weight must be attributed to the influence of the mothers, but factors other than milk production may also be important. In one study[2] calves grew more rap-

[1] Gifford, W., *Ark. Agr. Expt. Sta. Bul.* 531, 1953.
[2] Drewery, K. J., *Jour. Anim. Sci.*, **18**:938–946, 1959.

Table 12-2. Heritability Estimates for Several Economically Important Traits in Beef Cattle

Character	Approximate average heritability*	
Calving interval	low	0 to 15
Birth weight	medium	35 to 40
Weaning weight	medium	25 to 30
Weaning conformation score	medium	25 to 30
Maternal ability of cows	medium	20 to 40
Steers or bulls fed in dry lot from weaning to final age of 12–15 months:		
Feed-lot gain	high	45 to 60
Efficiency of feed-lot gain	high	40 to 50
Final weight off feed	high	50 to 60
Slaughter grade	medium to high	35 to 40
Carcass grade	medium to high	35 to 45
Area rib eye per cwt. carcass weight	medium to high	30 to 50
Fat thickness over rib per cwt. carcass weight	medium to high	25 to 45
Tenderness of lean	high	40 to 70
Summer pasture gain of yearling cattle	medium	25 to 30
18-month weight of pastured cattle	high	45 to 55
Cancer eye susceptibility	medium	20 to 40
Mature cow weight	high	50 to 70

* Summarized from published sources. Wider ranges indicate characters for which fewer estimates have been made and for which probable average heritability is less precisely known. References at end of chapter will guide student to the scientific literature on heritabilities as well as other aspects of beef-cattle breeding.

idly if their mothers were scored as being more "protective" or if their mothers were older—in both cases independent of milk production.

Heritability estimates of weaning weights have been quite variable, but they are predominantly moderately high as are the few available estimates of heritability of maternal ability of beef cows (Table 12-2). This leads us to believe that consistent selection of herd sires and replacement females which were themselves heavy at weaning should result in improvement. Whether this improvement comes largely from improved ability of the calves to grow or from improved maternal abilities of cows is immaterial, since selection is automatically for whichever (or both) of these is most important. The following selection procedures for weaning weight have been suggested:

(1) Retain a high percentage of heifers for one or two calf crops and select those with demonstrated ability to wean heavy calves for further use in the herd; (2) select sires from among sons of cows which have repeatedly

demonstrated their ability to wean heavy calves and which are grandsons of bulls whose daughters have on the average produced heavy calves at weaning; and (3) where possible, use sires whose daughters have proven to have good maternal abilities.[1]

As with other characters, weaning weights can be evaluated accurately only if all animals are managed as much alike as possible. Some factors, including sex of calf and age of dam, cannot be held constant, and adjustment must be made for them.

The magnitude of sex differences has varied widely from herd to herd, but in most cases bulls and steers have exceeded heifers. Unless data from the herd concerned indicate otherwise, an upward adjustment of about 10 per cent for heifer weights and 5 per cent for steer weights is needed to make their records comparable to those of bulls.

Age of the dam is related to weaning weights of calves. Weaning weights gradually increase until a cow is five or six to nine or ten years old and then tend downward.[2] The actual amount by which calves from three-year-old cows are lighter than those from mature cows has varied from about 20 lb to over 70 lb in different studies, with other ages showing proportional differences. Probably the magnitude of age differences depends to a large extent on management conditions, although the possibility of genetic differences in the rates of maturity in cows cannot be overlooked. The breeder must make allowance for age of dam according to the actual observed differences in his herd or by the use of adjustment factors which have been shown to be generally applicable to herds similar in breed and general environmental levels to his. Average upward adjustments required to convert 205-day weights of calves to a mature-dam basis are approximately as follows: 2-year-old dams—15 per cent; 3-year-old dams—10 per cent; 4-year-old dams—5 per cent; and dams over 10 years of age—5 per cent. Calves from cows 5 to 10 years of age would not be adjusted. These standard adjustments can be used in many herds.

Both additive adjustments in which a certain number of pounds is added to an animal's weight to adjust for sex or to a mature age-of-dam basis, and multiplicative factors in which weights are increased or decreased by a given percentage, have been suggested for both sex and age-of-dam adjustments. A multiplicative adjustment for sex appears preferable.[3] Neither additive nor multiplicative factors have been entirely satisfactory under all conditions for age-of-dam adjustments. In one herd

---

[1] Dawson, W. M., et al., *Jour. Anim. Sci.*, **13**:556–562, 1954.
[2] See Hamann, H. K., et al., *Jour. Anim. Sci.*, **22**:316–319, 1963, for one study and references to earlier work.
[3] Koch, R. M., et al., *Jour. Anim. Sci.*, **18**:738–744, 1959.

in which differential effects of environmental differences from year to year made either simple additive or simple multiplicative adjustments inaccurate, a modified method was found to fit the situation.[1] In this herd, differences between young and old cows were greater in years of poor environment than in years when faster average growth rates indicated environment had been better. On the assumption that changes required in a multiplicative adjustment are linearly related to changes in mean weights, the following adjustment formula was devised:

$$y_A = y_i + b_1\bar{y}_1 + b_2(y_i - \bar{y}_1)$$

where y_A = calf weight adjusted to a standard age of dam

y_i = weight of an individual calf from a given age of dam such as 3 years old

$\bar{y}_1$ = average unadjusted weight of calves from the given age-of-dam group

$b_1 = (\bar{y}_2 + \bar{y}_1)/\bar{y}_1$, where $\bar{y}_2$ is average weight of calves from standard age-of-dam group

b_2 = regression coefficient which gives amount of increase in adjustment to be made per unit increase in y_i

b_2 can be estimated by dividing the data into "high" and "low" groups and applying the following formula:

$$b_2 = \frac{(H_S - H_A) - (L_S - L_A)}{H_A - L_A}$$

where H and L represent high and low, and S and A represent standard age of dam and age of dam to be adjusted, respectively.

In making any adjustment for sex or age of dam it should always be remembered that at best they are estimates of how a given animal would have performed had it been of the other sex or had its dam been of a given age. Adjusted data are useful for making general comparisons, but can never be completely accurate. Thus, they should be used with some caution in individual cases.

Weaning weights must be taken at the standard age or some adjustment must be made to a standard-age basis if calves of different ages are to be compared fairly. Since in usual herd management it is impossible to weigh each calf as it reaches standard weaning age, some provision must usually be made for adjusting weights. A normograph[2] was devel-

[1] Vernon, E. H., et al., *Jour. Anim. Sci.*, 23:21–27, 1964.
[2] Koger, Marvin, and Knox, J. H., *Jour. Anim. Sci.*, 4:285–290, 1945.

oped for use in New Mexico range herds and has been used elsewhere. Growth was not linear in the herd in which it was developed, and it is strictly applicable only in herds with similar growth curves.

In most locations, calves tend to gain at a rather constant rate per day during the period up to six or eight months of age, which covers the usual weaning ages. Calves can therefore be weighed in groups when their ages average near the standard age. The difference between this weight and the actual or assumed birth weight can be used to calculate the average daily gain for each calf, and this figure can be used to adjust weights to a constant age basis. Since birth weight makes up a larger proportion of the total weight at younger ages, younger calves would be favored if no allowance were made for it. If, as is often the case, birth weights are not taken, only a small error will be introduced into the calculations by assuming an average birth weight for all calves in a herd.

For example, if a calf had a birth weight of 70 lb and weighs 360 lb at 195 days, he has gained 1.49 lb for each day of age. If the standard weaning age is 210 days, an approximation (close enough for usual purposes) of what the calf would weigh at 210 days is obtained by multiplying 1.49 by 15 and adding the product, 22.35 (better round this to 22 lb), to the actual weight at 195 days. Thus, the estimated 210-day weight is $360 + 22 = 382$. Slightly more precision would be attained if all calves in a herd were weighed two or more times at intervals of two to four weeks as they approached the standard age, and the average daily gain made during the interval used to estimate weight at the standard age. This procedure may be advisable in herds where growth rates are known to increase or decrease as calves approach weaning age. In most cases, however, a single weight should provide a reasonably good basis for estimating weight at a standard age.

Fewer calculations will be necessary, and comparisons of preweaning growth rates between calves will be nearly as valid, if use is made merely of the average daily gain without going to the trouble of converting to a standard age. This simpler method is being used in several performance testing programs.

There is little uniformity in relationships between rates of growth in the pre- and postweaning periods. In some herds there appears to be no relationship, but in others positive correlations of $+0.20$ or higher have been observed. This is probably related to preweaning nutritive level. If an animal is underfed during one period, he will make greater-than-average gains during a subsequent period of more liberal feeding. This phenomenon of *compensatory growth* has long been recognized in beef cattle and other animals. The reverse is also true under certain conditions. Thus, a calf which is so poorly nourished before weaning as to gain far below his genetic capability will usually gain relatively faster later. Opposed to this strictly environmental circumstance is a tendency to gain in accordance with inherited ability. Under conditions of optimum nutrition, there should be a definite positive correlation between gain in consecutive periods.

These two tendencies, environmental and genetic, apparently offset each other under some circumstances and result in no correlation or even a negative relationship between pre- and postweaning rate of gain. In other cases, the hereditary factors are important enough in both periods to result in a positive relationship. Under most circumstances, this relationship will not be strong enough to give weaning weight a high value in predicting subsequent gain of individuals. However, averaging the preweaning growth records of several progeny by a sire tends to average out the effects of different nutritional levels associated with milk production of their dams. In cases in which two or more sires have been bred to equalized groups of cows in the same herd and the calves raised to weaning under similar environmental conditions, fairly high correlations between pre- and postweaning gains of progeny groups have been found. Thus, progeny average growth rates to weaning are of considerable value in estimating the transmitting ability for growth of two or more sires bred to cows in the same herd.

As an illustration of the need for postweaning gain evaluation, a Nebraska study[1] indicated that if selection is aimed at increasing 550-day weight, 200-day and 386-day weights would be only 52 and 81 per cent, respectively, as efficient in selecting for it as would 550-day weight itself. The studies were made with bulls, steers, and heifers weaned at about 200 days of age and fed in dry lot at low levels for 196 days after weaning. Thereafter, for 154 days the bulls were full fed on rations high in concentrates while the steers and heifers were grazed without supplementation.

Weaning weight is usually negatively correlated with efficiency of gain during a subsequent time-constant finishing period. The lower efficiency is due to the fact that the calves with heavy weaning weights are heavier when they go on feed and are heavier throughout the time-constant period. Thus, the lower efficiency of calves with heavier weaning weights is due to the fact that they have reached a more expensive part of the growth curve rather than to inherent lower efficiency.

Post-weaning Rate and Efficiency of Gain. Although beef calves are sometimes sold for slaughter at weaning, especially in the South, they are more often fed in dry lot on concentrates, pastured, or fed according

[1] Swiger, L. A., et al., *Jour. Anim. Sci.*, **22**:514–510, 1963.

to some system involving both pasture and dry-lot feeding for a period of time prior to being marketed. The animal which gains rapidly during this period has to be fed fewer days to reach a given weight, with a resulting saving in labor and either a more rapid turnover of capital or an increased volume of business. The rapid-gaining animal also tends to be the most efficient.

Since no other characters seem to be highly enough related to post-weaning performance to be of predictive value, feeding tests are needed to measure this important character. How such feed tests should be conducted is not a settled question. Individual feeding is necessary if estimates of efficiency of feed utilization for each individual are to be obtained. Group feeding is satisfactory if only rate of gain is measured.

The earliest proposals called for feeding individually for a fixed number of days after weaning. This is by far the most convenient method, but it has the disadvantage that animals varying in weaning weight may be in very different portions of the growth curve during the period of similar chronological age. This method is therefore not well suited for measurement of efficiency, but since there is little or no relationship between weaning weight and subsequent rate of gain, it does provide a convenient and acceptable method of evaluating rate of gain for use in selection. Within rather wide limits, the age at which cattle go on feed after weaning has little effect on subsequent rate of gain. Therefore, a convenient, satisfactory, and widely used means of conducting gain tests is to put all calves on feed on a given date after weaning and feed them for a predetermined number of days.

It has been proposed that animals on test be fed through a given weight range—the suggested range often being from 500 to 900 lb live weight. A very high relationship exists between rate and economy of gain when this method is used.

Animals differing in rate of maturity may differ greatly in composition at a given weight. Since the energy required to produce fat is greater than that required to produce lean meat and bone, it is apparent that the early maturing, rapidly fattening calf would be at a disadvantage in efficiency evaluation. Therefore, feeding to a definite degree of finish in an effort to control the composition of the gains has been recommended.

The most comprehensive study[1] to date of relationships between rate of gain, rate of feed consumption, and gain in relation to feed consumed utilized data on 1,324 bull and heifer calves individually fed to limit to appetite at three locations on rations differing in proportion of concentrates and thus in total digestible nutrient levels. Genetic parameters

[1] Koch, R. M., et al., *Jour. Anim. Sci.*, **22**:486–494, 1963.

differed at the three locations but were apparently not related to concentrate levels in the rations fed. The findings indicate that selection for rate of gain should result in correlated responses in both efficiency and feed consumption with about 60 per cent of the genetic increase in gain due to increased efficiency and 40 per cent to increased feed consumption. Selection for gain would be expected to give 81 per cent as much improvement in efficiency as direct selection for efficiency itself. Other studies with smaller numbers of animals have also shown positive relations between rate and efficiency of gain.

In view of the expense involved in individual feeding, the uncertainty in interpreting the results, and the apparent gross relationship between rate of gain and efficiency of gain in cattle of reasonably uniform type, the practice of group rather than individual feeding is usually recommended.

The feeding period should involve full- or self-feeding either the entire ration or one or more components of it. Genetic differences are not fully expressed if a limited feed program puts a ceiling on rate of gain. Length of test period needed for most efficient tests of gaining ability is not known with certainty. There is a tendency for heritability of gain to increase somewhat at later stages of feeding periods exceeding 140 days, but in most cases selection for final weight would be almost as effective after 140 days on test as after a longer period. In one study with farm-raised calves weaned at about 230 days of age and fed in dry lot for 140 days, selection for weight at a year of age would have been almost as efficient with 84- or 112-day feeding periods as with the full 140 days.[1]

The question needs to be studied in additional herds. Pending additional information it is generally recommended that tests be 140 days or more.

Most postweaning gain testing has been under dry-lot conditions. Heavy grain rations have been fed in some tests, and in others rations with only 30 to 40 per cent concentrates have been fed. Hereditary differences have been found under both regimes. Since in the entire production process some 70 to 80 per cent of the nutrients consumed by beef cattle come from pasture and other forage, it might seem logical that they be gain tested on pasture rather than in dry lot on high concentrate rations. However, the fact that ability to finish at weights required to meet market demands and the ability to produce types of carcasses in demand are important in the beef production process makes it seem necessary that at least some of the testing be done under conditions simulating commercial finishing practices.

[1] Swiger, L. A., and Hazel, L. N., *Jour. Anim. Sci.*, 20:189–194, 1961.

If there were no genetic-environmental interactions, animals would have the same rank regardless of the regime under which they were tested. They could then be tested in whatever manner was most economical for individual breeders. Information is not conclusive on the importance of genetic-environmental interactions. Available data strongly suggest, but are not extensive enough for drawing definite conclusions, that genotypes leading to good performance under feed-lot conditions also lead to high gains on pasture.[1,2] In another study[3] significant pair $\times$ ration interactions were observed when monozygotic twins were split with one member fed a high-roughage and the other a high-concentrate ration. This suggests, but does not prove, the existence of hereditary types predisposed to proportionally better performance on one type of ration than on another.

The entire question of how best to test gaining ability and feed utilization warrants further studies.

With heifers it is usually not economically feasible to full-feed high-concentrate rations. There is also a possibility of harming their future reproductive and maternal qualities and longevity by such feeding, although detrimental effects have not been definitely proven. As a general rule, it appears that gain evaluations of heifers should be made on high-roughage rations and/or on pasture. The fall of the yearling year at 18 to 20 months of age after a summer pasture season is often a convenient time to evaluate size in heifers.

Conformation as an Item in Beef Improvement. The factor of human judgment enters into the usual methods of evaluating conformation. It is thus subject to more error than the measurement of weaning weight or growth rates. Because conformation has historically been the principal criterion of selection in beef cattle, we will discuss its possibilities and limitations in some detail.

Legend tells us that the great pioneer English breeder Robert Bakewell gave attention to economy of production in his breeding operations, and various writers on animal husbandry have stated that the master breeders of the past insisted on high and economical production of meat before animals found favor with them. In view of the fact that satisfactory objective measurements of efficiency, in a modern sense, were lacking, it appears more probable that most selection during the pioneer stages of livestock improvement was based on visual appraisals

[1] Miquel, Cesar, and Cartwright, T. C., *Jour. Anim. Sci.*, **22**:821, 1963 (abstract).
[2] Brown, C. J., and Gacula, Maximo, *Jour. Anim. Sci.*, **21**:924–926, 1962.
[3] Warwick, E. J., et al., *Jour. Anim. Sci.*, **23**:78–83, 1964.

of body form or conformation, just as has been the case up to the present time.

To a certain extent, conformation reflects production quality in beef cattle and other meat animals, being related especially to values which markets place on carcasses. The ability of trained judges to predict carcass grades from an examination of the conformation and finish of live animals just prior to slaughter is quite high if a considerable range of variation is present. The ability to predict small differences in carcass grades or scores of animals of uniform type from visual appraisal of the live animals is much more limited. The frequent disagreement of placing between "on-foot" classes and the carcasses of the same animals in livestock shows illustrates this fact.

The development of animal form is an exceedingly complex process, and the relationships of type or form to productive characters and carcass desirability are incompletely understood. A great deal of our present thinking regarding animal form is based purely on empirical reasoning and tradition. Intensive research on these problems would appear to offer great promise in (1) directing future efforts toward changing type in constructive directions and (2) freeing animal improvers from the necessity of adhering to standards which may be unrelated to, or negatively related to, economic value.

It has long been recognized that the body proportions of animals change with age, the young animal being largely head and legs but gradually changing with age so that the proportions of the more valuable parts are increased in relation to those less valuable, such as the head, neck, and shanks. In general, the tissues develop in the order of skeleton, muscle, and fat. There seem also to be additional growth gradients running from the cranium backward and from the tail forward. Thus, for example, fat is the last tissue to develop and the loin region is the last to fatten. In times of reduced nutrition, fat deposition in a growing animal may virtually cease and muscle development be slowed, but skeletal growth will be slowed only if the plane of nutrition approaches starvation levels. Undernourished animals resemble unimproved types in conformation.

On the basis of such observations, the idea was developed[1] that the principal change brought about by selection for improved meat type has been to increase the speed of changes in body and tissue proportions normally associated with age so that they occur at earlier ages and continue to a greater degree than in unimproved types. Obviously both

[1] Hammond, J., "Farm Animals: Their Breeding, Growth and Inheritance," Edward Arnold (Publishers) Ltd., London, 1952.

the genetic make-up and the plane of nutrition influence the speed of these changes, especially the rate of fat deposition.

Percentages of High-priced Cuts.   It has been generally believed and stated that rearing on a low plane of nutrition increases the proportion of low-priced cuts in the carcass. Similarly, animal-husbandry textbooks for many years stated unequivocally that improved beef-type animals have higher percentages of the high-priced cuts than unimproved or dairy types. Stress on "maximum development in the regions of the high-priced cuts" has long been a part of livestock judging, especially in beef cattle. Both these generalizations appear to require examination.

Missouri studies many years ago[1] showed that carcasses of cattle raised on a high plane of nutrition had higher percentages of two high-priced cuts, the loin and the rump, than those of animals raised on lower planes of nutrition. However, the percentages of two low-priced cuts, the flank and the plate, also increased, while the rib changed little. The percentage of round, a high-priced cut, decreased, as did the lower-priced cuts of chuck, neck, shin, and shank. Collectively the four high-priced cuts, rib, loin, rump, and round, totaled 50.2 per cent of carcass weight in six high-plane steers slaughtered at 20 months or more of age, and they totaled 51.4 per cent in four low-plane animals of similar ages. Changes with age generally paralleled those with fattening. Thus, although fattening and aging resulted in increased proportions of some high-priced cuts, the proportion of some low-priced cuts also increased.

The historically accepted assumption that improved types have higher proportions of the high-priced cuts is not well supported by experimental data. In fact, data on the question, so far as the authors are aware, are all negative. As long ago as 1903[2] it was reported that the percentages of the various wholesale cuts were virtually the same for carcasses of beef and carcasses of dairy animals. Several later reports confirm this generalization.[3] Texas work[4] in which carcasses of more than 200 steers varying greatly in conformation and finish were cut out showed no consistent differences in percentage of high-priced cuts associated with conformation. This work has confirmed the earlier Missouri work in indicating no increase in proportion of high-priced cuts in highly finished animals.

Dangers in Selection for Type or Conformation.   Selection for body form and breed trademarks becomes dangerous when the items selected

[1] *Mo. Agr. Expt. Sta. Res. Bul.* 54, 1922.
[2] *Iowa Agr. Expt. Sta. Rpt.*, 1903.
[3] Cole, J. W., et al., *Jour. Anim. Sci.*, **23**:71–77, 1964.
[4] Butler, O. D., Jr., et al., *Jour. Anim. Sci.*, **15**:93–96, 1956.

for have no relation to productive value. When this occurs, selection pressure which can be put on the really important things is reduced. Selection for certain characteristics becomes doubly dangerous when the factors selected for actually lower performance. The history of animal breeding has been marred by several cases of this kind. Some examples will be reviewed in the chapters on Sheep and Swine. In these cases animals with certain characteristics became popular, and their germ plasm was widely distributed through the breed concerned before a realization of the true nature of the fad became apparent.

The beef industry has long been plagued by divergent ideas on the ideal type, especially as related to mature size. Historically, selection in beef cattle was for thick, compact, low-set animals with the ability to fatten at light weights. Changes toward this type were apparently accompanied by some loss of size and ruggedness, with the result that by the 1940s and 1950s many producers were questioning the usefulness of some modern-type cattle for economical production, at least under range conditions. To add to the confusion, ultra-compact types known as *comprests* or *compacts* arose a number of years ago in two breeds, and for a time, especially in the years immediately following World War II, they were quite popular in American show rings. Their impact on thinking in American beef-cattle circles is typified by the following statement:

Then out of nowhere came "Comprest" Herefords—and they really were a sight to behold. Although of popular bloodlines, the new type differed from anything thereto seen. Extremely lowset on ideal underpinning with smooth, bulging quarters and deep wide middles to match, with short thick necks and beautiful heads, the calves and yearlings were sensational.[1]

Although many questioned the usefulness of such types from the standpoint of economical beef production, it was apparently generally assumed that they represented the ultimate in carcass quality.

As a result of the many questions which cattle of these types raised in the minds of the cattlemen, two state experiment stations (Colorado and Oklahoma) began in the late 1940s to make comparisons between such small-type cattle and conventional types. In these two studies cow herds were maintained, and performance records from birth to maturity or slaughter were studied. In addition, three stations undertook the work of comparing the feed-lot and grazing performance of steers sired by bulls of three types, small, medium, and large. This work was supported in part by the American Hereford Association. A major breed associa-

[1] Forbes, Rank C., *Breeder's Gaz.*, October, 1946.

Fig. *12-3.* *Above,* normal and "comprest" steers showing differences in type. *Below,* two dwarfs of the extreme type produced by mating of comprest × comprest. (*Courtesy of Animal Husbandry Research Division, U.S. Department of Agriculture, and Dr. H. H. Stonaker, Colorado State University.*)

tion's sponsorship of research to aid its members in evaluating a specific type of animal represented a new and commendable approach in the meat-animal industry at that time.

At each of the three stations cooperating in the steer work, representatives of each of the three types were fed out under three different regimes, namely,

1. Fattening by full feeding as calves
2. Grazing as yearlings, followed by a dry-lot fattening period
3. Fattening on grass as two-year-olds

The results showed a definite tendency for the steers sired by large-size bulls to gain more rapidly than those sired by medium-size bulls and for the latter to gain faster than those sired by small-size bulls. This tendency was more pronounced during the wintering and grazing phases than during the full-feeding phase. Over-all differences in economy of gain were small, but there was a tendency for the progeny of large- and medium-size bulls to gain more efficiently on high-roughage rations. The large-size steers showed less finish in the phases of the test where full feeding was deferred or omitted. Differences in finish were small in the immediate full-fed groups. In summary, it was said: "The results of these tests indicate that medium-size cattle tend to combine the gaining ability of large cattle and the finishing ability of small cattle without sacrifice of efficiency of gain."[1]

Comprest and conventional steers produced in the Colorado study were put on full feed at weaning and fed to a constant finish of low choice. There were no appreciable differences between types in efficiency of gain, ages at which a low-choice finish was attained, percentages of the various wholesale cuts, nor in percentages of lean, fat, and bone in the 9-10-11 rib sample. The conventional steers ate more, gained faster, and weighed about 20 per cent more when they reached a low-choice finish. Feed required to winter heifers and cows was proportional to body weight, with no apparent differences between types in basic efficiency of feed utilization.

Matings of comprest × comprest produced about ¼ conventional-type calves, ½ comprests, and ¼ dwarfs, of which most were an extreme type. The comprest type was apparently due to the action of a single semidominant gene which, when homozygous, produced dwarfism. Comprest calves out of conventional dams averaged about 40 lb lighter than conventional calves at weaning and calves from comprest cows averaged about 50 lb lighter than calves from conventional

---

[1] Weber, A. D., *Amer. Hereford Jour.*, May 15, 1951.

dams. Comprest cows raised eight fewer calves per 100 cows than conventional cows.[1]

These test results are of interest from two standpoints. First, the experiments failed to demonstrate any carcass superiority for the extreme comprest types except ability to finish at very light weights. The results thus drive home to us the danger of assuming that extreme types have virtues before they are proved beyond all reasonable doubt to exist. Secondly, the comprest type had several very serious production disadvantages in that they produced a high percentage of dwarfs, calves with low weaning weights, and low-percentage calf crops. As a result of these experiments and the observations of breeders, cattle of these extreme types lost their popularity and have now virtually disappeared.

Conclusions about Conformation. In the foregoing material we have attempted to point out the difficulty of evaluating conformation, some of the ways in which long-accepted ideas regarding the values of certain kinds of conformation have not been borne out by experimental approaches, and the dangers of selecting for new or extreme types without prior data verifying their real economic value.

We should not, however, leap to the conclusion that conformation is of no consequence or importance in improving beef cattle. First, it is an important factor in selling both slaughter and breeding cattle. Unless a breeder had unlimited financial resources, he might well be unable to survive as a breeder if he strayed too far from accepted type at a particular time. Changes in preference for type or conformation tend to occur in the industry as an evolutionary process rather than by abrupt changes.

A second fact is that structural soundness, especially in feet and legs, is essential for beef cattle under commercial conditions. Scientific proof of the relation of generally accepted indicators of structural soundness in young cattle to durability under commercial conditions is lacking, but observation suggests some relationships.

A third reason for including conformation in the evaluation of beef cattle is that there are real differences in some aspects of carcass quality which favor beef types. Percentage bone and ratio of lean to bone in some cuts are two of these. Attractiveness and shape of several retail cuts is another important factor.

Further, it appears that past failures of conformation evaluation to identify beef types with ability to cut out high percentages of favored cuts is not an indication of inability to relate external traits to carcass

---

[1] Stonaker, H. H., *Colo. Agr. Expt. Sta. Bul.* 501-S, 1958. See also Chambers, D., et al., *Jour. Anim. Sci.,* 13:956–957, 1954 (abstract).

characters but rather to faulty standards based on what people "thought" should be true rather than on established relationships.

Studies of relationships between live-animal measurements and yield of preferred cuts,[1] together with studies in which yields of these cuts and other carcass characters are related to visual scores, given specific parts and characters of the animals before slaughter, are gradually modifying our views on what constitutes most desirable beef type and providing leads for future improvements. It appears that emphasis in finished cattle should be on trimness of middle and brisket, on width of shoulders, loin, and round, since width in these areas indicates muscling, and on other indicators or muscling including depth of twist, and visual evidence of muscling in the forearm and lower round. Greater length and less depth of body than formerly thought ideal seem to be desirable, and moderate length of legs has no adverse effect on carcass quality. The extreme smoothness sometimes favored seems not to be compatible with high proportions of lean. In other words, if an animal is well-muscled, the muscles can be expected to bulge.

Application of the above criteria has resulted in correlations of about +0.30 to +0.60 between live estimates of cutability and carcass yields of trimmed, boneless, preferred retail cuts.[2]

Instruments emitting ultrahigh-frequency sound waves and capable of indicating depth of different tissues by recording echoes as the waves pass from one tissue to another have been studied extensively for their ability to enable breeders to "look under the skin" of beef animals. Commonly known as ultrasonic devices, they have shown considerable promise for estimating thickness of fat over the back and a potential for estimating thickness of lean and area of loin muscle. Other mechanical and electronic devices have also shown promise for aiding in estimating fat thickness of living animals.

Carcass Evaluation. In specific details there is lack of agreement as to what constitutes desirability in beef carcasses. However, there is general agreement in the essentials, namely, that carcasses should have (1) a maximum of lean combined with tenderness and other factors giving maximum palatability and (2) a minimum of excess or waste fat.

As shown in Table 12-2, many carcass traits have medium to high heritabilities. Selection should be effective in bringing about changes if sufficient selection pressure can be attained. This is extremely difficult

[1] Green, W. W., *Jour. Anim. Sci.*, **13:**61–73, 1954.
[2] Gregory, K. E., et al., *Jour. Anim. Sci.*, **23:**1176–1181, 1964; and Wilson, L. L., et al., *Jour. Anim. Sci.*, **23:**1102–1107, 1964.

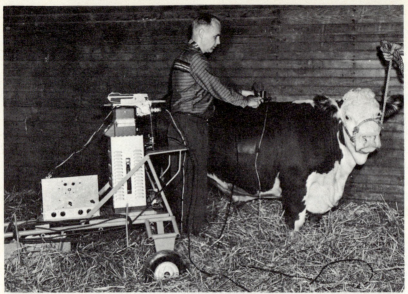

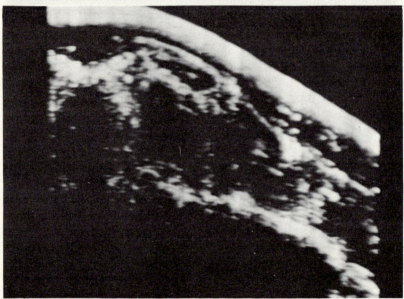

Fig. 12-4. *Above,* using an ultrahigh-frequency sound device to estimate fat thickness and, with less precision, to estimate thickness and area of muscles in a live animal. *Below,* a photograph of the "B" scan of an ultrasonic device. This particular scan was taken at the 12th rib of a 1,210-lb steer with several undesirable characteristics. From the picture taken while the animal was alive, external fat thickness was estimated at 1.5 inches and rib eye at 11 square inches. Actual carcass measurements were 1.54 inches and 11.1 square inches, respectively. Note the pocket at the right of the backbone. This was found to be an unusual fat deposit when examined in the carcass. (*Courtesy of Dr. J. R. Stouffer, New York State College of Agriculture at Cornell University.*)

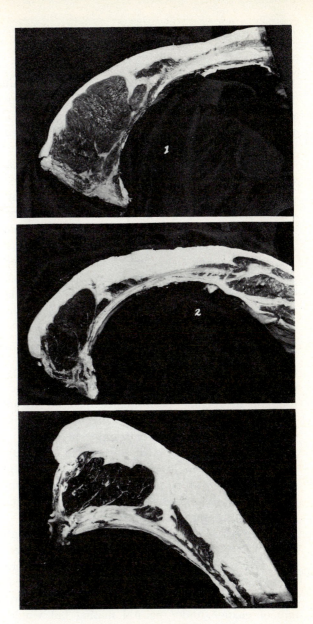

Fig. 12-5. Photographs of 12th-rib cuts from three beef carcasses illustrating extreme variation in proportions of fat and lean. The ideal is to develop cattle with no more external or waste fat than in the top photo but with lean of desired palatability. The two lower cuts have excessive waste. Fat thickness at this point on the carcass is a good indicator of the cutout of trimmed round, loin, rib, and chuck—the preferred cuts in a beef carcass. (*Courtesy of Dr. Val. H. Brungardt, University of Wisconsin.*)

if complete reliance must be placed on sib evaluation and/or progeny testing. These are the only available methods if potential carcass quality cannot be satisfactorily evaluated in live animals. Due to the long generation interval and low reproductive rates it has been estimated[1] that even with optimum progeny testing and sib evaluation, annual progress for traits which can be evaluated only in the carcass will be only about one-third as great relative to the size of the standard deviations as it can be for traits measured directly in the living animal. In the latter case, animals identified as superior can themselves be used for breeding.

This accounts for the great interest in developing methods of evaluating potential carcass quality in living animals. As pointed out in the preceding section, some success has been attained on this in estimating potential cutability. Little progress has been made on evaluating potential palatability. Until methods for this are available, breeders have no alternative but to use progeny tests in spite of their slow and expensive nature.

Selection Plans and Genetic Correlations among Traits. The evaluation of economically important beef-cattle traits depends upon a careful program of keeping performance records. All animals in a herd must be fed and managed as nearly alike as possible so observed differences will have a maximum genetic component.

Estimates[1] indicate that if selection in *large populations* of beef cattle were solely for single traits, *annual genetic improvement* under mass selection plans with natural service could approximate 4.3 lb for weaning weight, 0.043 lb in average daily post-weaning gain, and 8.4 lb in average total digestible nutrients required per 100 lb of gain. These figures are large enough in relation to usual standard deviations of approximately 40 lb, 0.3 lb, and 60 lb, respectively, for the three traits, to suggest that in a 10-year period beef-cattle population averages could be shifted by a standard deviation or more. Using progeny-testing systems to supplement individual selection would give little or no increase in annual rate of change under natural mating but would increase progress by 50 to 100 per cent if artificial insemination were used to permit superior progeny-tested sires to have large numbers of offspring.

Progress in relation to sizes of standard deviations will be much less for carcass characters. It is estimated that they could average only 0.029 square inch increase annually in rib-eye area and —0.088 lb of force required to shear a 1-inch core of cooked meat—an indication of increase in tenderness. Usual standard deviations for these traits are approxi-

[1] Warwick, E. J., *Natl. Acad. Sci., Natl. Res. Coun. Publ.* 751, pp. 82–92, 1960.

mately 1.0 square inch and 3.0 lb, respectively. Progeny testing and artificial insemination would be proportionally much more effective tools than for traits which can be measured in the same individuals used for breeding.

The foregoing will serve to illustrate possibilities. The estimates are unrealistic in that selection for single items would seldom if ever be practical for beef cattle. Also, the assumption was made that all animals in a population would be evaluated in the case of weight and efficiency items and that random samples of all sib groups would be slaughtered for carcass evaluations. This would seldom if ever be possible.

The practical breeder must select simultaneously for several traits and should weight them in his selection program according to their heritabilities, genetic correlations, and economic importance. Since relative economic importance of different traits varies greatly depending upon type of production practiced and upon herd levels already attained, universally applicable selection indexes for beef cattle seem improbable.

Further, knowledge of genetic correlations, both as to their general levels of magnitude and the degree to which they vary from breed to breed or herd to herd, is too incomplete to make the development of indexes feasible under most circumstances at present. Variability in indexes suggested to date[1] is illustrative of the effects of varying economic values and genetic parameters on weightings.

Weighting deviations from herd averages (in terms of standard deviations) by economic values should not lead to serious errors for most growth, conformation, and carcass traits since they all tend to be medium to high in heritability. Because of its low heritability, fertility should not be given as much emphasis in a selection program as its economic importance would suggest.

Estimates made to date of genetic correlations among traits of beef cattle[2] are subject to large sampling errors, but there appear to be general patterns:

1. There are positive genetic relationships between all measures of size and growth. Thus, selection for any one or a combination of birth weight, weaning weight, postweaning gain, 18-month weight, mature weight, etc., can be expected to have indirect effects on the others even though no direct selection is practiced for them.
2. Genetic relationships between measures of size and growth and conformation scores have varied considerably from study to study but average

---

[1] See Wilson, L. L., et al., *Jour. Anim. Sci.,* **22**:1086–1090, 1963, for one index and references to earlier studies.
[2] See *ibid.;* Shelby, C. E., et al., *Jour. Anim. Sci.,* **22**:346–353, 1963; and Brinks, J. S., et al., *U.S. Dept. Agr. Tech. Bul.* 1323, 1965.

near zero. The variation is likely due to the fact that scoring is subjective and people are probably influenced either positively or negatively by size in evaluating conformation with the magnitude and direction of the tendency varying from one group of research workers to another.

3. Genetic relationships are low between measures of growth and carcass characteristics not related to size. This, taken together with item 2, above, constitutes a strong indication that a wide range of type and conformation is compatible with rapid growth.

4. Genetic relations between maternal qualities of cows and growth and carcass characteristics have been insufficiently studied to permit drawing of conclusions, but there is some evidence of negative relationships. At present it does not appear that these are of a magnitude sufficient to make it impossible to combine good mothering ability with other desired traits, but they may be large enough to retard progress.

It will be necessary to await the results of long-time selection experiments with beef cattle to determine with certainty the genetic relationships among traits and the indirect effects of selection for those desired. The fairly high heritabilities for most traits other than fertility, the apparent absence of serious negative relations among traits of economic importance, the results of two studies of the effectiveness of selection,[1] and observation of past changes in the industry, particularly for conformation, all indicate that beef cattle may be one of the more plastic types from a breeding standpoint.

Breeding Systems for Seedstock Herds. The owner of a seedstock herd has the options of (1) following an outbreeding system within his breed, (2) following a mild linebreeding or inbreeding program, or (3) inbreeding intensely. No one of these can be said to be preferable under all circumstances.

With the medium to high heritability estimates for most traits except fertility in beef cattle, an outbreeding system should be effective in developing the kinds of cattle desired. It has the disadvantage that in all except large herds a breeder will have to go outside his own herd for herd sires. Often he will not know as much about the genetic potential of purchased bulls as he will about bulls raised in his own herd.

The other two options involve two rather different degrees of inbreeding. Inbreeding effects have been somewhat variable in different beef-cattle studies[2] but on the average it appears that weaning weight will

---

[1] Flower, A. E., et al., *Jour. Anim. Sci.*, **23**:189–195, 1964; and Brinks, J. S., et al., *U.S. Dept. Agr. Tech. Bul.* 1323, 1965.

[2] Clark, R. T., et al., *Oreg. Agr. Expt. Sta. Tech. Bul.* 73, 1963.

be depressed about 1 to 2 lb for each 1 per cent inbreeding in the calf and an additional 1 to 2 lb for each 1 per cent inbreeding of the dam. There is some evidence that inbreeding effects may be more severe in heifer calves and that bull calves are more affected by inbreeding of dam.[1] Postweaning growth and conformation scores at all ages tend to be lowered by inbreeding, although trends in these regards have varied markedly between studies.

Fertility and calf viability are reduced more importantly than other traits. Colorado results suggest a reduction of 10 to 12 per cent in calf crop raised in herds averaging approximately 20 per cent inbreeding in cows and 30 per cent in calves.[2]

These depressions in performance, although too much to be tolerated in a commercial herd, are not great enough to preclude the use of mild inbreeding in seedstock herds in which ultimate value of the herd depends upon the breeding value of animals from it when used in other herds. Closed lines have been maintained for periods of 15 to 30 years with mild inbreeding at several different experiment stations. In some of these cases selection has apparently been intense enough to more than offset inbreeding depression for all growth and conformation traits studied. Data on fertility and viability of calves are not conclusive, but even if reduced somewhat, net reproductive rates have remained at or near normal levels.

Breeding within a closed herd has the advantages of (1) enabling a breeder to make selections on the basis of records made under conditions with which he is familiar, and (2) permitting maintenance of herds with high relationships to outstanding animals (usually sires) used or bred in the herd. There is no reason to avoid this if it is impossible to find potential sires outside the herd which are equal or superior to those raised in it. An additional potential advantage of a mild inbreeding or linebreeding program is that there should be some increase in predictability of breeding behavior when animals go to other herds.

The formation of inbred lines for use commercially in top crosses or rotation linecrossing programs is being explored experimentally. Inbreeding levels increase rather slowly in beef cattle even when inbred as rapidly as possible. In spite of some rather drastic fertility and survival problems in some lines, most of those started can be maintained for at least 15 years or more. Tests of Colorado lines in rotation crosses[3] have resulted in excellent productivity. Unfortunately, no control populations were established when the lines were started, and it is now impossi-

[1] Brinks, J. S., et al., *U.S. Dept. Agr. Tech. Bul.* 1323, 1965.
[2] Stonaker, H. H., *Colo. Agr. Expt. Sta. Bul.* 501-S, 1958.
[3] *Colo. Agr. Expt. Sta. General Series* 789, 1963.

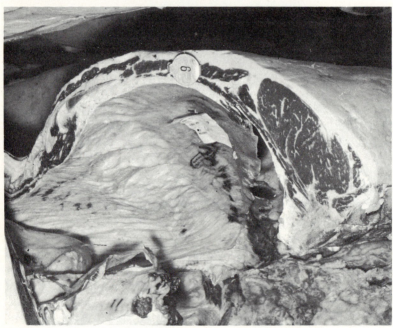

Fig. 12-6. Above, Emulous 7000, the first Certified Meat Sire to qualify under the program of Performance Registry, International. *Below,* a 12th-rib cut from one of the progeny on which his certification was based. To qualify for the Certified

ble to know with certainty whether the crosses exceed the productivity of the original foundation stocks or are more productive than cattle which could have been developed during the same period of time and with the same expenditure of time, money, and effort, without inbreeding, but with selection for improved performance. The average of all crosses has about equaled the productivity of a small herd established a few years ago with a group of commercial heifers and since bred each year to a different, purchased bull representative of available non-inbred commercial bulls. Crossline cattle by sires from the superior lines have excelled the control herd by approximately 10 per cent in pounds of calf weaned per cow bred.

Preliminary results of linecrossing at other locations are similar in showing that crossline animals are more productive than inbreds, but very few data comparing them to cattle produced by other breeding systems are available.

Data are insufficient at present to determine with any certainty whether heterosis from crosses of selected inbred lines within a breed is sufficient to pay for the rather large cost of their development. Development costs of the few superior lines must include the effort expended in developing many lines which ultimately prove inferior and must be discarded.

With beef cattle, as with other species, one of the most important detrimental effects of inbreeding is on the lowly heritable traits affecting reproduction. Inbred bulls produced lower-quality semen than outbreds at young ages.[1] Their slower attainment of puberty and the possibility of lower later fertility is a problem to be considered in producing inbred bulls for sale to commercial producers.

Organized Performance Testing of Beef Cattle. From what has been said about the various characters of economic importance in beef cattle, it is obvious that carefully kept records on traits of economic importance are a vital step in improvement of seedstock herds.

Meat Sire award, it is necessary that at least 10 progeny be fed to carcass weights of 475 lb or more. For certification, the average of all test progeny must be above for each criterion, and at least 50 per cent of the carcasses must meet or exceed all the following: (1) Carcass weight per day of age of not less than 1.25 lb for steers and 1.15 for heifers; (2) A minimum adjusted rib-eye area of 2.0 square inch per 100 lb of carcass; (3) A maximum adjusted fat thickness of 0.13 inch over the rib eye per 100 lb of carcass; and (4) a minimum of a "small" amount of marbling. There are alternate standards for (2) and (3) of a minimum of 48 per cent boneless, closely trimmed preferred cuts and for (4) of a maximum of 8.0-lb shear force on ½-inch cores of cooked rib steak. (*Courtesy of the owner, Carlton W. Corbin, Fittstown, Oklahoma.*)

[1] Harris, Laura A., et al., *Jour. Anim. Sci.*, **19:**665–673, 1960.

Beef-cattle breeders in general were slow to realize this, and in spite of scattered privately owned herds in which records were kept, usually at the suggestion of or under supervision of research or extension personnel, it was not until 1953, in Virginia, that the first organized program of performance testing was established. During the next few years, more than 40 state extension services developed programs. In more than 20 states these are wholly or partially under the direction of organizations of breeders. Procedures vary somewhat between states, but in most cases include evaluation of weaning weights and conformation scores and postweaning gains and scores. Some programs are sponsoring central tests for evaluating postweaning gains of bulls and are beginning to supervise progeny tests for carcass quality.

In 1955 an organization was chartered at Amarillo, Texas, under the name American Beef Cattle Performance Registry, for the purpose of

Fig. 12-7.   Sam 951, a Charolais bull rated 100 as a Certified Meat Sire by Performance Registry, International. His steer progeny out of Angus, Hereford, and Angus-Hereford cross cows averaged 673 lb carcass weight at 13 months of age; the carcasses had rib eyes 2.32 square inches in average area and fat thicknesses averaging 0.06 inch per 100 lb carcass weight, graded choice, and ½-inch cores of cooked meat averaged 6.43 lb in shear force. (*Courtesy of the owner, Litton Charolais Ranch, Chillicothe, Missouri.*)

recognizing outstanding performance in beef cattle. The name was later changed to Performance Registry, International. It is now headquartered at 1140 Delaware St., Denver, Colorado. It offers recognition and herd-record services on many factors important to beef producers, including weaning weights, postweaning gains, and carcass evaluation through progeny testing.

In the late 1950s and early 1960s most breed associations developed programs in one or more phases of performance, conformation, or carcass evaluation. Details can be obtained from the association of the breed in which a person is interested.

MAXIMIZING PRODUCTIVITY OF COMMERCIAL HERDS

Commercial beef producers usually buy bulls from seedstock herds. It is essential that they obtain animals with the greatest possible potential for transmitting improved weaning weight, postweaning growth, conformation and carcass quality obtainable for the money which can be justifiably expended on sires. This can be accomplished through buying from seedstock herds in which records are kept and made fully available to bull buyers.

The type of operation in a commercial herd will determine how extensively records can be kept on individual animals. In the long run, the productivity of a commercial herd in which all replacement females are raised will depend upon the transmitting ability to the sires used. Thus, if sires are purchased, and if breeding is in multiple-sire units so that progeny of individual sires cannot be identified, keeping the complete records needed in a seedstock herd cannot be justified. The commercial herd owner should strive to select replacement heifers from dams which have been the best producers and which are themselves above average in growth rate and conformation. A simple record system in which identity of dams and offspring are maintained, birth dates are recorded, and weights and conformation scores taken at weaning and at a later age (usually around 18 months) when final selections are made has been found useful by some commercial cattlemen and is to be recommended where mode of operation makes it possible. If taken, these records are subject to the same type of adjustments for sex and age of dam as in seedstock herds.

A commercial cattleman can also do some culling in his cow herd on the basis of calf weight and grade, but this will usually be justified only if production is relatively low. Even if she is a low producer relative to her contemporaries, a cow will be approaching maturity before this is known. In the ensuing few years she will likely raise calves as heavy as those from an inherently better young heifer which might replace her.

Commercial cattlemen should always follow an outbreeding or crossbreeding system. As discussed previously, even mild inbreeding on the average depresses performance to a degree for all traits, but proportionally the most for the traits of low heritability related to fertility and viability. Since the income of a commercial cattleman depends more upon per cent of calf crop born and raised than on any other one factor, he must do everything possible to maximize it.

Using an outbreeding system is one thing which on the average will help in this regard. If he produces grade animals of a single breed, he should insofar as possible obtain unrelated bulls so that daughters of bulls of one bloodline will be bred to bulls of other bloodlines. The possibilities of heterotic responses from use of bulls of inbred lines in planned rotation systems have been mentioned in an earlier section.

Summaries of available data on crossbreeding with three British breeds (Angus, Hereford, Shorthorn) of beef cattle are given in Table 12-3. Not all aspects of desired data from all experiments have been published, but it appears from the average results of these experiments that crossing gives results as follows as compared to pure or straight breeding with the same foundation stocks:

1. An increase of 4 per cent or more in net calf crop raised due in part to higher calving rates and in part to reduced calf mortality
2. An increase of from 3 to 5 per cent in weaning weight
3. A slight increase in postweaning rate of gain
4. Little effect on efficiency of postweaning gain
5. Little effect on carcass grades or yield of preferred cuts

The foregoing results are from matings of straightbred bulls to straightbred cows of another breed compared with matings of the same bulls to cows of their own breed. Only limited data have been published on the productivity of crossbred British-type cows as compared to that of straightbreds, but results from Montana[1] in which Shorthorn × Hereford cows were bred to Angus bulls showed an advantage of 9.8 per cent in calf crop and advantages of 13 and 8 per cent in weaning weight and postweaning gain, respectively, and carcass grades were slightly higher as compared to straightbred Herefords. Unfortunately, this experiment had no straightbreds of the other two breeds to determine how much of the advantage was due to crossbreeding per se. In preliminary

[1] Knapp, Bradford Jr., et al., *U.S. Dept. Agr. Cir.* 810, 1949.

Table 12-3. *Crossbreeding Studies with British Breeds—Advantages of Crossbreds over Straightbreds*

(See list at end of chapter for references given)

	Ft. Robinson, Nebr., Gregory et al. (1965) and (1966a, b, c) Wiltbank et al. (1965)*	Virginia, Gaines, et al. (1961) and (1966)*	Ohio, Gerlaugh et al. (1951)†	Miles City, Mont., Knapp et al. (1949)‡	Louisiana, Damon et al. (1959, 1960)†
No. straightbred matings	447	144	195	139	54
No. crossbred matings	470	142	196	119	53
% diagnosed pregnant	3.0	.....	.....	.....	.....
% calving	0	7.6	.....	.....	.....
% calf death loss	.....	6.0	3.6	.....	.....
% calves weaned of cows bred	3.6	12.3	.....	5.1	.....
% weaning weight (both sexes)	4.6	3.8	5.1	3.5	−2.8
% postweaning gain (both sexes)	.....	.....	2.4	.....	.....
% TDN per 100 lb-gain	−0.4	.....	−0.7	1.6	.....
% yearling or slaughter weight:					
Steers	3.3	4.4	.....	7.8	−4.1
Heifers	8.1	3.0	.....	6.9	.....
Carcass grade:					
Steers	slight	0	0	⅓ grade	0
Heifers	.....	slight	0	.....	.....
% carcass cutability§	−0.3	.....	.....	.....	.....

* Angus, Hereford, Shorthorn in all possible crosses.
† Angus, Hereford, and reciprocal crosses. In both studies, figures of matings are for cows actually calving.
‡ Straightbreds were Herefords, crosses were Shorthorn × Hereford.
§ Per cent of closely trimmed, boneless retail cuts from round, loin, rib and chuck from steer carcass.

Louisiana and Nebraska data,[1] crossbred cows have had substantially higher calving rates than straightbreds and have weaned heavier calves.

Data in Table 12-4 summarizing results on Brahman-British crosses from several state and federal stations cooperating in the Southern Regional Beef Cattle Breeding Project show marked heterosis of approximately 10 per cent for preweaning growth rate for both Brahman × British and British × Brahman crosses. There is an additional 4 to 5 per cent increase in growth of calves from crossbred cows bred to either type of bull. Carcass data from similar crosses (Table 12-5) show heterosis for growth rate as expressed by carcass weight per day of age and

[1] *La. Agr. Expt. Sta., Proc. 4th Livestock Producers' Day,* Jan. 24, 1964; and *Nebr. Agr. Expt. Sta. Beef Cattle Progress Rpt.,* 1967.

Table 12-4. Studies with Brahman- and Charolais-British Crosses—Advantages of Crossbreds over British Types in Average Daily Gain, Birth to Weaning, Expressed as Percentages*

		Dams	
Sires	*British*	*Brahman-British Cross*	*Brahman*
British	0	15.0	11.4
Brahman	10.8	15.6	1.2
Charolais	8.4	18.8	22.2

* Data from experiment stations in several Southern states. Adapted from Kincaid (1962).

some heterosis for dressing per cent. For other carcass traits the crosses tend to be intermediate between the parental types.

In Texas[1] straightbred Hereford and Brahman cows were compared with crossbred cows backcrossed to bulls of both types. The crossbreds

[1] Cartwright, T. C., et al., *Tex. Agr. Expt. Sta. Monograph* 1, 1964.

Table 12-5. Some Growth and Carcass Characteristics of British and Brahman Steers, Their Crosses, and Crosses with Charolais*

Characteristic	*British*	50% British 50% Brahman	*Brahman*	50% British 50% Charolais	50% Brahman 50% Charolais
Slaughter age, days	429	422	405	429	405
Carcass wt. per day of age, lb	1.03	1.15	0.98	1.15	1.12
Dressing per cent	57.2	60.1	59.1	58.9	60.2
Carcass grade†	11.1	10.2	8.4	9.3	7.7
9-10-11 rib cut:					
% fat	30.6	28.5	20.6	26.2	17.5
% lean	52.1	53.7	58.3	56.0	60.2
% bone	17.3	17.8	21.1	17.8	20.3
Warner-Bratzler shear‡	13.8	15.6	20.2	13.6	15.5

* Data from experiment stations in several Southern states Adapted from Kincaid (1962).

† On scale in which 13 = Choice; 10 = Good; 7 = Standard.

‡ Pounds of force required to shear a 1-inch core of cooked meat. Smaller figures indicate more tender meat.

Fig. 12-8. Above, a registered Brahman bull. *Below,* an F_1 Brahman $\times$ Hereford cow with a ¾ Brahman calf at side. This cow was a regular producer until 21 years of age in a herd of the Texas Agricultural Experiment Station and weaned 17 calves with an average 205-day adjusted weight of 535 lb. (*Courtesy of American Brahman Breeders Association.*)

showed a superiority of over 20 per cent in pounds of calf weaned per cow bred as compared to the better parent breed. Part of the advantage was due to higher calf crop raised and part to higher calf-weaning weights.

Available evidence thus strongly indicates that, regardless of the types or breeds involved, there are heterotic effects on many traits of prime importance in the economical production of beef. Probably the most important of these is calf crop—dependent upon both fertility and viability of calves. Both these things are low in heritability and thus unlikely to be improved to an important extent by selection. Estimates from various sources indicate that only 75 to 80 per cent of the beef cows bred annually in the United States raise calves. Thus, improvement in net calf crop is sorely needed. It appears that systematic crossbreeding systems can be helpful.

Possible systematic crossbreeding systems with beef cattle vary greatly with size of herd and whether natural mating or artificial insemination is used. They include but are not limited to the following:

1. One of the possible systems is based upon maintenance of a base herd of straightbred cows. About ⅓ of them are bred to bulls of their own breed and about ⅔ to bulls of another breed. All males from both types of mating are sold commercially. The straightbred heifers are used for replacements in the straightbred herd. The crossbred heifers are bred to bulls of a third breed and all offspring sold commercially. The crossbred cows are used as long as they remain productive.

   This general system has been successfully used in Southern States with base herds of Brahman cows. Breeding them to bulls of a European breed results in daughters which are highly productive (see Table 12-4) and long-lived. When bred to bulls of a second European breed, the resulting three-breed cross calves are highly productive and of very acceptable market quality.

   The system has the disadvantages of (*a*) requiring that 25 to 50 per cent (or even more depending on reproductive rates) of the total cow herd be straightbreds and thus of less than maximum productivity, (*b*) not permitting much selection among heifers since virtually all are required for use, (*c*) being somewhat complicated, and (*d*) being possible under natural service only in large herds which have enough cows to justify the use of three or more bulls.

   It has the advantage of a uniform breed composition in the commercial offspring of the crossbred cows and of a maximum degree of heterosis in the three-breed cross calves. It would appear to be most valuable in situations in which some proportion of inheritance of one breed is needed for high productivity of brood cows, but a minimum of the inheritance of this breed is desired in market animals.

2. Another possibility is crisscrossing between two breeds. In this system,

bulls of two breeds are used and daughters of bulls of one breed are bred to bulls of the other. As will be seen on page 394, this system results on the average in about ⅔ as much heterozygosity in the calves as would be true of single crosses between two breeds. All the cows are crossbreds and should exhibit heterosis in reproductive performance. The system is simple and under natural mating is adaptable to any herd using two or more bulls in which matings can be controlled.

The system has the disadvantage of producing two rather distinct calf types. If one were less in demand for slaughter than the other, total income could be less than maximum.

3. A third possibility is the rotation crossing of three breeds. In this system (see page 395) a higher degree of heterosis is maintained than with crisscrossing. On a continuous basis, under natural mating, it can be applied only in herds large enough to use three bulls and in which matings can be controlled. Here too, there will be distinct calf types depending upon the breed of bull siring calves.

From the foregoing it is obvious that utilization of crossbreeding systems with beef cattle is not as easy as with swine because of slower reproductive rates. Artificial insemination, in management situations in which it is practical, eliminates most of the problems, since semen of two or three breeds can easily be stocked and each cow can be individually inseminated with the semen required to maintain the systematic crossing program being followed.

Under natural mating, difficulties are rather great for small one-sire herds. The same difficulties apply in large herds in which it is impossible to maintain controlled breeding pastures. A further problem sometimes encountered is that serious prejudices against crossbreds exist in some markets. Even if this is unjustified, a producer can lose more than he gains from crossbreeding if the selling price of his product is seriously reduced. Thus, before embarking on a crossbreeding program, a cattleman should ascertain whether there are markets available which will not discount his product.

Although predictions are dangerous in a business in which tradition and breed loyalties are as deep as in the American commercial beef industry, it appears likely that systematic crossbreeding will increase in the United States. It is difficult to see how a major portion of an industry can ignore potential improved efficiencies of 5 to 10 per cent or more. These appear possible from proper utilization of crossbreeding.

FORMATION OF NEW BEEF-CATTLE BREEDS FROM CROSSBRED FOUNDATIONS

Formation of new beef-cattle breeds from crossbred foundations has been active in the United States in this century. Much of this activity has been in the Gulf Coast areas where abundant experimental evidence, as well as observation by cattlemen, indicates that the three British breeds of beef cattle commonly used elsewhere in the United States lack the heat and insect resistance, the ability to live on sparse feed, or perhaps some other quality necessary for maximum production.

Zebu-type cattle (originally imported from India and known in the United States as *Brahman*) on the other hand do not produce carcasses as well suited to market demands in the United States as those of European breeds. Further, as straightbreds they tend to grow rather slowly and to have low reproductive rates. As we have seen in the last section, crosses of Brahman and European breeds result in a high degree of heterosis with the result that the crosses are considerably more productive than either type of straightbred. The fact that crossbreds seem to represent much of what was needed for efficient beef production in the area has led to the development of several new breeds.

The honor of developing the Santa Gertrudis, the first breed of this type (and incidentally the first American breed of cattle), goes to the King Ranch, Kingsville, Texas. Established in 1851, the King Ranch was originally stocked with *Texas Longhorns*—animals able to survive under the heat and sparse forage conditions typical of the huge ranch, but lacking in carcass quality. The herds were graded up to Shorthorns and Herefords from 1880 to about 1900. Carcass qualities improved,

Fig. 12-9. A group of Santa Gertrudis cows and calves on a South Texas pasture. (*Courtesy of King Ranch, Inc., Kingsville, Texas.*)

Fig. 12-10. A three-year-old Beefmaster cow. (*Photo by Darol Dickinson, Cashan, Colorado, supplied courtesy of the Lasater Ranch, Matheson, Colorado.*)

but ability to live and grow on the range decreased. Limited crosses with Brahman bulls were made during a test period (1910–1918). These were so promising that the development of a new breed containing about ⅜ Brahman and ⅝ Shorthorn inheritance was started. After more than 40 years of development, the breed is now in wide use. A breed association was established in 1951.[1]

There are several other new strains or breeds based upon Brahman-European crosses. Among these are the Beefmaster, developed by the Lasater Ranch of Falfurias, Texas, and Matheson, Colorado, from Shorthorn, Hereford, and Brahman stock; the Brangus, developed from Brahman-Angus crosses; and the Charbray, developed from the Brahman and the French Charolais breed.

These breeds, and others, all attempting to combine the qualities of different types to produce new strains adapted to beef production under rigorous environmental conditions, are now undergoing extensive trials both in the areas of origin and in other regions of the United States. Their ultimate usefulness will depend upon how nearly their performance comes to that of the initial crossbred. If the new strains can

[1] For more detailed breed history see Rhoad, A. O., "Breeding Beef Cattle for Unfavorable Environments," pp. 203–210, University of Texas Press, Austin, Tex., 1955.

be made to maintain as high a level of productivity as crossbreds between the parent breeds and to further improve upon the crossbreds through selection for specifically desired traits, their future use in grading programs is assured. If, however, as seems more probable, subsequent generations lose considerable proportions of the hybrid vigor of the initial crosses, they will find their greatest use in crossbreeding programs in which a relatively small fraction of Brahman inheritance is desired.

COOPERATIVE BEEF-CATTLE BREEDING RESEARCH PROGRAM

A cooperative beef-cattle breeding research program was initiated in 1946 between the U.S. Department of Agriculture and state agricultural experiment stations to provide for a larger-scale research program than any one agency could develop alone. Thirty-five states are cooperating in the program. Work is organized into three regional projects including states in the Western, Southern, and North Central regions. A large fraction of the research evidence upon which this chapter is based has come from this cooperative research program.

DUAL-PURPOSE CATTLE AND BEEF PRODUCTION FROM DAIRY BREEDS

With increasing specialization in both the beef and dairy industries, the relative importance of the traditional dual-purpose cattle breeds has been declining. They undoubtedly still have a place in certain types of agriculture and it appears that further improvement will depend upon concurrent selection for dairy traits using procedures recommended for dairy herds, and for beef traits using procedures recommended for beef breeds. Heritabilities of dairy and beef traits in dual-purpose breeds are comparable to those found in dairy and beef breeds. Concurrent selection for both types of characters would be expected to result in reduced progress as compared to selection for either one alone. Whether the fact that dual-purpose breeds in general have attained neither the average milk and fat production of strictly dairy breeds or the early maturity and fattening ability of beef breeds depends solely upon the reduced selection intensity possible or whether genetic antagonisms are involved is not definitely known. In one study[1] only low and unimportant genetic

[1] Mason, I. L., *Anim. Prod.*, 6:31–45, 1964.

correlations were found between milk production and growth in two dual-purpose breeds. Fairly large, but not statistically significant, negative genetic correlations were found between milk production and dressing per cent—an indicator of fatness.

Numerous European studies indicate very low genetic relationships between milk production traits and growth. Since most European countries depend to a large extent upon beef produced as a by-product of the dairy industry, emphasis is being put upon some selection for beef traits in dairy breeds. It is felt that beef potential can be improved with little or no harm to dairy production.

Studies in both Europe[1] and the United States[2] show that animals of dairy breeds produce surprisingly useful beef carcasses when raised by proper methods. Undoubtedly, as population increases and pressures on land become greater in the United States, we will be making better use of surplus dairy animals for beef production.

SUMMARY

Goals of seedstock breeders with all classes of meat animals are to develop qualities in their herds which will result in maximum productivity of quality products in the commercial herds to which they directly or indirectly supply breeding stock. Most growth, efficiency, and carcass characters in beef cattle are medium to high in heritability, and there appear to be no negative genetic correlations among them large enough to seriously interfere with improvement.

Thus, selection for those of economic value should be effective in changing beef populations in desired directions. Growth is moderately heterotic, but carcass traits are little affected by heterosis. External indicators of potential carcass quality are needed since direct selection for carcass traits through progeny and sib testing are slow and expensive. Beef-cattle conformation is related in some degree to carcass quality, but more knowledge of these relationships is needed to serve as a guide to selection. Ultrasonic and other devices which will enable breeders to "look under the skin" show promise for some carcass traits.

Fertility and viability are low in heritability but are considerably affected by heterosis, thus indicating that commercial production should be based on outbreeding systems and, where possible, on crosses of breeds or lines under systems designed to maximize heterosis. An additional use of crossbreeding is to incorporate desired characters of two

[1] Brooks, A. J., and Latham, J. O., *Empire Jour. Expt. Agr.*, **25**:339–355, 1957.
[2] Cole, J. W., et al., *Jour. Dairy Sci.*, **47**:1138–1144, 1964.

or more breeds in a commercial operation. Several new breeds have been formed in this century on crossbred foundations in efforts to stabilize such desired combinations.

Dual-purpose cattle breeds have never attained maximum productivity for either beef or dairy characters and their relative importance in the industry is decreasing. Carcasses of properly fed young animals of dairy breeding are surprisingly palatable, and further augmentation of beef supplies from this source is probable as population increases.

REFERENCES

Brinks, J. S., Clark, R. T., and Kieffer, N. M. 1965. Evaluation of Response to Selection and Inbreeding in a Closed Line of Hereford Cattle, *U.S. Dept. Agr. Tech. Bul.* 1323.

Cartwright, T. C., Ellis, G. F., Jr., Kruse, W. E., and Crouch, E. K. 1964. Hybrid Vigor in Brahman-Hereford Crosses, *Texas Agr. Expt. Sta. Tech. Monograph* 1.

Clark, R. T., Brinks, James S., Bogart, Ralph, Holland, Lewis A., Roubicek, Carl B., Pahnish, O. F., Bennett, James A., and Christian, Ross E. 1963. Beef Cattle Breeding Research in the Western Region, *Oregon Agr. Expt. Sta. Tech. Bul.* 73.

Cunha, T. J., Koger, M., and Warnick, A. C., eds. 1963. "Crossbreeding Beef Cattle," University of Florida Press, Gainesville, Fla.

Damon, R. A., Jr., McCraine, S. E., Crown, R. M., and Singletary, C. B. 1959a. Performance of Crossbred Beef Cattle in the Gulf Coast Region, *Jour. Anim. Sci.*, **18**:437–447.

———, ———, ———, and ———. 1959b. Gains and Grades of Beef Steers in the Gulf Coast Region, *Jour. Anim. Sci.*, **18**:1103–1113.

———, Crown, R. M., Singletary, C. B., and McCraine, S. E. 1960. Carcass Characteristics of Purebred and Crossbred Beef Steers in the Gulf Coast Region, *Jour. Anim. Sci.*, **19**:820–844.

———, Harvey, W. R., Singletary, C. B., McCraine, S. E., and Crown, R. M. 1961. Genetic Analysis of Crossbreeding Beef Cattle, *Jour. Anim. Sci.*, **20**:849–857.

Gaines, J. A., McClure, W. H., Vogt, D. W., Carter, R. C., and Kincaid, C. M. 1966. Heterosis from Crosses among British Breeds of Beef Cattle: Fertility and Calf Performance to Weaning, *Jour. Anim. Sci.*, **25**:5–13. (See also *Jour. Anim. Sci.*, **20**:906, 1961.)

Gerlaugh, Paul, Kunkle, L. E., and Rife, D. C. 1951. Crossbreeding Beef Cattle, *Ohio Agr. Expt. Sta. Res. Bul.* 703.

Gregory, Keith E. 1961. Improvement of Beef Cattle through Breeding Methods—Summary of Results from Regional Project NC-1, *North Central Regional Publ.* 120, published as *Nebraska Agr. Expt. Sta. Res. Bul.* 196.

———. 1964. Beef Cattle Breeding, *U.S. Dept. of Agr., Agr. Information Bul.* 286.

————, Swiger, L. A., Koch, R. M., Sumption, L. J., Rowden, W. W., and Ingalls, J. E. 1965. Heterosis in Preweaning Traits of Beef Cattle, *Jour. Anim. Sci.,* **24**:21–28.

————, Swiger, L. A., Koch, R. M., Sumption, L. J., Ingalls, J. E., Rowden, W. W., and Rothlisberger, J. A. 1966a. Heterosis Effects on Growth Rate of Beef Heifers, *Jour. Anim. Sci.,* **25**:290–298.

————, Swiger, L. A., Sumption, L. J., Koch, R. M., Ingalls, J. E., Rowden, W. W., and Rothlisberger, J. A. 1966b. Heterosis Effects on Growth Rate and Feed Efficiency of Beef Steers, *Jour. Anim. Sci.,* **25**:299–310.

————, Swiger, L. A., Sumption, L. J., Koch, R. M., Ingalls, J. E., Rowden, W. W., and Rothlisberger, J. A. 1966c. Heterosis Effects on Carcass Traits of Beef Steers, *Jour. Anim. Sci.,* **25**:311–322.

Kincaid, C. M., 1962. Breed Crosses with Beef Cattle in the South, *Southern Cooperative Ser. Bul.* 81. (Published by *Texas Agr. Expt. Sta.*)

Knapp, Bradford, Baker, A. L., and Clark, R. T. 1949. Crossbred Beef Cattle for the Northern Great Plains, *U.S. Dept. Agr. Cir.* 810.

Mason, I. L. 1966. Hybrid Vigor in Beef Cattle, *Animal Breeding Abstracts,* **34**:453–473 (a comprehensive review).

Swiger, L. A., Gregory, K. E., Sumption, L. J., Breidenstein, B. C., and Arthaud, V. H. 1965. Selection Indexes for Efficiency of Beef Production, *Jour. Anim. Sci.,* **24**:418–424.

Warwick, E. J. 1958. Fifty Years of Progress in Breeding Beef Cattle, *Jour. Anim. Sci.,* **17**:922–943.

————. 1960. Genetic Aspects of Production Efficiency in Beef Cattle, *Nat'l. Acad. Sci., Nat'l. Research Council, Publ.* 751, pp. 82–92.

Wiltbank, J. N., Gregory, K. E., Swiger, L. A., Ingalls, J. E., Rothlisberger, J. A., and Koch, R. M. 1966. Effects of Heterosis on Age and Weight at Puberty in Beef Heifers, *Jour. Anim. Sci.,* **25**:744–751.

————, J. N., Rothlisberger, J. A., Ingalls, J. E., Gregory, K. E., and Rowden, W. W. 1965. Reproduction in Straightbred Cows Bred for Crossbred and Straightbred Calves, *Nebraska Agr. Expt. Sta. Beef Cattle Progress Rpt.,* pp. 8–9.

CHAPTER THIRTEEN

Improving Swine

The popular name "mortgage lifter," often given the hog, is indicative of the fact that hog production is, over a period of years, likely to be a moderately profitable enterprise with fewer speculative hazards than are often encountered in other forms of livestock production. The hog is a more efficient converter of nutrients into human food than either sheep or beef cattle. It should be remembered, however, that the greater part of the hog's diet must be in the form of concentrates which are ordinarily more expensive than the roughages consumed by cattle and sheep.

IMPROVING SEEDSTOCK HERDS

From a breeding standpoint the following items are of primary importance in determining the efficiency and profitability of swine:

1. Litter size
2. Livability
3. Weight per pig and per litter at weaning time
4. Daily gains between weaning and marketing
5. Efficiency of feed conversion
6. Body type and carcass desirability

Production Characters. Although greatly influenced by herdmanship, the number of pigs farrowed and raised and their weight at weaning time together form the best measures we have of prolificacy and mothering ability of sows. They are particularly important since they have a tremendous influence on profitability. On the average, a sow will have consumed a total of 1,500 to 2,000 lb of feed during the period between breeding and the date her litter is weaned. If this all has to be charged against a litter of only four or five pigs, the chance of eventual profit is small. In addition to number of pigs farrowed and raised, the weaning weight of the pigs is important since pigs which are the heavier at weaning time retain their weight advantage and therefore reach market weights sooner.

The eventual impact on the swine industry of developments in the fields of nutrition and management which permit weaning pigs during the first to third week of life instead of at the customary eight weeks cannot be estimated with any certainty at present. Since sows ordinarily do not return to estrus or conceive while nursing a litter, early weaning permits earlier rebreeding and the production of an increased number of pigs per year by the sow. Also, the sow is an inefficient milk producer, and it is possible that nutritionally adequate milk substitutes can be provided at lower cost than can sow's milk. Apparently there are disease-control and management complications which have thus far discouraged widespread use of artificial pig-rearing systems at ages of less than about five weeks. If the system of very early weaning should become general, however, the necessity of selecting sows for nursing ability would be lessened or eliminated, and more attention could be given to other traits. It is not impossible that artificial rearing of pigs could go as far as it has in poultry, in which natural rearing is virtually unknown and selection against broodiness has been so effective as almost to eliminate maternal instincts from many strains and breeds.

Daily gain from weaning to market weight is important because the fast-growing pig is on hand fewer days, thus requiring less labor and fewer days of shelter and exposure to risk of disease. Another important consideration is that rate of gain and efficiency of feed conversion are related. The strength of this relationship apparently varies considerably with differences in breed and/or management and feeding conditions. Whether breeders should depend entirely on the indirect effects of selecting for rate of gain or whether they should keep and use efficiency data in selection is not entirely clear. Most testing stations both in the United States and abroad feed pigs either individually or in litter groups and keep feed records. Relatively few breeders are following the practice.

Conformation in Swine. Conformation as it can be evaluated externally in swine is of importance to the seedstock breeder if it is related (1) to potential carcass characters or (2) to production characters such as litter size or rate of gain.

There are relationships between conformation and these characters.

Fig. 13-1. An artist's representation of Pet 2nd 68646, one of the leading show and producing Poland China sows of the early 1900s. She was an ideal of the so-called "medium type" of that era. Compare to modern Poland China in Fig. 13-13. (*Reproduced at suggestion of the Poland China Record Association from Davis and Duncan, "History of Poland China Swine" published in 1921 by the Poland China History Association.*)

Under some circumstances they can be of considerable usefulness in a breeding program. However, in most cases the relationships are not close enough that conformation evaluation can effectively substitute for objective measurements of carcass and performance characters. Rapid changes in swine type or conformation are possible. As a result it has been the focal point of drastic changes in the industry. Fashions in hog type have undergone more changes during this century than has been true of any other class of livestock. These are of interest both from a historical standpoint and as demonstrations of the fact that undesirable changes as well as improvement can occur in animal breeding.

Through 1920 lard was in demand as an export item. Emphasis in hogs was on production of thick-fat types yielding large amounts of lard. This emphasis sometimes led to extremes, and during the period around 1900, particularly during the few years immediately following the turn of the century, the Poland China breed of hogs was swept by the so-called "hot-blood" craze in which selection was for an extremely small, short-legged, refined, fat, early-maturing type of hog. The position an individual breeder should take if some fad such as this should sweep his breed is a question for thought. Should he follow the fad in order to produce animals for which there is an immediate market, or should he make the present financial sacrifice of staying with an unpopular but basically better type? That this is not a hypothetical situation is illustrated by the story of the action of Peter Mouw, of Orange City, Iowa, during the hot-blood craze. Mr. Mouw and a few other breeders believed this to be a basically unsound type and continued to produce a more rugged "big type" which was often the subject of ridicule at the shows. The tide finally turned, however, and almost overnight these hogs became the most popular in the breed. In this case, adherence to an ideal paid off with a small fortune to Mr. Mouw.

During the period between the First and Second World Wars, export demand for lard declined greatly and hydrogenated vegetable fats began to compete successfully with it. Coincidental with this lessening demand, popular ideas of swine type swung toward a long, lean, upstanding type. This extreme type was losing its popularity prior to World War II allegedly because it came into disfavor with feeders. During and immediately following World War II, demand for lard was again high, and for a number of years breeders tended to select for extreme thickness and fatness. Since shortly after World War II demand for lard has been low, and selection is again for swine types producing less lard.

Relationships of conformation in swine to both carcass and production characters were demonstrated in experiments by the U.S. Department of Agriculture and by the Illinois Agricultural Experiment Station. Large, intermediate, and small types were compared. Large- and intermediate-type sows averaged about one more pig raised per litter than small-type sows. The large and intermediate types also grew somewhat more rapidly and at a given slaughter weight produced much leaner carcasses.[1]

It should be emphasized that these studies were with animals representing rather extreme differences in type. Within the narrower ranges of type usually found in a population, relationships between conformation and production and carcass characters are much less marked.

Although changing hog types have been, to a certain extent, based upon economic trends in the prices of the various products, it is true that, once a shift has started, there has been a tendency for it to become a fad and go too far. There is no doubt that the extreme shifts have been harmful to the entire industry. In recent years many studies have been made on relationships between conformation and cutout percentages of preferred cuts. These studies have exerted much influence on present concepts of what constitutes desirable meat type. It is hoped that continued reliance on research of this kind will prevent the development of unrealistic fads in the future. The hog-marketing system has

[1] *U.S. Dept. Agr. Cir.* 698, 1944; and *Ill. Agr. Expt. Sta. Bul.* 321 and 322, 1929.

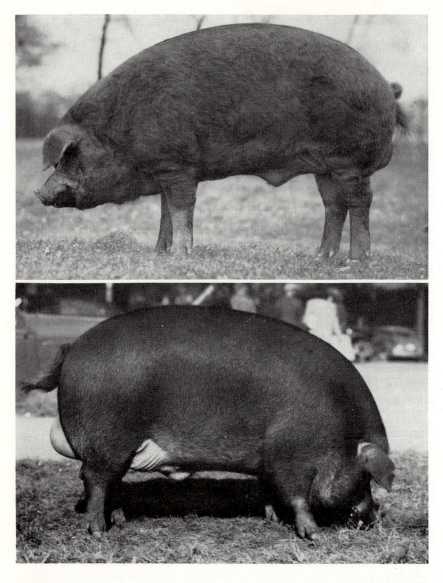

Fig. 13-2. Durocs showing how ideals of type change. *Above,* Wavemaster Stilts, 1931 Grand Champion. *Below,* Top Set, first place aged boar, Iowa State Fair, 1947. Compare to current ideal type as shown in Fig. 13-8. (*Courtesy of United Duroc Record Association.*)

been somewhat slow in translating demonstrated differences in carcass value into price differentials for live hogs. Progress is being made in this direction, however, and there appear to be good possibilities of continued constructive development. Even if individual producers do not receive adequate price incentives for the production of superior-type hogs, the general production of meat-type animals should put the entire industry in a stronger competitive position and strengthen the position of all producers.

External appearance is of value as an indicator of potential carcass characters, but it is far from perfect. Therefore, research has continued to find more accurate means of estimating potential carcass characters of live animals.

The development of a "probe" technique[1] for measuring back-fat thickness in living animals represented a first in animal breeding—actually measuring an item of potential carcass value in an animal and still using the animal for breeding purposes. The technique is being used widely in swine improvement programs. It is relatively simple, involving cutting a small slit in the skin in the back and thrusting a thin metal ruler down through the soft fat to the firmer lean and reading fat thickness directly. Subsequently Andrews and Whaley, of Purdue University, developed a device employing the difference in electrical conductivity of fat and lean to accomplish the same purpose. In using this device, a special needle with two electrodes at the end is thrust slowly into the back until a difference in flow of electricity indicates that lean tissue has been reached. This method is somewhat quicker and involves less tissue damage than the probe method, but it has the disadvantage of requiring a moderately expensive piece of equipment.

More recently, devices employing the principle of reflection of ultra-high-frequency sound waves by tissue interfaces have been successfully used to predict back-fat thickness and, with somewhat less accuracy, to predict size of the *longissimus dorsi* or loin muscle. These devices have the advantage of inflicting no tissue damage, but the equipment is expensive.

In present-day swine breeding, the objective is to produce animals with a maximum percentage of the high-priced primal cuts—ham, trimmed loin, shoulder, and bacon—and a minimum of lard, while maintaining desirable eating quality in the lean. In general, there is a relationship between outside fat thickness and yield of primal cuts with minimum outside fat being most desirable. This is the relationship which leads to usefulness of the probe and other devices for estimating fat thickness in living animals. This relationship is, however, far from the

[1] Hazel, L. N., and Kline, E. A., *Jour. Anim. Sci.,* 11:313–318, 1952.

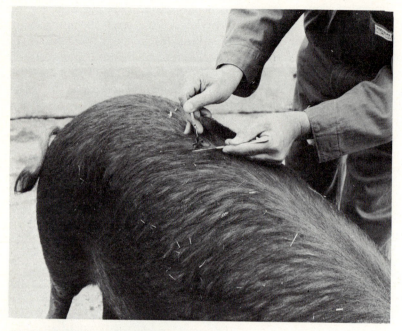

Fig. 13-3. Illustrations of two methods of estimating back-fat thickness in live swine. *Above,* the probe. *Below,* the Leanmeter. (*Courtesy of Dr. H. O. Hetzer and Mr. J. W. Stevenson, Animal Husbandry Research Division, U.S. Department of Agriculture.*)

Fig. 13-4. Grand Champion swine carcass, 1964 International Livestock Exposition. The pig from which this carcass came weighed 215 lb. The carcass was 30.9 inches long, had a back-fat thickness of 1.03 inches, a loin eye area of 7.09 square inches and yielded 46.2 per cent of ham and loin. (*Courtesy of Prof. E. A. Kline, Iowa State University.*)

complete answer to breeding for carcass quality since fat deposits between muscles (so-called "seam" fat) and variations in lean characteristics may also provide problems.

Thus, even with the usefulness of visual conformation evaluation and methods of estimating fat thickness and lean areas in live animals, there is a continuing need for slaughter and carcass evaluation to guide breeding programs.

Effects of Environment on Swine-carcass Development. In the chapter on Beef Cattle it was pointed out that environmental differences exert differential effects on the development of the various tissues in the ani-

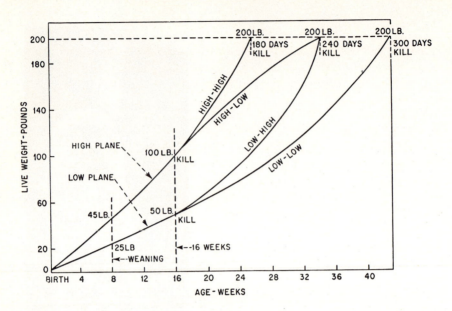

Fig. 13-5. Plan of experiment to determine the influence of plane of nutrition at different periods with respect to age upon growth rate and carcass composition of bacon hogs. (*Courtesy of Dr. C. P. McMeekan and the late Dr. John Hammond, Cambridge University, Cambridge, England.*)

mal body. One of the most vivid experimental demonstrations of this fact comes from a study[1] of the growth and carcass composition of pigs of an inbred Yorkshire strain when fed from birth to a slaughter weight of 200 lb on various feeding regimes. Pigs were divided at birth into two groups; one group was fed on a high plane and the other on a low plane for the first 16 weeks of life. At that time, half the pigs on each plane of nutrition were shifted to the other, so that four nutritional regimes resulted, namely, groups fed on

1. A high plane from birth to slaughter (high-high)
2. A high plane from birth to 16 weeks and a low plane from that time to slaughter (high-low)
3. A low plane from birth to 16 weeks and a high plane thereafter (low-high)
4. A low plane from birth to slaughter (low-low)

It was shown that if the pig has grown very slowly early in life, the early developing parts of the carcass, such as the head, legs, and

[1] McMeekan, C. P., *Jour. Agr. Sci.*, 30:276–343, 387–436, and 511–568; 31:1–49; 1940–1941.

bone, and, to a lesser extent muscle, are permanently reduced in size. When these animals are later put on a high plane of nutrition, the later developing parts, particularly fat, grow rapidly and result in carcasses too fat for bacon. Conversely, pigs fed from birth on a high plane have a rapid development of the early growing parts, and if later put on a low plane of nutrition, produce carcasses with a high content of lean.

Later experiments[1] in which nutritional limitations imposed early in life were not so severe or of such long duration have not duplicated the earlier results, suggesting either that very severe retardation is necessary to permanently inhibit lean growth or that different strains or breeds of swine react differently.

The question may be of only academic interest in the United States, since the most profitable type of swine production is usually that of full feeding from birth to market weight. Thus, we need swine bred to lay down only the desired amount of fat under this kind of regime. However, if market demand should be for leaner hogs than can be produced under full-feeding regimes with available breeds and strains, McMeekan's work shows that limitation of feed intake late rather than early in life will be helpful in producing leaner carcasses.

Heritability of Economic Characters in Swine. Estimates of heritability of swine characters show considerable variation from study to study but do show general trends. Figures representing best current estimates are given in Table 13-1. The heritabilities of litter size at birth and weaning and of weaning weight of the litter are all low.

All estimates of the heritability of individual pig weights at 56 days of age are low, indicating that weight at that age is largely a function of the nursing ability of the sow rather than of the pig's own genes. After weaning, however, the pig's own genes evidently exert their influence and heritability increases, probably to 20 to 30 per cent at 154 to 180 days of age.

Heritabilities of various type and conformation scores average around 30 per cent in studies made within breeds and when only animals of basically similar types were studied. In one study in which strains of distinctly different types were included, heritability of type score was estimated at 92 per cent. The latter figure probably explains why breeders have been able to alter hog types rather rapidly upon occasion in the past, and the figure of 30 per cent within strains indicates why further improvement often seems to come slowly in a herd already fairly near desired characteristics. Another reason for progress often being

[1] Nielson, Henning E., *Anim. Prod.*, 6:301–308, 1964.

Table 13-1. Heritability of Economically Important Traits of Swine

Character		Approximate average heritability*
Litter size at birth	low	5 to 15
Litter size at weaning	low	5 to 15
Weaning weight of litter	low	10 to 20
Pig weight at weaning	low	10 to 20
Pig weight at 154 to 180 days	medium	20 to 30
Daily postweaning gain	medium	25 to 40
Feed per unit gain	medium to high	30 to 50
Conformation and type scores:		
Within strains	medium	25 to 35
Between strains	high	92†
Live animal back-fat thickness (probe)	high	40 to 60
Carcass characteristics:		
Back-fat thickness	high	40 to 60
Thickness of belly	high	40 to 60
Yield of lean cuts	medium to high	30 to 50
Area of *longissimus dorsi*	high	40 to 60
Length of body	high	40 to 60
Dressing per cent	medium	25 to 35
Number of nipples	medium	20 to 40

* From many published sources. Publications listed at end of chapter and as footnotes will give references on heritability as well as on other aspects of swine breeding.
† One study only and it on small, intermediate, and large Poland Chinas.

discouragingly slow in such herds is that selection may be for a genetic intermediate. If we simplify our concept of type by considering it for a moment as being merely a matter of length of body and leg, it is easy to see that improvement in an already acceptable herd would have to be made by reducing the frequency of animals that are either too long or too short rather than by changing average length. Regardless of the number of pairs of genes influencing a character, selection for an intermediate is virtually powerless to reduce variability unless accompanied by intense inbreeding. Even then as discussed in a previous chapter, reduced variability may not be marked.

In order to put the foregoing heritability estimates in more practical terms, let us consider a hypothetical example. If it is assumed that the inheritance of several important characters is due to additive gene action, that litter size weaned and litter weaning weight are each 10 per cent heritable, that 180-day weight and conformation are each 30 per cent heritable, and that we have herd averages for the various items as indicated, then the progress to be expected from one generation of selection would be as shown herewith:

Trait	Av. of herd	Selected individuals	Av. selection differential	Heritability, %	Expected genetic improvement next generation
Av. no. of pigs weaned	6.0	7.2	1.2	10	+0.12
Av. litter weaning wt., lb	180.0	203.0	23.0	10	+2.3
Av. individual 180-day wt., lb	180.0	198.0	18.0	30	+5.4
Av. type score*	5	7.5	2.5	30	+0.75

* Graded on a scale from 1 to 9, with 1 being inferior and 9 excellent.

The selection differentials indicated for each character are approximately the size observed[1] in herds of the Regional Swine-Breeding Laboratory in which selections were made concurrently for several traits utilizing production records.

The gains expected for any one character are relatively small, and even if they were all attained, environmental differences from year to year are great enough to make it difficult to know whether progress was being made until records had been kept for several years.

It is also uncertain from studies to date whether the estimates of possible progress for individual traits are really attainable.[1,2] Studies of data from some stations in the Regional Swine-Breeding Laboratory suggest that selection was effective in improving litter size and weaning weight and postweaning growth rate or at least in overcoming the depressing effects of inbreeding. On the other hand, more extensive studies incorporating data from several stations showed that the expected effects of selection for these traits had not been attained. These studies are complicated by the fact that most of the lines studied were being inbred, and this may have depressed performance more than was realized. Also, the studies covered only a few years. A discouraging item in considering probable effectiveness of selection for litter size comes from a realization that continuous automatic selection for larger litter size may have been going on for centuries without raising average litter size above its present modest average.

[1] Dickerson, G. E., et al., *Mo. Agr. Expt. Sta. Res. Bul.* 551, 1954.
[2] Craft, W. A., *Jour. Anim. Sci.*, 17:960–980, 1958.

It is probable, however, that past natural selection may have been for an optimum rather than large litter size. This could be the case since death rates are usually higher in larger than in smaller litters farrowed. This is particularly true under conditions of poor husbandry and probably even more acute under wild conditions. Thus, with improved husbandry there may be more opportunity for genetic improvement than would appear at first thought to be the case.

The automatic feature of selection for litter size is something which has often received insufficient emphasis in the past and thus warrants a word of explanation. To understand this it is necessary to differentiate between *average litter size raised* per sow farrowing and *average size of litter in which the pigs were raised*. To make this clear, let us consider an example of a group of 12 litters in a herd in one season.

Size of litter (no. of pigs raised)	No. of litters	Total no. of pigs raised in litter of this size	Litter size × no. of pigs raised
8	1	8	64
7	1	7	49
6	5	30	180
5	2	10	50
4	1	4	16
2	2	4	8
Total	12	63	367

Av. litter size raised per sow farrowing = $63/12$ = 5.25
Av. size of litter in which the pigs were raised = $367/63$ = 5.82

If breeding stock were selected from this group of litters entirely at random so far as litter size was concerned, there would be a positive selection of $5.82 - 5.25 = 0.57$ pig per litter over the average of that generation. This is due to the fact that the larger litters have more pigs available. A high percentage of the selection practiced for litter size is of this automatic type.

It is possible that in breeders' herds animals from small litters are better grown than pigs in larger litters and therefore have a greater probability of being selected for breeding purposes. This could result in a lack of positive selection for litter size. The fact that there is by no means an absolute negative correlation between litter size and rate of development, together with the fact that many breeders have consciously selected for large litters, makes this seem unlikely.

If, as seems likely, litter size is failing to respond to selection as rapidly

Fig. 13-6a. *Left,* a 10th-generation high-fat line gilt from the U.S. Department of Agriculture fat-thickness selection study. She had a probe-estimated back-fat thickness of 2.53 inches at 173 lb. *Right,* a 10th-generation low-fat line gilt with an estimated back-fat thickness of 0.88 inch at 178 lb. (*Courtesy of Dr. H. O. Hetzer, Animal Husbandry Research Division, U.S. Department of Agriculture.*)

as heritability estimates would lead us to think it should, the reason is probably due to one or a combination of the following:

1. Even the present low heritability estimates may be higher than is actually the case, thus leading us to overestimate potential progress.
2. Prolificacy may depend to a certain extent upon overdominance, with the best performers being heterozygous for a larger-than-average number of gene pairs.
3. There may be negative genetic correlations between prolificacy and other desired traits, so that when we select for some of these, such as rate of gain, we indirectly select for smaller litters and override any actual positive selection we have made for litter size.

No positive evidence for overdominance has as yet been advanced with swine, although strong circumstantial evidence for its existence has been presented.[1]

Rapid changes which breeders have been able to make in swine type in the past constitute evidence of the effectiveness of selection for specific aspects of conformation. A U.S. Department of Agriculture study[2] shows that response to selection for back-fat thickness is rapid. In this study all animals are probed for back-fat thickness at 175 lb live weight. Two lines of Durocs have been selected for ten generations—one for high

[1] Dickerson, G. E., "Heterosis," pp. 330–351, The Iowa State University Press, Ames, Iowa, 1952.

[2] Hetzer, H. O., et al., *Jour. Anim. Sci.,* **24**:849, 1965 (abstract).

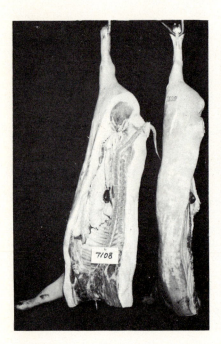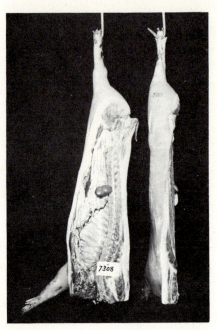

Fig. 13-6b. Carcasses from 8th-generation animals of the U.S. Department of Agriculture fat-thickness selection study. *Left,* high-fat line gilt slaughtered at 208 lb. She had a back-fat thickness of 2.73 inches and a carcass length of 27.6 inches. *Right,* a low-fat line gilt slaughtered at 211 lb with corresponding measurements of 1.18 and 30.3 inches. (*Courtesy of Dr. H. O. Hetzer, Animal Husbandry Research Division, U.S. Department of Agriculture.*)

back fat and the other for low. The average back-fat thickness after ten generations of selection was 2.09 inches for the high line and 1.05 for the low. A control random-bred line remained at about the original population average of 1.50 inches. Eight generations of selection in the Yorkshire breed has resulted in somewhat less rapid progress, but the high and low lines are distinctly different.

Genetic Relationships between Traits. Estimates of genetic correlations are subject to large sampling errors, and it is also possible that the true relationships differ among breeds of swine and perhaps also with the same breeds under different nutrition or management regimes.

In some cases different studies have given generally similar results and we have considerable confidence in the general applicability of the results. In other cases this has not been true.

All known studies have shown positive genetic relationships between rate of gain and gain per unit of feed consumed, but the magnitude of the relationships has varied considerably.[1] Relationships between growth at different ages are positive but often rather low. There are positive correlations between carcass length and measures of leanness. As would be expected, genetic relations among different measures of fatness are high as are those among different measures of leanness. Thus, for the foregoing characters, positive selection for one trait can be expected to result in correlated desirable responses in related traits. The magnitude of the correlated responses will differ.

A considerable degree of uncertainty exists regarding genetic relationships (1) between carcass characteristics and the production traits of growth rate and feed efficiency and (2) between litter size and weight and both postweaning performance and carcass traits.

In the U.S. Department of Agriculture study of selection effectiveness for back-fat thickness, the low-fat lines have consistently had bigger litters, faster growth, and greater feed efficiency than the high-fat lines, but litter size has decreased somewhat in both selected lines as compared to a random-bred control line.

Some reports indicate that rapid gains may be strongly correlated genetically with rapid fat deposition and that the same genes leading to rapid growth may be responsible for poor nursing ability. Other work indicates that genetic relationships between growth and fattening ability are low with the associations in many cases being desirable rather than undesirable.

Quite aside from the experimental data on the subject, a priori reasoning (always a dangerous procedure!) might indicate a negative relationship between maternal and fattening abilities. Bacon hogs are commonly reputed to raise larger litters than most other breeds. Likewise in cattle, no one has yet succeeded in combining maximum milking ability with the propensity to fatten at early ages. Some studies with laboratory mice suggest negative associations between maternal performance and fattening.

Regardless of whether negative associations exist among desirable traits, they are far from absolute. Thus, they may have the effect of slowing but not preventing, progress if selection is made simultaneously for two or more negatively related traits.

The costs of producing breeding stock makes it probable that a line extremely deficient in maternal ability but exceptionally good in growth rate, or vice versa, could not be profitably used. Thus we shall have to continue selecting animals for reasonable excellence in all qualities.

[1] Vogt, D. W., et al., *Jour. Anim. Sci.*, **22**:214–217, 1963; and Jonsson, Per, and King, J. W. B., *Acta Agr. Scandinavica*, **12**:68–80, 1962.

It is true, however, that existing swine breeds differ significantly in certain characteristics, some being strong in one character and some in another. These differences may depend upon chance fluctuations in gene frequency or upon the type of selection which has historically been practiced in the various breeds. As inbred lines are developed, they likewise differ from each other. Judicious crossing takes advantage of these differences, and evidence to date strongly suggests that maximum performance may be attained only by this method. An extension of this idea is the formation and development of lines, solely or almost solely, on the basis of their performance in crosses with each other. This is commonly termed *selection for combining ability* (see chapter on Outbreeding).

In spite of the admittedly slow rate of progress usually made in selecting swine, there is ample evidence that our present hogs show marked improvement in many characters over their wild ancestors, thus showing that selection can be effective in bringing about improvement. Likewise, the rather drastic shifts in type of American swine during the past half-century indicate that breeders have been able to select effectively for the type popular at a given time.

Danish Swine Testing.   The best data showing improvement in swine over a long period of years come from the records of animals fed in the testing stations in Denmark as part of the swine progeny-testing system in that country.

The Danish system of swine breeding[1] is based upon a group of state-recognized breeding centers where a high percentage of the boars used for commercial production in the nation are produced. The breeding centers are visited twice annually by a committee representing farmers' organizations and cooperative bacon factories and scored for numerous items, including type and prolificacy of breeding stock, management and general appearance of farm, and health of animals. Each year the breeding centers are required to send to the testing centers enough test litters of four pigs each to average two pigs per sow in the herd. The test litters are fed out under standard conditions and slaughtered at about 200 lb live weight. Records are kept of rate of gain, efficiency of feed conversion, and several carcass measurements and scores.

Annual reports are issued which include the above information for each center. These reports are designed to serve as guides for the purchase of breeding stock and the establishment of breeding programs.

The first of the testing stations was established in 1907, and, with

---

[1] Lush, J. L., *Iowa Agr. Expt. Sta. Bul.* 204, 1936; and Jonsson, Per, *Ztschr. f. Tierzücht. u. Züchtungsbiol.*, **78**:205–252, 1963.

the exception of an interruption during World War I, testing has been continuous since that time. The same basic ration has been used for the entire period. Litters were fed as groups for many years, but individual feeding was instituted beginning with 1951–1952 in hope of increasing effectiveness of selection for one or more traits.

The following data[1] indicate some important changes which have taken place in the performance of Danish Landrace pigs fed out at the testing stations:

Trait	1909–1910	1924–1925	1947–1948	1953–1954	1962–1963
Av. daily gain, lb	1.18	1.32	1.46	1.49	1.48
Av. feed units required per unit gain	377	357	319	303	298
Av. age at 20 kg live wt., days	....	69	74	77	79
Av. age at 90 kg live wt., days	....	186	180	181	184
Class 1 bacon sides, %	....	40	90	90	97

Detailed data indicate that progress in lowering feed requirements per unit of gain was slight during the period 1939–1950. New testing stations were built in 1950, and apparent increases in efficiency since then may be due at least partially to improved environment. The data show that increases in rate of gain while on test have been accompanied by increases in age at 20 kg live weight. This suggests that selection for rapid gains on test may have resulted in indirect selection for lowered milking ability of sows and thus for lower gains in pigs during the nursing period. Although rate of gain on test has increased about 12 per cent since 1924–1925, the decrease in age at 90 kg live weight has been small. It has increased somewhat in recent years and in 1962–1963 was four days greater than in 1947–1948.

Figure 13-7 shows that progress has been made in increasing body length, increasing belly thickness, and decreasing back-fat thickness. For several years prior to 1952–1953 the ideal bacon carcass was considered to be 93 to 95 cm long and to have a back-fat thickness of 3.4 cm—dimensions which had on the average been attained. Thus, little selection pressure had been put on these items. In recent years, however, customers have preferred leaner bacon, and selection is now for carcasses 95 or 96 cm long with minimum back fat. Carcass length has averaged between 95 and 96 cm since 1957–1958, and the average

[1] See Lush, J. L., *Iowa Agr. Expt. Sta. Bul.* 204, 1936; and Clausen, H., et al., *Beret. Forsøgslab.* 336, (Copenhagen), 152 pp. 1964 (English summary).

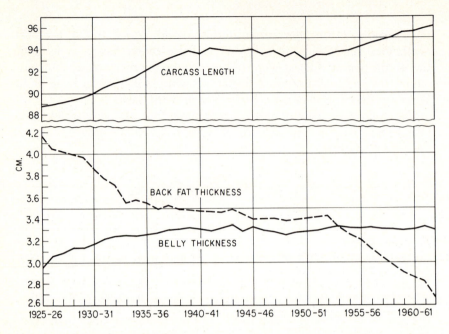

Fig. 13-7. Graph showing trends in carcass length, back-fat thickness, and belly thickness in carcasses of Danish Landrace hogs fed out and slaughtered in testing stations in the Danish swine progeny-testing program. (*Redrawn from graph presented by Clausen et al., 1964.*)

is now considered ideal. Efforts are now concentrated on discarding the too-long or too-short pigs. Back-fat thickness has continued to decrease, with the reduction of 0.16 cm in 1962–1963 being the largest since 1952–1953. Part of the decrease may be nongenetic,[1] but if so the specific environmental factors which may be responsible are not known. Thickness of belly (or "streak") has been about 3.30 cm for several years—a thickness considered ideal.

Regardless of whether progress is becoming slower with some traits or whether some traits may be somewhat antagonistic to one another, real progress has been made in improving the Danish Landrace hog while this program has been in operation. Since the reports serve only as guides to breeders, and since management of the test litters has undoubtedly improved over the years (although known changes in rations or management other than the shift to individual feeding are slight), it is impossible to determine precisely how much of the improved performance can be attributed to genetic improvement resulting from the program.

[1] Smith, C., *Anim. Prod.*, **5**:259–268, 1963.

Several other European countries and Canada have progeny-testing programs which are carried on under plans more or less similar to the one in Denmark.

Swine Performance-testing Programs in the United States. In the United States, production testing on an organized basis was slower in

Fig. 13-8. *Above,* CC Long Trend, the Duroc breed's first Superior Certified Meat Sire. He has certified 26 litters. Fifteen head of progeny on test reached 200 lb at an average age of 157 days with a feed conversion ratio of 1-lb gain for each 2.77 lb of feed eaten. Twelve head slaughtered at an average weight of 203 lb averaged 29.4 inches in length, 1.21 inches in back-fat thickness, and 4.75 square inches in loin eye area. *Below,* a Duroc boar of ideal type by 1965 standards. Compare with animals in Fig. 13-2. (*Courtesy of United Duroc Record Association.*)

Fig. 13-9. HT2 Pecatonica 5–8, an outstanding Yorkshire Certified Meat Sire. Progeny in his certified litters had a 175-day adjusted weight of 232 lb. At an average slaughter weight of 209 lb, they averaged 31.0 inches in length, 1.50 inches back-fat thickness, and 4.71 square inches in loin eye area. (*Courtesy of American Yorkshire Club.*)

developing, but participation in programs initiated since 1938 indicates that American breeders are coming to a realization of the need for such procedures. Development of programs which can be used practically with the large hog populations and under the management systems used in swine production here has required time and experience.

In 1938 the United Duroc Record Association initiated a program of Production Registry (PR) which is designed to give recognition to sows having the ability to farrow and raise large litters and nurse them well to weaning. Recognition is also given to boars. Most of the other breed associations developed somewhat similar programs during the following years under rules which varied in some particulars. In 1946 all the major American breed associations adopted a single set of rules which provided that both gilts and older sows must raise litters of at least eight pigs to 56 days of age and that the total weight of the litter must be at least 275 lb for gilt litters and 320 lb for litters from older sows. Provision was made for additional recognition for sows producing more than one qualifying litter. Boars received recognition both for siring qualifying litters and for siring daughters that produced qualifying litters.

In subsequent years some associations changed their rules, in some cases going to ages younger than 56 days for taking official weights.

Fig. 13-10. *Above*, a Hampshire boar, King Edward, the first Certified Meat Sire of any breed in the United States. *Below*, Greenlite, a descendant of King Edward. As shown in Fig. 13-11, five of the seven boars in the first three generations of his pedigree were Certified Meat Sires and four of the first seven sows were Production Registry sows. Greenlite is himself a Superior Certified Meat Sire. Fifteen of his offspring fed on test reached 200 lb at an average age of 155 days and required only 2.52 lb of feed per lb gain. Thirteen offspring slaughtered had carcass data as follows: length, 30.3 inches; back fat, 1.3 inches; and loin eye area, 5.11 square inches. (*Courtesy Hampshire Swine Registry.*)

At least two associations recognize PR litters on the basis of number farrowed alive and birth weights. This may well be a backward step since birth weights are related to future growth to only a low degree. Further, live pigs farrowed and their weight at birth provide no measure of the ability of a sow to raise and nurse her litter.

Heavy weaning weights of individual pigs are important to the producer since the heavier pig at weaning will have attained a greater percentage of its final weight and can therefore reach market weight in fewer days and with less feed required than if lighter at weaning. There are low positive correlations between weaning weights and post-weaning rates of gain but these are thought to be due principally to the fact that heavier weaning pigs are in a more advanced phase of the growth curve at weaning and grow more rapidly thereafter due to this rather than to a genetic correlation between growth in the two phases. Heritability of individual pig weaning weight is low and is greatly influenced by maternal qualities of dams. Selection for it, if effective, may be more largely selection for maternal qualities of dams than for genetic growing ability in pigs.

Some breed associations now have programs of herd testing in which weights are taken on all litters in a herd. This is a constructive step and gives a better appraisal of a herd than merely maintaining records on litters which are thought to have a chance of qualifying for PR.

In 1954 and 1955 most major breed associations in the United States developed more inclusive performance-testing programs aimed at the efficient production of modern meat-type swine. The rules of the various associations differ in some details but in general provide for the certification of litters which meet the following standards:

1. The litter must qualify for Production Registry.
2. Two pigs from the litter must be slaughtered at not over 220 lb live weight and must have 175-day equivalent weights of 200 lb or more.
3. Carcasses of these two pigs must meet the following standards:

Maximum slaughter weight	220 lb
Minimum loin-eye area	4 square inches
Minimum body length	29 inches
Maximum back-fat thickness	1.6 inches

Some associations do not require that the litter qualify for Production Registry. Sires of at least five certified litters from different sows are given recognition as certified sires, provided the dams are not too closely related.

Some associations have initiated on-the-farm production-testing pro-

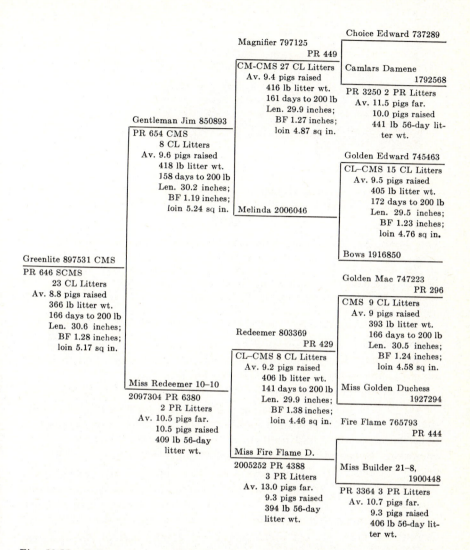

Fig. 13-11. Pedigree of the Hampshire boar, Greenlite, showing an unusual concentration of animals with superior performance information among his ancestors. In the 4th edition of this book, published in 1950, we printed a swine pedigree with the greatest amount of performance information then known to us to be available for a privately owned animal. The information was limited to two generations and included only number and weights of pigs to 56 days of age. The pedigree of Greenlite is indicative of the increased use of records by seedstock breeders since that time. (*Courtesy of Hampshire Swine Registry.*)

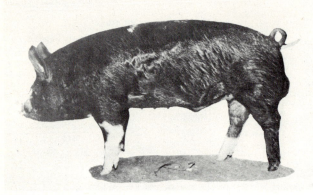

Fig. 13-12. Berkshire Certified Meat Sire, Velvet's Mixer 27th. Certification data on his progeny showed a live weight of 209 lb at 171 days of age; carcass length 30.6 inches; back-fat thickness 1.40 inches; and loin eye area of 5.99 square inches. (*Courtesy of American Berkshire Association.*)

grams in which 15 or more pigs by a sire are fed as a group for slaughter. For recognition of the sire, the pigs must meet prescribed standards for gain, efficiency of gain, and carcass measurements.

Another significant development since 1950 has been the organization and growth of performance-testing programs in a number of states under the auspices of state extension services or local or state breeder organizations. Some of these aid both in on-the-farm testing programs and the operation of central test stations while some limit their activities to testing stations. In most of these, samples of litters under test (usually two pigs per litter) are put on feed under standard conditions at weaning and fed to market weight with gain and efficiency records kept in much the same fashion that the Danish stations operate. In some programs barrows or gilts are fed and slaughtered at the end of the tests for carcass evaluation. In these cases the tested animals serve as tools for estimating genetic worth of their sires, dams, and sibs. In other tests boars are being fed and evaluated for gain, efficiency, and usually for back-fat thickness. Those with acceptable records are used for breeding. In some tests both boars and litter-mate barrows or gilts for slaughter are included.

As mentioned, some of the organized performance-testing programs are including back-fat determinations of the live hogs through the use of the probe technique, thus obtaining an estimate of probable carcass quality in addition to the strictly production factors, even though the animals are not slaughtered.

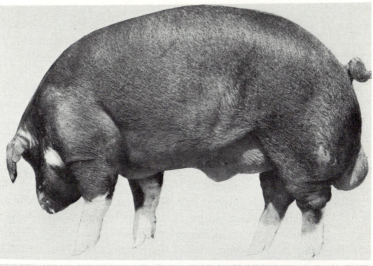

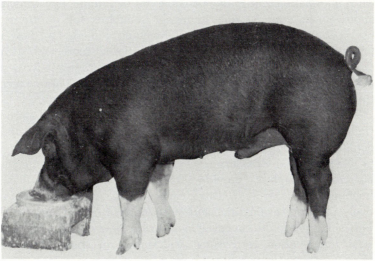

Fig. 13-13. Above, the Poland China boar Supersire PR24 SMS, Grand Champion at the 1963 Illinois and Iowa State Fairs. This "Superior Meat Type Sire" had tested progeny which averaged 200 lb at 143 days of age and gained a pound for each 2.71 lb of feed consumed. When slaughtered at an average weight of 210 lb they produced carcasses averaging 29.6 inches in length, 1.37 inches in back-fat thickness, and 4.90 square inches in loin eye area. *Below,* Grand Champion barrow at the 1963 National Barrow Show. His carcass had only 1.13 inches back-fat thickness, had 5.84 square inches of loin eye area, and produced 16.2 per cent of live weight in trimmed ham. (*Courtesy of Poland China Record Association.*)

Production Registry programs have never included a high proportion of registered American swine, and no critical studies have been made of their effectiveness. The general feeling is that, due to the low heritabilities of the traits measured, effectiveness has probably been low in improving litter size and weaning weight. Continued maintenance of positive selection pressure for these traits may result in improvement over long periods.

Most testing stations have shown only minor changes in average daily gain and efficiency of gain over the 6 to 10 years since they became operational, but most have shown important reductions in back-fat thickness and (in tests where carcass evaluation is part of the program) in yield of primal cuts. From current interest in performance records among breeders, there is reason to believe that the Certified Meat Type programs and testing stations are having an important impact on swine-breeding practice. Observation, backed up by marketing data, indicates that important advances in improving meatiness of hogs (with reduced lard production) have occurred in the period since 1950. It is impossible with information currently available to determine how much of this improvement may be due to higher standards for meatiness applied by breeders to visual selection of breeding animals and how much to formal testing procedures and the application of more sophisticated aids to selection such as the back-fat probe. Doubtless all have played a part. Continued application of these techniques with improvements when they are technically proven should lead to further progress.

With seedstock swine populations as large as are required in the United States to provide boars for commercial producers, it appears doubtful that publicly supported or cooperative central testing stations to which several breeders send animals will ever be large enough or numerous enough to provide facilities for testing more than a small fraction of animals in seedstock herds. Thus, by far the greater proportion of testing, evaluation, and selection will be in herds of breeders. This situation could change if new technical developments caused artificial insemination to become widespread in commercial herds and greatly reduced the number of boars needed.

Unless artificial insemination becomes much more widely used, it appears that central test stations will have their greatest usefulness

1. As a method of comparing the productivity of animals from the herds of several breeders, thus giving the individual breeder an unbiased appraisal of how his animals compare with those of other breeders
2. As an educational tool for demonstrating potential rates of improvement and setting goals for the industry
3. To some degree serving as a testing place for animals from elite or nucleus herds

Fig. 13-14. Junior Champion Chester White boar at the 1964 Iowa State Fair. From a Meat Certified Litter whose littermates weighed 200 lb at slightly over five months, cut well over 4.0 square inches of loin eye, and averaged 29.0 inches in carcass length and 1.4 inches in back-fat thickness. (*Courtesy of Chester White Swine Record Association.*)

The stations will have relatively little value for the latter purpose unless it becomes possible for individual breeders to enter fairly large numbers in the tests.

Programs of on-the-farm herd testing will thus remain the basis for most improvement. Participation in an organized effort is very helpful in expediting the necessary record keeping in a testing program and is particularly valuable to the smaller breeder. Also, organized programs sponsored by breed associations or the Agricultural Extension Service have an official standing which increases their value somewhat over and above the value of private records.

Whether official or private, testing programs will do the breeder and his breed the most good if they are conscientiously used for selecting the most productive stock with highest potential carcass quality. To accomplish this it is necessary (1) that entire herds, not just selected individuals, be tested and (2) that real use be made of the records in determining which animals are to be retained for breeding purposes. The use of selected records on a few individuals purely as promotional tools cannot be expected to be an effective agent for breed or industry improvement.

The keeping of personal records is to be encouraged if organized programs suitable for individual situations are not available. These need not be elaborate but should include pedigree records on each litter, records of the number of pigs farrowed and weaned, and weights of the individual pigs at weaning and at some later date, preferably about five months of age. Factors have been developed for adjusting individual pig weights to the usual weaning weight of 56 days and to 154 days if actual weights are taken a few days before or after the actual ages. These make it possible to weigh several litters at one time and convert all records to a standard age for use in comparing the records of various individuals.

Factors for adjustment to the two ages are given in Table 13-2 for ages nine days each way from the standard ages. The formulas can be used to calculate factors for ages further from the standards, although

Table 13-2. Factors for Converting Pig Weights to Standard Ages of 56 and 154 Days

Factors for converting individual pig weights to 56 days*		Factors for converting individual pig weights to 154 days†		
Age, days	*Multiplicative factor*	*Age, days*	*Multiplicative factor*	*Additive factor*
47	1.281	145	1.047	7.3
48	1.242	146	1.042	6.5
49	1.206	147	1.036	5.6
50	1.171	148	1.031	4.8
51	1.139	149	1.026	4.0
52	1.108	150	1.020	3.2
53	1.079	151	1.015	2.4
54	1.051	152	1.010	1.6
55	1.025	153	1.005	0.8
56	1.000	154	1.000	0
57	0.976	155	0.995	0.8
58	0.954	156	0.990	1.5
59	0.932	157	0.985	2.3
60	0.911	158	0.980	3.0
61	0.891	159	0.975	3.8
62	0.872	160	0.971	4.5
63	0.854	161	0.966	5.3
64	0.837	162	0.961	6.0
65	0.820	163	0.957	6.7

* Calculated from formula:

$$\text{Adjusted weight} = \text{actual weight} \times \frac{41}{\text{actual age} - 15}$$

† Calculated from formula:

$$\text{Adjusted weight} = \frac{(\text{actual weight} + 154)(199)}{\text{actual age} + 45} - 154$$

it must be recognized that the accuracy of adjustment will decrease. To use the 56-day adjustment factors, it is necessary only to multiply the actual weight by the proper factor. To use the 154-day adjustment factors, it is necessary to multiply the actual weight by the multiplicative factor, then add the additive factor. For example, a weight of 140 lb taken at 146 days of age would be multiplied by 1.042 and 6.5 lb added to the product to give 152.3 lb as the adjusted 154-day weight.

In addition to the foregoing records, a back-fat probe or ultrasonic estimate of back-fat thickness at a standard weight is highly desirable as are slaughter tests of sample animals. Probes are usually made at three locations and averaged. As is the case with weighing, it is usually not practical to probe each pig when it reaches a standard market weight. Studies with different breeds of swine have shown that back-fat thickness increases with weight by an average of about 0.006 inch per pound increase in weight for at least 30 to 40 lb either above or below the usual standard weight of 200 lb. If, for example, a pig is probed at 180 lb and has 1.30 inches of back fat, his probable back-fat thickness at 200 lb would equal $1.30 + (20 \times 0.006)$, or 1.42 inches. The increase per lb of weight increase varies with breed and nutritional level. In different studies it has usually ranged from about .004 to 0.0075. Thus, for most accurate adjustment a breeder should use a factor applicable to his particular breed. However, if he probes within a range of not more than 30 lb either way from the standard weight, only small errors will be introduced by using a standard or average adjustment factor. Adjustment tables based on average figures such as the one in Fig. 13-15 have been developed to make the adjustment simple.

Swine Selection Indexes. Assuming production records are being kept, the breeder is faced with the problem of how to evaluate them in selecting breeding stock for retention in the herd or for making selections from other herds for use in his. How much emphasis should be put on weight for age, how much on litter size, how much on litter weaning weight, how much on conformation, how much on back-fat thickness, etc.?

If economic values of different traits were uniform with the passage of time and in different regions, and if phenotypic and genetic parameters were constant in all herds, breeds, and management systems, use of a uniform index would be feasible. However, these things are all unlikely. Further, at least to date in the United States, test procedures and exact items of performance measured have not been unified enough to permit use of a single index.

Several indexes have been proposed or are in use under specified

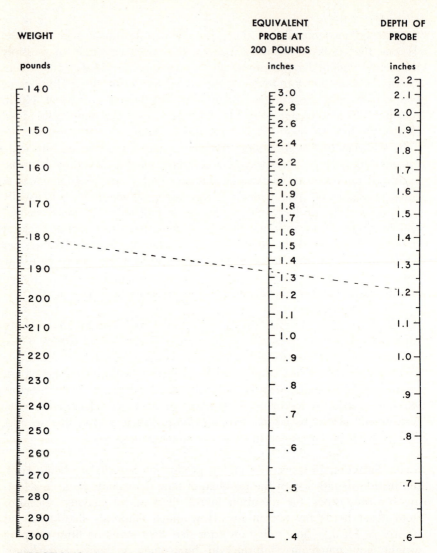

DIRECTIONS: Lay a ruler, or other straightedge, from the figure representing the weight of the pig, as read on the left-hand scale, to the figure for
the depth of the probe, as read on the right-hand scale. The point where
this line intersects the center scale shows the equivalent probe at 200
pounds. EXAMPLE: In the above example the depth of the probe was 1.2
inches for a pig weighing 180 pounds. This is equivalent to a probe of
nearly 1.35 inches for a pig weighing 200 pounds.

Fig. 13-15. Normograph for adjusting back-fat probes to 200 lb liveweight
basis. (*Developed by H. W. Bean and published by T. R. Greathouse in Illinois
Agric. Ext. Cir. No. 868, 1964.*)

conditions. We will present a few of these to illustrate the possibilities rather than as recommendations for all situations.

The following index, covering performance of a pen of four pigs by one sire during the test period and back-fat thickness of the individual at its end, has been used in the Iowa Swine Testing Station for evaluating boars for which no slaughter information on littermate barrows is available:

$$\text{Index} = 240 + (50 \times \text{gain}) - (50 \times \textit{EFF}) - (50 \times \textit{BF})$$

where gain = individual average daily gain on test, lb
    EFF = pen average efficiency of gain, lb of feed per lb of gain
    BF = individual back-fat thickness at the loin as determined
         by probe expressed in inches

For boars with a littermate barrow which has been slaughtered, the index is modified as follows:

$$\text{Index} = 117 + (50 \times \text{gain}) - (50 \times \textit{EFF}) - (40 \times \textit{BF}) + (3 \times \textit{H\&L\%})$$

where *H&L%* is the per cent of ham and loin in the littermate barrow.

In the indexes the figures 240 and 117 are merely constants chosen for convenience to make the indexes positive and average 100 or above. Both indexes were developed to put approximately equal emphasis on rate of gain, efficiency of gain, and meatiness.

In practice the index has been supplemented by independent culling levels by requiring that individuals gain at least 1.70 lb per day on test, have no more than 1.35 inches of back fat as determined by probe, and come from a pen of four all by the same sire having a feed requirement no higher than 3.15 lb of feed per 1 lb of gain in a fall and winter test, or 3.05 in a summer test. This has been done since it was felt that, for example, animals occasionally got acceptable indexes because of outstanding performance in other traits, even though they had too much back fat to be acceptable. To a degree this negates the index approach, but in practice relatively small percentages have been eliminated for failure to measure up in one respect. For example, in the fall of 1963 no boars were eliminated for failing to meet the back-fat requirement, 4.7 per cent were eliminated for failing to meet the efficiency standard, and 12.2 per cent for failing to meet the rate-of-gain standard. The reports have not stated how many of those eliminated for failure to meet standards would have had low indexes, but doubtless a good many of them would have had.

The following index was developed for use in the Wisconsin on-the-farm testing program in which animals are born and raised on the same farm:[1]

$$\text{Index} = 164 + 5X_1 - 4X_4 - 43.3X_6 + 1.2X_7 - 2X_8$$

where X_1 = individual weight at 154 days
X_4 = length of foreleg
X_6 = back-fat depth at loin
X_7 = number in litter at farrowing
X_8 = number in litter at 154 days

In the data upon which this index was based, shorter forelegs were associated with higher carcass yield if other factors were held constant, and it was included in the index since the measurement was one which could be taken.

As a general rule it would appear that major emphasis in swine selection programs should be given to rate of gain (or alternatively, weight at a given age or age at a given weight), efficiency of gain when available, and factors, such as back-fat thickness, which are indicative of carcass leanness. Slaughter data when available in sufficient volume to be a reliable test on a half- or full-sib group should have strong emphasis. Lesser emphasis should be put on litter size and weaning weight—not because they are unimportant, but because their heritability is so low as to make progress slow and uncertain.

Progeny and Sib Testing. Some of the testing programs discussed in the foregoing material have been for testing performance of the individual with the view of selecting those with superior performance for breeding use. Information from these tests is also useful for estimating breeding values of sires, dams, and full or half sibs. Information from tests involving slaughter is useful only for the latter purpose.

Much remains to be learned about optimum use of the different types of tests and of information from them for maximum improvement of swine populations. A weakness of many testing programs to date has been the tendency of some breeders to put undue reliance on carcass information from the small number of animals slaughtered. As is apparent from the material on family selection in the chapter on Principles of Selection, slaughter information on two pigs from a litter is of only limited value in providing estimates of the genotypic values of the dam or of full or half sibs. Carcass characters related to yield of lean cuts are medium to high in heritability. Back-fat probe is a good indicator of them. Therefore, a probe measurement on an individual is likely

---

[1] Robison, O. W., et al., *Jour. Anim. Sci.*, **19**:1013–1023, 1960.

to be a better indicator of its genotype for carcass yield of lean cuts than would the actual carcass cutout of several full or half sibs.

In seedstock herds under natural mating systems, plans for sire selection in which all prospective herd sires are progeny tested on females outside the regular population before being used in the seedstock herd itself appear not to be effective in attaining maximum rate of genetic improvement because of the lengthening of generation interval in the seedstock herd.[1] This is true for all production characters which can be measured in the individual himself. It is also true for carcass traits in which reasonably good estimates of potential carcass quality can be made from the live animal on the basis of conformation, back-fat probe, and/or ultrasonic estimates of back-fat thickness. Average progeny performance including slaughter characteristics of a sample should, however, be a major factor in determining which boars to retain for use in subsequent breeding seasons.

For sows, it appears that maximum progress can be made if intense culling is practiced on the basis of the first litter performance with only approximately one-third to one-half kept for second litters.[2] Only a very few of outstanding productivity should be kept for three, four, or more litters. Selection among sows having produced one or more litters is really partially on individual and partially on progeny performance. The number of pigs farrowed and raised and their growth to weaning are largely maternal characters and culling sows on the basis of these factors is really individual selection. Although some maternal influences remain to normal slaughter weights, postweaning and carcass characters depend to an increasing extent upon the genotypes of the individual pigs. Thus, selection of sows on the basis of these traits of her pigs is largely for transmitted genetic influences and is progeny testing as it is usually considered.

Tests of samples of half sibs by different sires and full sibs from the same sires and dams are useful in estimating potential breeding values of others in the sibling groups whether alive at the time of the test or to be born in the future. Theoretical studies[3] indicate that for maximum genetic improvement in a population it is desirable (1) that as large a number as possible be tested relative to the size of the population being improved and (2) that if testing facilities are limited, tests be restricted to a *nucleus* group of breeders or portion of the population. The first item is important for both individual and sib testing. The probability of finding individuals which are truly superior increases with the number tested. Similarly, locating families of half or full sibs with

[1] Dickerson, G. E., and Hazel, L. N., *Jour. Agr. Res.*, 69:459–476, 1944.
[2] Dickerson, G. E., and Hazel, L. N., *Jour. Anim. Sci.*, 3:201–212, 1944.
[3] Smith, C., *Anim. Prod.*, 1:113–121, 1959.

superior merit increases with the number of families from which animals have been tested. The desirability of restricting testing to nucleus groups, if the total which can be tested is limited, depends upon the hierarchical structure of most seedstock and commercial breeding systems. A small or nucleus group of seedstock breeders provides most of the sires used by a lower layer of *multiplier* seedstock herds which in turn provide sires for commercial herds. Another way of looking at the situation is that, if testing facilities are limited, tests should be restricted to those herds or population segments in which previous breeding practices are believed to have resulted in improvement to levels well above population averages. Tests among these already improved groups will have higher probabilities of discovering superior individuals and/or families than would tests of random samples of the entire population.

From a practical standpoint, limitations on testing facilities are important only for certain traits. Every individual seedstock breeder can and should obtain information on every animal in his herd for litter size and growth rate. He can also make systematic ratings of conformation. Any person worthy of being a seedstock breeder can learn to make back-fat probes or secure the services of someone who can. Thus, measurements of back-fat thickness can be made on every animal in seedstock herds. Having records on every animal in the herds for the foregoing important traits would provide maximum information by which different progeny and sib groups could be compared as well as permit individual comparisons. This should make for maximum selection intensities if the records are intelligently used.

Limitations on test facilities are likely to be important only for evaluating efficiency of gain and carcass characteristics. Information is not currently complete enough that specific recommendations can be made to seedstock breeders on (1) what proportions of their pigs should be fed out either individually or by litters or samples of litters to obtain data on efficiency of feed utilization or (2) what proportions of their pigs should be fed out for slaughter in order to obtain direct carcass information. The questions are partly economic and partly genetic. How much can a breeder afford to spend to obtain the additional information on efficiency not supplied indirectly by rate-of-gain information? Similarly, how much can he afford to spend to obtain more specific information on carcass characteristics than can be inferred from back-fat probes? Unfortunately, specific information is not available to answer the questions at the present time, but indications are that expenditures for individual feeding and slaughter data can be justified for only relatively small proportions of the total populations. Obtaining detailed observations for these characters on relatively small samples of a herd under central test-station conditions should enable a breeder to better evaluate his herd in relation to those of others. Further, detailed efficiency and slaughter tests on small samples would enable breeders to quickly spot problems which might arise as a result of imperfect predictions of efficiency or carcass quality if entire reliance were placed on the indicators, i.e., rate of gain and back-fat probe. Looked at in this way, test stations would not be a major factor in selecting breeding stock, but rather serve as a check on adequacy of selection practices being followed.

The foregoing discussion has assumed central test stations, but most of the same arguments would apply to individual breeders developing their own test facilities.

Effects of Inbreeding in Swine. As with other species, on the average, inbreeding swine is detrimental. From extensive Regional Swine Breeding Laboratory data it has been established that on the average each 10 per cent increase in inbreeding is accompanied by reductions of 0.3 to 0.6 pig per litter at birth, 0.5 to 0.7 pig per litter at weaning, and of 3 to 6 lb per pig at 154 days of age.[1] The reduction in litter size is much more important economically than reduced growth rate.

Efforts to find the physiological bases for reduced reproductive efficiency in inbred swine have shown that in some lines inbreeding delays testicular development, delays puberty in both sexes, reduces the number of ova shed by females, and increases early embryonic death rates. Observations indicate that the percentage of boars which refuse to breed and those that are slow breeders increases with inbreeding. In spite of the rather drastic average effects of inbreeding on reproductive performance in swine, some lines have performed reasonably well for several generations of inbreeding.

Carcass characters have been affected to only minor degrees by inbreeding, with different studies giving somewhat different trends.[2]

The experimental results are clear in indicating that the average effects of inbreeding in swine are detrimental and that the detrimental effects cannot *on the average* be overcome by selection in herds in which inbreeding is increasing very much—probably as little as 3 to 5 per cent per generation. Thus, inbreeding in even mild forms should be avoided in commercial herds.

However, the effects of inbreeding are not so drastic but that some lines can be maintained to very high levels of inbreeding and a majority of lines can be maintained to inbreeding levels of perhaps at least 30 to 50 per cent. Formation and use of inbred lines is thus a possible

[1] Craft, W. A., *Jour. Anim. Sci.*, **17**:960–980, 1958; and Dickerson, G. E., et al., *Mo. Agr. Expt. Sta. Res. Bul.* 551, 1954.
[2] King, J. W. B., and Roberts, R. C., *Anim. Prod.*, **1**:123–127, 1959.

breeding system for developing stocks to be used commercially in crosses and in combinations for future breed improvement as proposed by Wright many years ago. There has been little investigation of the latter use. Use of lines in crosses will be discussed in a subsequent section.

MAXIMIZING PRODUCTIVITY OF COMMERCIAL SWINE

As with other classes of livestock, the genetic potential of a commercial swine herd depends largely on the merit of the sires used during the immediately preceding generations. However, a commercial producer can and should exert as much pressure as possible in the selection of replacement females by (1) selecting replacements only from litters above average in number and weight of pigs weaned, and (2) selecting as replacements those gilts with superior growth rates and meat-type conformation. Simple records including litter identification are necessary as selection aids.

Sires should be obtained from seedstock herds on the basis of the best available evidence on inherent productivity and meat type. Physical characteristics, probable transmitting ability, and potential combining ability with the sow herd should all be considered in evaluating individual boars for use in a herd or for deciding which breed, line, or strain from which boars should be obtained.

All evidence indicates the desirability of maintaining maximum heterosis in commercial swine herds. Doing this is most beneficial for the traits of low heritability—litter size, litter weaning weights, and vigor as measured by livability—but heterosis is also expressed in growth rate and efficiency of feed utilization to a lesser degree.

Crossbreeding Hogs. Crossbreeding for commercial production has been practiced more extensively with hogs than with any other class of farm animal. It is estimated that 80 to 90 per cent, or even more, of the market hogs in the principal corn-belt states are crossbred. Much experimental crossbreeding work has been done with hogs, and most of it shows economically important advantages for crossbreds. Most workers in the field have recommended the practice.

Crossbreeding is an old practice[1] with swine, and experiment stations as long ago as 1902 published results generally showing advantages for crossbreds. The early experiments were based on small numbers,

[1] Craft, W. A., *Jour. Anim. Sci.*, 17:960–980, 1958.

Table 13-3. Advantages of Crossbred Swine over Purebreds in Minnesota Work*

	First cross	Three-breed cross	Back cross
Farrowing results:			
No. of live pigs	0.93	1.66	−0.19
Birth wt. per live pig, lb	0.05	0.01	0.37
Wt. of live pigs per litter	2.83	4.38	2.53
Nursing period:			
No. lost per litter	−0.60	0.39	0.87
Litter size at weaning	0.33	2.05	0.68
Wt. per pig at weaning, lb	5.00	5.00	7.00
Total litter wt., lb	39.00	96.00	63.00
Feed-lot period:			
Daily gain, lb	0.12	0.11	0.14
Less feed per 100-lb gain	12.68	16.21	12.15
Birth to 220-lb wt.:			
Fewer days to reach 220 lb	17	17	22
Fewer lb feed per 220-lb pig	27.90	35.66	26.73

* From *Minn. Agr. Expt. Sta. Spec. Bul.* 180, 1936.

and experimental designs were often not adequate to permit clear-cut and dependable results. Proponents of pure breeding were inclined to ignore or down-grade the early results. In the 1920s and 1930s experiments were conducted and reported which demonstrated conclusively the value of crossbreeding systems.[1] The Minnesota results are summarized in Table 13-3. The results in the column headed "First cross" are typical of results secured in many experiments on crossing straightbred animals of two breeds, except that the death rates of the crossbreds in most other experiments have been lower than for straightbreds. In most such experiments, litter size (presumably a result of lower prenatal death rates) at birth, litter size at weaning, growth rate both before and after weaning, and efficiency of feed utilization have favored crossbreds over the averages of the parental breeds. Advantages tend to be relatively small for any one item, but they are multiplicative, i.e., more pigs raised per litter that are also heavier and have a greater effect on pounds of pigs raised per litter than for either component separately.

Although crossing two breeds results in increased production, systems

[1] Winters, L. M., et al., *Minn. Agr. Expt. Sta. Bul.* 320, 1935; and Lush, J. L., et al., *Iowa Agr. Expt. Sta. Bul.* 380, 1939.

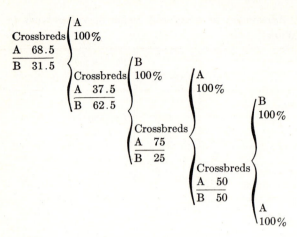

Fig. 13-16. Diagram of crisscrossing.

which involve the use of single- and multiple-cross sows are much more practical and widely used. The practice of using crossbred sows was formerly frowned upon, but experimental work since 1920 has shown conclusively that crossbred sows exhibit heterosis in their ability to raise pigs. Using them as mothers thus permits the producer to take advantage of heterosis in this important character as well as in rate of gain, survival, and efficiency of gain. Furthermore, no one has demonstrated an undesirable decrease in uniformity of type in progeny of crossbred sows.

To take advantage of the heterosis both in the growing pig and in the sow without having to start over periodically with purebred females, workers at the Minnesota Experiment Station suggested systems known as *crisscrossing* and *triple,* or *rotational, crossbreeding.* In crisscrossing, sows of breed A are mated to a boar of breed B. Crossbred gilts from these matings are then selected and bred back to a boar of breed A, selected females from these matings are bred to boars of breed B, etc. In getting started on a crisscrossing program, a backcross, that is, mating crossbred gilts of boars of the breed of one of their parents, is necessary.

It is seen from Fig. 13-16 that the crossbreds in a crisscrossing system soon come to have about ⅔ of their inheritance from the breed of their immediate sire and ⅓ from the other breed being used in the system.

Another system of continuous crossbreeding involves the use of three pure breeds and is known as triple or rotational crossing. In this system boars of breed B are bred to sows of breed A, their selected crossbred daughters to boars of breed C, the female offspring of these matings to boars of breed A, with boars of breeds B and C being used in turn. As seen in Fig. 13-17, the crossbreds soon come to have about $\frac{4}{7}$ of the inheritance of the breed of their immediate sire, $\frac{2}{7}$ from the breed of their maternal grandsire, and $\frac{1}{7}$ of the hereditary material of the other breed being used.

Theoretically, after the first few generations, the pigs from a criss-crossing program will exhibit about $\frac{2}{3}$ as much heterosis as first crosses between the two breeds. This follows from the fact that crosses will be between purebred sires and sows carrying about $\frac{2}{3}$ of their genes from the other breed. Similarly, in rotational crossing, the crosses will be between boars of one breed and sows carrying about $\frac{6}{7}$ of the inheritance of other breeds, giving about $\frac{6}{7}$, or about 86 per cent, as much heterosis as first crosses.

In both these systems, heterosis in both the offspring itself and the dam is utilized, thus leading to the expectation that total production will be as great or greater than in first crosses. Experimental results from these systems have been good.

Data in Table 13-4 from the Ohio Agricultural Experiment Station are illustrative of results from a rotation system of crossbreeding. In the experiments from which these data were taken, work was done on three farms with the original sow herd and all the purebreds on each being of one breed. Each of the three farms had a different breed. Boars of the three breeds represented by the sows were used in a rotational breeding system on each farm, and purebreds were raised each season for comparison with the various types of crossbreds. Thus, in spite of the fact that all purebreds and all crossbreds were not raised

Fig. 13-17. Triple, or rotational, crossing.

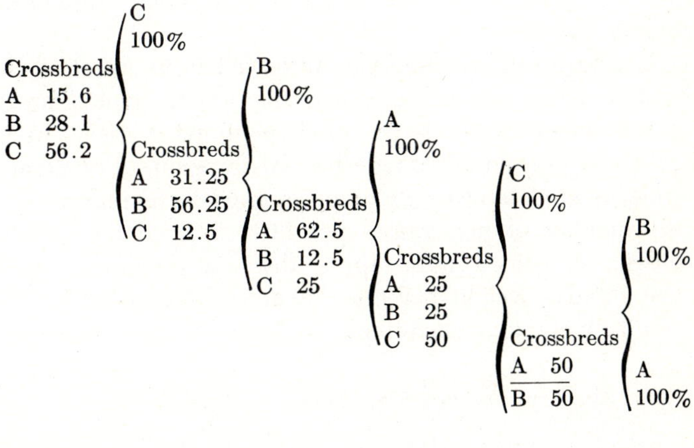

Table 13-4. Average Litter Performance of Purebreds, Crossbreds, Three-breed Crosses and Advanced Generations of a Rotation Crossing Program*

Type of litters	No. litters		Av. no. pigs raised		Av. wt. per litter at 180 days	
	Gilt	Sow	Gilt	Sow	Gilt	Sow
Purebred	56	46	5.7	6.7	931	1,319
Crossbred	16	23	5.0	7.2	936	1,463
Three-breed cross	8	6	7.6	7.4	1,344	1,580
Advanced generations†	33	14	7.1	8.0	1,272	1,722

* Adapted from *Ohio Agr. Expt. Sta. Bul.* 675, 1948.
† The third, fourth, and later generations of pigs produced by a rotation crossing program.

on the same farm, approximately equal proportions of purebreds and crossbreds were raised in each season on each farm, and the data are useful for comparative purposes.

As a general conclusion of the Ohio work it was stated:

Although the data are not regarded as providing the final answer, they suggest that unless or until a better plan of breeding for the production of market hogs is developed, rotating purebred sires of three or more breeds on successive generations of sows selected from the herd is worthy of consideration.[1]

The usefulness of both two-breed and multiple-breed crosses involving crossbred sows are illustrated in Table 13-5 from results of an extensive study conducted in the 1950s at the Iowa Agricultural Experiment Station.

Certain cautions should be observed in interpreting the material presented on crossbreeding. First, the results quoted are average figures based on relatively large numbers in most cases. In individual cases there may be no advantage for crossbreeding. For example, in the Ohio work quoted, above, 19 crops of pigs were produced. In 15 of these the number of pigs raised per litter favored the crossbreds, in 16 the crossbreds gained more rapidly than the purebreds, and in 18 the average 180-day weight per litter favored the crossbreds. Thus, the average results were overwhelmingly in favor of the crossbreds. This is what

[1] *Ohio Agr. Expt. Sta. Bul.* 675, 1948.

a commercial breeder would expect to obtain, but there would be some rare exceptions.

A second consideration is that results from crossbreeding depend as much on quality of sires used as they do in any other breeding system. With a given quality of breeding stock, heterosis from crossbreeding will be expected to give higher productivity than straightbreeding. However, the consistent use of sires bred for performance could easily result in a grade or purebred herd with higher average productivity and carcass value than in a herd where mediocre sires had been used in a criss-crossing or rotational breeding scheme.

To obtain maximum benefits, producers who intend to use a cross-breeding system must follow a definite plan which fits in with their hog-management system. There are no practical difficulties in producing replacement sows and at the same time following a systematic crossing program with hogs. Quite commonly commercial hog producers keep sows for only one, two, or perhaps three pig crops, then market them all and replace the entire herd with gilts. The high reproductive rate of hogs permits selecting the necessary number of replacement gilts from one pig crop. Boars of the proper breed can be purchased to use on each group of gilts saved.

The foregoing represents our interpretation of experimental results on crossbreeding swine. As a matter of historical interest we believe it desirable to bring an alternative viewpoint, once seriously espoused, to the attention of students.

Using only data from crossbreeding experiments where both parental breeds and their crosses were involved, workers at the Illinois Experiment Station summarized the results of numerous experiments which in total included more than 50,000 animals. They took the following

Table 13-5. *Comparative Performance of Purebred and Crossbred Swine for Litter Size and Growth Rate*[*,†]

Characteristic	*Purebred*	*First cross*	*3-breed cross*	*4-breed cross*
Litter size at birth	100	101	111	113
Litter size at 8 weeks	100	107	125	126
Weight at 8 weeks (per pig)	100	108	110	109
Weight at 154 days (per pig)	100	114	113	111
Pork production per litter	100	122	141	140

* All data expressed as percentages of the purebred performance.
† From Hazel, L. N., *Proc. World Conf. on Anim. Prod.*, **2**:189–198, European Association for Animal Production, Rome, 1963.

Table 13-6. Summary of Results from Many Swine Crossbreeding Experiments *

Character	Better purebred†		Poorer purebred†		Crossbred†		Av. of purebred parents	Per cent by which crossbreds exceed av. of purebreds
Size of litter	10.1	(1,538)	9.4	(1,081)	9.5	(1,515)	9.75	−2.5
Av. birth wt., lb	2.9	(1,728)	2.65	(4,176)	2.79	(6,137)	2.775	+0.6
Per cent survival	80	(8,288)	72	(15,874)	80	(9,935)	76	+5.3
Av. weaning wt., lb	33.4	(15,522)	31.7	(8,133)	33.3	(9,519)	32.55	+2.3
Av. daily gain, lb	1.43	(489)	1.32	(574)	1.43	(794)	1.37	+4.3
Feed per cwt. gain, lb	346	(366)	382	(274)	368	(591)	374	+1.6

* Adapted from *Ill. Agr. Expt. Sta. Bul.* 489, 1942.
† Figures in parentheses show number of pigs or litters involved in each case.

as their point of reference: "For crossbreeding to be judged beneficial, the performance of the crossbreds must exceed the performance of the better of the two parental strains of purebreds."[1] They considered the various items of productive importance shown in Table 13-6 one by one. In only one item, average daily gain, did the crossbreds exceed the better purebred, and this by only 0.006 lb per day.

Although agreeing with the basic premise that crossbreds must exceed the better of the parental breeds to be of value in commercial production, we believe the Illinois summary is logically faulty for the following reasons:

1. When two breeds, for example, Poland Chinas and Durocs, were being compared with their crosses, the crossbreds of each year were compared with the breed that was best in *that year*. If the two parental breeds were actually equal in productivity, chance variations would make one better in some years and the other in other years. The crossbreds, to show superiority in this type of analysis, thus have to be not only *better* than the purebreds but *enough better to offset this bias in the selection of the best purebred.*

2. In the Illinois summary, crossbreds were compared with the best-performing purebred separately for each of the six traits studied. The crossbreds were thus compared with one parent for some traits and with the other parent for other traits. In effect, they were compared with a "composite purebred" which did not exist.

3. The individual items considered are cumulative in their effects. Even if the crossbreds were not superior to the best purebred in any one item, still in a measure of total productivity they might be distinctly better. This is illustrated by a hypothetical example in Table 13-7.

[1] *Ill. Agr. Expt. Sta. Bul.* 489, 1942.

In the hypothetical example of Table 13-7 the crossbreds are assumed not to exceed the best purebred in any of the single items. In litter weight at 180 days of age, however, the cumulative effect of the crossbreds being above the average of the purebreds for the individual items would result in their having an advantage of 8.9 per cent.

In view of the above considerations, we believe it is fairer to compare the crossbreds with the average of the purebreds. The latter has been done in the last column of Table 13-6. This summary indicates small average advantages for the crossbreds in all items except number of pigs at birth. The advantage of 5.3 per cent in survival more than offsets this, so that a larger average number of crossbred pigs were raised. This, together with the advantages of the crossbreds in weaning weight, in average daily gains after weaning, and in efficiency of gain, means that there would have been an appreciable advantage for crossbreds in litter weight and cost of production.

Taking advantage of breed differences in crosses. Thus far we have discussed crossbreeding largely as it applies to crossing hogs of the same general kinds for the purpose of utilizing heterosis. Hog breeds do differ in qualities of economic importance. These differences are often particularly marked between European bacon breeds and American swine breeds. The former often have advantages in litter size, the latter in growth rate.

Crosses of American-breed boars on sows of bacon breeds are often made for the purpose of taking advantage of the best qualities of each type. The results shown in Table 13-8 of work conducted a number of years ago at the Miles City, Montana, station of the U.S. Department of Agriculture illustrate the possibilities. The Yorkshire dams, on the average, raised more pigs per litter than the Chester Whites, and the pure Chester Whites had faster growth rates. Crosses of Chester White boars on Yorkshire females gave litters as large as the purebred York-

Table 13-7. Hypothetical Crossbreeding Results with Swine

	Purebred A	Purebred B	Crossbreds
Av. no. pigs farrowed	10.0	6.0	8.5
Per cent surviving	70	90	80
Av. no. pigs alive, 180 days of age	7.0	5.4	6.8
Av. wt. per pig, 56 days of age, lb	30	40	35
Av. daily gain, 56 to 180 days, lb	1.2	1.6	1.5
Av. wt. per pig, 180 days, lb	178.8	238.4	221.0
Av. wt. per litter, 180 days, lb	1251.6	1287.4	1402.8

Table 13-8. A Comparison of Two Types of Purebred Swine and Their Crossbreeds°

Matings	No. of litters	No. of pigs far-rowed	Birth wt.	No. weaned	Wt. 70 days	Daily gain	Feed, 100-lb gain
Purebred Yorkshires	38	10.6	2.40	7.6	38.4	1.21	375
Purebred Chester Whites	36	9.8	2.38	6.6	39.4	1.30	403
Yorkshire male × Chester White female	29	9.9	2.49	7.4	42.7	1.35	371
Chester White male × Yorkshire female	29	10.1	2.42	8.0	42.8	1.33	370

* From *Iowa Agr. Expt. Sta. Bul.* 380, 1939.

shires (actually slightly larger) and growth rates slightly higher than those of purebred Chester Whites.

On the basis of an extensive study in which it appeared that there was probably a strong negative genetic relationship between maternal qualities and postweaning rate and economy of gain, the following suggestion was made:

. . . that maximum performance can be secured only through judicious crossing of different strains of swine. For example, sows of a cross that has exceptionally good milking ability and prolificacy mated to boars from a strain that excels in rate and economy of post-weaning gains would give maximum litter performance.[1]

Although the relationships are probably not generally as strong in most breeds and strains of swine as suggested by this study, it does seem that commercial swine producers would utilize existing breed differences to the fullest extent possible for producing the most profitable crossbreds.

Unfortunately, the breeds most universally used in the United States have not been fully enough characterized for us to be able to state with a high degree of precision their best use in crosses.

As a general principle, breeds most diverse in origin will give the greatest amount of heterosis in crosses. In the United States the breed least related to the traditional American breeds is the American Landrace. As discussed in an earlier section, the Danish Landrace has

[1] Dickerson, G. E., *Iowa Agr. Expt. Sta. Bul.* 354, 1947.

been subjected to a systematic selection program since shortly after 1900. In 1934 the U.S. Department of Agriculture and the Iowa Agricultural Experiment Station imported a small number of Danish Landrace hogs under an agreement with the Danish government that they would not be bred pure for commercial use in the United States. In 1949 this agreement was modified. Since then private breeders in the United

Fig. 13-18. Highly productive representatives of the American Landrace breed. *Above,* Certified Meat Sire GV-R17-5 Majestic Hi-Lite. Animals in the 8 Meat Certified litters he sired had carcass data as follows: length, 30.6 inches; back-fat thickness, 1.35 inches; loin eye area, 4.48 square inches; and per cent ham and loin, 39.2. *Below,* LM Ann who has farrowed 264 and raised 193 pigs in seventeen litters. (*Courtesy of American Landrace Association.*)

States have combined descendants of the 1934 importation, some high-grade Landrace animals tracing mainly to the 1934 importation but carrying small fractions of inheritance of some American breeds, and Landrace animals of types similar to the Danish Landrace imported from Norway and Sweden, to develop the *American Landrace* breed. Animals from the 1934 importation and new strains based in part on it proved to cross exceptionally well with American breeds both from the standpoint of contributing meat type and for the heterosis apparent in productive traits.

Carcass traits exhibit little heterosis. Crossbreds tend to have carcass characteristics averaging about the same as the averages of the parental breeds. Thus, to produce carcasses with maximum lean and minimum fat it is necessary to have parental breeds both of acceptable meat type.

Crossing Inbred Lines for Commercial Production. As this is being written in the middle 1960s, the potential place of inbred lines produced for use commercially in crosses is uncertain. During approximately the first 20 years after the founding of the Regional Swine Breeding Laboratory in 1937, a considerable number of inbred lines were developed both within existing breeds and among the descendants of crossbred foundations. Some of the latter were never highly inbred and will be discussed later as new breeds.

Most of the lines formed within established breeds by experiment stations were started from foundation stocks selected from breeders' herds during the lard-type hog era. After emphasis began being put on meat-type hogs in the 1940s and subsequently, breeders were able to move fairly rapidly in changing type to meet new demands. The inbreds were less flexible, and after a few years they and their crosses were found to be producing more lard than contemporary animals from breeders' herds. This occurred largely as a consequence of the lines maintaining the basic types in vogue when they were split off from the parent breeds. The inflexibility of inbred lines in meeting changing conditions must be counted as one of their disadvantages. Cost of formation and maintenance of lines is another factor.

This inflexibility in ability to meet changing conditions was a major factor in most of the lines being discarded and research facilities being turned to use for other studies which appeared to have more potential for developing swine improvement methods.

Neither the general nor specific combining ability (i.e., usefulness in cross combinations) of individual lines or the utility of formation and crossing of inbred lines as a breeding system were studied as thoroughly as would have been desirable. However, the following conclusions seem justified from the results obtained:

1. As an average of all results, crosses among lines resulted in restoring performance by an amount about equal to performance losses incurred in the formation of the lines. In other words, there was no evidence of an over-all improvement in productivity from crossing lines as compared to what would have been expected from intermating the foundation stocks from which the lines were initially derived.
2. Crosses of some lines appeared to result in performance superior to that of outbred or crossbred populations of swine. This suggests that a breeding program based on inbred lines should include the formation of many lines. If a few of these proved superior in crosses, they would be used for parent stocks in commercial production. As illustrations of the productivity of certain linecrosses as well as for their historical interest, some results of using lines in crosses and topcrosses are given in subsequent sections.
3. Reproductive potential of most inbreds is low. As a result, inbred boars are not reliable breeders when attempts are made to use them routinely to breed sows in farmers' commercial herds.

Research on the formation and use of inbred lines is continuing at a few locations in the United States and in at least one other country. Thus, further research information will be forthcoming.

Although the use of inbred lines in crosses for commercial use must be considered an experimental technique, several individual breeders and a few large companies producing several hundreds or thousands of boars annually are using inbred lines in their breeding plans, usually on a somewhat flexible basis. To reduce breeding problems among boars sold for commercial use, they are often produced by crossing two or more related lines. Further, inbred lines are not necessarily maintained as completely closed lines. Rather, test introductions of outside inheritance are made from time to time. If successful, closed line matings are subsequently made to achieve uniformity of breeding behavior.

Some of the firms in this business operate with an open book, so far as the pedigrees of the animals they sell are concerned. Others sell animals (which are often called *hybrids*) on a closed-pedigree basis, considering the parentage a trade secret. Some of these companies not only sell boars to the producer, but they also offer him a breeding program whereby they supply boars from year to year that, on the basis of tests conducted by the company, combine well with stock used previously. Several commercial herds in which the plan has been followed for a number of years have been cited for their excellent production records.

The production of inbred lines is a long and expensive process, and the determination of their usefulness requires much experimentation. Thus, only persons or firms with adequate financial resources, a clear understanding of the problems involved, and access to competent technical help should consider entering the business of producing and selling inbred lines or their crosses.

The use of inbreds could develop more rapidly than can now be predicted. Each individual livestock producer should therefore keep abreast of current recommendations of his state experiment station regarding the use of inbred animals in crosses.

At the risk of eventually being proved wrong, the authors venture to predict that inbreeding and the use of inbred lines will be used in the future somewhat as follows:

1. Further extensive inbred line formation will be undertaken when performance levels appear to have reached plateaus beyond which further progress is slow or impossible and carcasses produced are *on the average* ideal for length, back-fat thickness, and perhaps other characters.
2. When situations such as the foregoing are attained, mild to moderate inbreeding will be used as a tool to promote uniformity and dependability of breeding behavior in specific lines rather than in expectation of obtaining greatly increased hybrid vigor from inbred linecrosses.
3. Lines will be selected partially on their ability to combine with other lines, and commercial production will be based on specific line combinations.
4. Boars sold for commercial use will probably be crosses of two separate but related lines having similar behavior in crosses with other lines and breeds.

When situations favoring extensive line formation and use in American swine production will develop is uncertain. There is reason to believe that a genetic equilibrium (or a plateau) has been reached for litter size in most if not all breeds and that further selection is not resulting in improvement. Failure of growth rate and efficiency of feed utilization to improve as rapidly as carcass characters in test stations during the past decade may be an indication that plateaus are being approached but could merely mean that breeders have been putting less emphasis on these characters in view of the urgent need for carcass improvement. Presently, it appears that attainment of ideal carcass characters is some time in the future.

From the results of Danish swine testing discussed earlier, it would appear that this breed could be at a point where it meets the criteria listed above. Perhaps formation of distinct lines and crossing them for commercial production should be considered in this breed.

Some results of crossing inbred lines. In many cases crosses of inbred lines have been compared only with parent lines and with each

*Table 13-9. Crossing Inbred Lines of Duroc Swine**

	Inbred line × same inbred line	Inbred line × different inbred line	Rotation line cross	Outbred
No. litters:				
From gilts	4	14	35	15
From sows	30	41	51	28
Av. no. pigs per litter:				
Birth	9.2	9.0	10.6	10.2
180 days of age	4.8	6.6	8.0	7.8
Wt. at 180 days of age:				
Per pig, lb	157	205	203	182
Per litter, lb	752	1,359	1,625	1,413
Performance on feeding test:				
Av. days of age at start	66	64	64	65
Av. initial wt., lb	35	41	41	41
Av. daily gain, lb	1.18	1.44	1.43	1.31
Feed per cwt. gain, lb	388	380	373	377

* Courtesy of the late W. L. Robison, Ohio Agricultural Experiment Station.

other. As mentioned previously, average results indicate that crossing restores the performance lost during inbreeding and that some selected crosses may go beyond this level.[1]

In the final analysis, however, the usefulness of inbred linecrosses for commercial production will depend on how animals produced in this fashion compare in productivity with those produced by conventional purebreeding or crossbreeding procedures. Therefore, we will limit our discussion of this subject to a few experiments in which conventionally bred animals were included. These experiments lack much of demonstrating conclusively whether the crossing of lines has a potential for commercial improvement, in part because conventionally bred animals differ so much in productivity that it is difficult to know whether representative samples have been included in the tests.

At the Oklahoma Experiment Station[2] rather extensive comparisons were made of inbreds, two-line crosses, three-line crosses, and conventionally bred purebreds, all within the Duroc breed. Because of small litters raised by inbred dams, the two-line crosses were inferior to outbreds in total 180-day litter weight. Three-line crosses (two-line cross

[1] Craft, W. A., *U.S. Dept. Agr. Cir.* 916, 1953; and *Jour. Anim. Sci.*, 17:960–980, 1958; Dickerson, G. E., et al., *Mo. Agr. Expt. Sta. Res. Bul.* 551, 1954.
[2] Chambers, Doyle, and Whatley, J. A., Jr., *Jour. Anim. Sci.*, 10:505–515, 1951.

sows mated to boars of a third line) exceeded outbreds in total 180-day litter weight by a substantial amount, with most of the difference being due to larger numbers of pigs raised per litter.

The results of somewhat similar tests in Durocs at the Ohio Experiment Station as given in Table 13-9 also indicate an advantage in 180-day litter weight for rotation linecrosses, but in this case most of the advantage was the result of faster growth of individual pigs.

At Purdue University, crosses between inbred lines of different breeds produced results superior to outbreds (Table 13-10), especially when linecross gilts were used as mothers.

Oklahoma results shown in Table 13-11 generally indicate advantages in important productive characters for linecrossbreds over hogs of other breeding. The trials included several linecrossbred combinations, and

Table 13-10. Comparative Performance of Conventionally Bred Swine and That of Crosses between Two Inbred Lines*

(White King, an inbred Chester White line; the Landroc, a line developed from a Landrace × Duroc foundation. Inbreeding 30 to 40 per cent in each line)

		Crossline		Backcrosses		W.K. × Landroc back-cross
	Conventionals	Landroc × W.K.	W.K. × Landroc	Landroc × crossline	W.K. × crossline	
Years tested	1947–1949	1947–1949	1947–1948	1948–1949	1949	1949
Total no. litters†	30‡	21	9	11	4	4
Av. no. pigs per litter:						
Farrowed	8.1	9.2	6.8	11.2	13.2	12.0
Raised to 56 days	4.9	6.5	4.1	7.6	6.5	8.5
Raised to 154 days	4.8	6.3	4.1	7.6	6.1	8.0
Av. wt. per litter, lb:						
At 56 days of age	167	263	153	274	266	355
At 154 days of age	877	1,288	854	1,405	1,016	1,542
Slaughter data:						
No. pigs slaughtered	68	53		28	14	15
Av. live wt., lb	215	218		215	215	216
Av. dressing per cent	69.1	68.6		69.1	70.1	69.3
Av. yield of 5 primal cuts as per cent of live wt.	43.8	43.9		44.0	44.8	45.0

* Adapted from *Purdue University Agr. Expt. Sta. Bul.* 552, 1950.

† All litters from gilts.

‡ Included litters as follows: 12 Duroc; 6 Chester White; 7 crossbred Duroc × Chester White; and 5 crossbred Poland China × Chester White.

Table 13-11. Comparative Performance of Crossbred, Linecross, Line-crossbred, and Outbred Hogs°

	Outbred Duroc	Crossbred	Linecross Duroc	Linecross-bred
Number of litters	11	17	22	61
Number of pigs per litter:				
Farrowed	8.48	8.56	9.72	10.08
At 56 days	6.94	7.70	7.94	8.29
At 154 days	6.81	7.52	7.83	8.16
Litter weight in lb:				
Birth	22.0	22.9	24.4	27.0
At 56 days	175	241	227	250
At 154 days	907	1,096	1,128	1,173
Postweaning feeding test:				
No. of pigs fed	77	129	172	497
Av. daily gain, lb	1.44	1.42	1.42	1.44
Feed per cwt. gain, lb	354	349	352	346
Per cent lean cuts†	35.7	38.2	35.0	36.7

* From *Okla. Agr. Expt. Sta. Bul.* B-415, 1954. Data from four trials at the Fort Reno Experiment Station, El Reno, Oklahoma.

† Weight of trimmed hams, loins, and shoulders as a percentage of the slaughter weight of the hog.

the advantage for this type of breeding was even more marked when the best linecrossbred combinations were considered. In summary of this work it was stated:

Although the development of inbred lines specifically for superior performance in crosses is a slow and costly process, the value of these lines once they are developed is considerable. Nearly all of the present inbred lines have certain deficiencies, yet many of them produce sufficient hybrid vigor to make them useful, particularly when applied in the most effective combinations. It seems likely that this type of program for commercial hog production will become more important in the future, particularly in herds large enough that a systematic crossing program can be economically handled.

The outline for future development seems clear. First, more extensive testing is needed to determine the best linecross combinations. In addition there must be improvement in existing lines, and the development of new and better lines, in order to get the maximum return from a systematic rotational linecrossing program. Finally, a more efficient procedure for developing new lines is needed, and current research is being aimed in that direction.[1]

[1] *Okla. Agr. Expt. Sta. Bul.* B-415, 1954.

Some results of use of inbred lines in topcrosses. Research information on the usefulness of inbred lines in topcrosses on non-inbred stock is limited. Little average difference was found[1] in the progeny performance to five months of age between inbred Poland Chinas of several lines and non-inbred purebred boars in farmers' herds. The lines differed in performance of progeny. Progeny of inbred Landrace boars had an advantage of 3.5 per cent in viability and 12 lb in five-months weight over progeny of non-inbred purebreds. This may well have been a manifestation of heterosis due to the wide genetic differences between Landrace boars and sows of American breeding rather than to the superiority of the Landrace line.

In a Wisconsin study[2] there were no significant differences in litter size or viability between progeny of inbred and non-inbred boars when all were bred to non-inbred sows under farm conditions. Four of the eight lines tested produced topcross progeny with superior growth rates to five months. Topcross gilts retained for breeding had an over-all advantage of 1.08 pigs raised per litter and 37 lb in weaning litter weight as compared to gilts in the same herds sired by non-inbred boars. Three of the lines produced topcross gilts significantly superior in pig-raising ability to nontopcross gilts, and all three of these lines were among the four lines giving superior growth rates. These three lines were derived, in part or all, from Landrace or Yorkshire foundation stock. This again suggests the importance of diversity of origin for parent stocks in crosses.

NEW BREEDS OF SWINE[3]

Most of the traditional American breeds of swine including the Duroc, Poland China, Chester White, and others, were developed in this country from various mixtures and crosses of swine breeds and types brought to this country by early settlers. These breeds have been well established for so long that many breeders overlook their crossbred origin.

A number of strains, some of which can now be classes as breeds, have been developed since 1934. Most of the new strains contain inheritance of the Danish Landrace and one or more American breeds.

The Landrace has been bred in Denmark for over 60 years under a systematic progeny-testing system which has already been described

---

[1] Hazel, L. N., et al., *Jour. Anim. Sci.*, 7:512, 1948 (abstract).
[2] Durham, R. M., et al., *Jour. Anim. Sci.*, 11:134–155, 1952.
[3] See *U.S. Dept. Agri. Farmers' Bul.* 1263, 1966, for more detail on new swine breeds.

in some detail. The breed appeared to have developed certain characters, including length of body, well-developed hams, prolificacy, and ability to gain rapidly and efficiently, which would make them useful in this country.

Partly because the original agreement under which they were imported prohibited the commercial use of the Danish Landrace in the United States, and partly because some characteristics of the breed, such as weak feet and legs, white color, sparse hair, and sagging backs, were considered undesirable for American conditions, efforts were begun to form new strains combining the good characters of the Landrace with the darker colors, stronger feet and legs, and other desirable characters of American breeds.

Several of the new strains have been released and are being used commercially. The Hamprace (also sometimes known as the Montana No. 1) was developed at the U.S. Range Livestock Experiment Station, Miles City, Montana, by the U.S. Department of Agriculture and the Montana Agricultural Experiment Station. Solid black unbelted segregates from purebred Hampshire herds and Danish Landrace animals provided the original stock. These breeds were crossed, and the new breed was developed by selection from the F_2 and backcrosses to each parent breed. The Hamprace is a solid black hog with much of the type of the Landrace.

The Minnesota No. 1 is a red strain developed by the Minnesota Experiment Station from Tamworth and Landrace foundation animals. It was the first of these new hog breeds to be used extensively. Another new strain developed by the Minnesota Station is the Minnesota No. 2, developed from Poland China and Yorkshire foundation animals. It does not carry Landrace inheritance and was developed primarily to use in crossbreeding with the Minnesota No. 1.

The U.S. Department of Agriculture developed seven new strains carrying Landrace inheritance at the Beltsville, Maryland, Research Center. Two of these, the Beltsville No. 1 (Landrace-Poland China foundation) and the Beltsville No. 2 (Landrace-Hampshire-Duroc-Yorkshire foundation) are in commercial use.

There are also several other new strains developed by experiment stations or private breeders.

The Inbred Livestock Registry Association of Augusta, Illinois, is registering animals of the new strains as well as a few inbred animals of longer-established breeds. The association registered 2,196 animals in 1963.

Most of the new strains have been inbred to a considerable extent as they have developed, and most of them now have average inbreeding

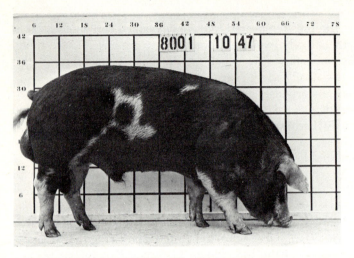

Fig. 13-19. Representatives of two of the new American breeds of swine. *Above,* a Beltsville No. 1 boar. *Below,* A Palouse gilt. This breed was developed at Washington State University on a Landrace-Chester White crossbred foundation. (*Courtesy of Animal Husbandry Research Division, U.S. Department of Agriculture and Washington State University.*)

coefficients of 25 to 40 per cent. Thus, they already have considerably more homozygosity than most established breeds. They have been developed primarily for commercial use in crosses with each other and with other lines and breeds of hogs. Their value depends upon the productivity of their crossbred offspring, not on their performance as pure strains.

REGIONAL SWINE BREEDING LABORATORY

General recognition of the problem inherent in breeding more productive swine and the need for cooperative effort because extensive facilities were necessary for such research led to the establishment of the Regional Swine Breeding Laboratory in 1937, with headquarters at Ames, Iowa. Ten Midwestern experiment stations have formal projects as part of the work of the laboratory. Other stations have cooperated informally. The program was organized on a broad basis, with all the cooperating stations having similar objectives but with each having broad latitude to investigate problems in which it was most interested and best prepared to undertake. The late Dr. W. A. Craft was director of the laboratory for many years. According to Dr. Craft,

The principal objectives of the program are to discover, develop, and test procedures of breeding and selection which may be used by hog producers to speed the improvement of hogs in performance; to investigate precisely the usefulness of inbred lines for improving breeding value of the pure breeds and for use in pork production; to enlarge knowledge concerning the genetic effects of inbreeding, and the inheritance of characters; and to evaluate and demonstrate application of such knowledge in swine breeding. Points emphasized in the investigations include productiveness of sows, vitality of pigs, growth rate, economy of gains, and desirability of carcasses.[1]

The laboratory has made extensive research contributions, and much of the material discussed in this book on effects of inbreeding and crossing of inbred lines, heritability of performance characters, selection, and the formation of new breeds from crossbred foundations is based on swine-laboratory results. Dr. C. E. Shelby became director in 1960.

SUMMARY

Characters of swine related to economic production of a high-quality product are numer of pigs raised per litter and their weight at weaning, postweaning rate and efficiency of gain and carcasses with a high ratio of lean to fat. Fertility and preweaning growth are low, postweaning traits medium, and conformation and carcass characters medium to high in heritability. There is a high genetic correlation between rate and efficiency of postweaning gain. Other genetic correlations among important traits are not well understood. Selection programs have been effective in changing some characters of seedstock herds both in the United

[1] *U.S. Dept. Agr. Cir.* 916, 1953.

States and abroad. Emphasis in selection to improve seedstock herds should be on traits of economic importance and medium to high heritability. They should utilize modern evaluation techniques including the back-fat probe and slaughter tests.

Heterosis or hybrid vigor is most apparent in traits of low heritability. For these traits it is so important that commercial producers should follow systems of crossing breeds or inbred lines to maximize total performance in their herds. Inbred lines are costly to form and evaluate, but their crosses may have an important place in future swine production if other breeding methods encounter plateaus beyond which further improvement by selection is difficult or impossible.

REFERENCES

Altman, Philip L., and Dittmer, Dorothy S. eds. 1962. "Growth, Including Reproduction and Morphological Development," Federation of American Societies for Expt. Biol., Washington, D.C. (On pages 115–126 this publication has extensive summaries of heritability estimates for swine and other farm animals.)

Comstock, R. E. 1960. Problems and Evidence in Swine Breeding, *Jour. Anim. Sci.*, **19**:75–83.

Craft, W. A. 1953. Results of Swine Breeding Research, *U.S. Dept. Agr. Cir.* 916.

———— 1958. Fifty Years of Progress in Swine Breeding, *Jour. Anim. Sci.*, **17**:960–980.

Dickerson, G. E., Blunn, C. T., Chapman, A. B., Kottman, R. M., Krider, J. L., Warwick, E. J., and Whatley, J. A., Jr. 1954. Evaluation of Selection in Developing Inbred Lines of Swine, *Mo. Agr. Expt. Sta. Res. Bul.* 551.

Jonsson, Per. 1963. Danish Pig Progeny Testing Results, *Ztschr. f. Tierzücht. u. Züchtungsbiol.*, **78**:205–252.

Smith, C., and King, J. W. B. 1964. Crossbreeding and Litter Production in British Pigs, *Anim. Prod.*, **6**:265–271.

————, ————, and Gilbert, N. 1962. Genetic Parameters of British Large White Bacon Pigs, *Anim. Prod.*, **4**:128–143.

CHAPTER FOURTEEN

Improving Sheep

Sheep are basically dual-purpose animals, producing both wool and meat. Although they are not known to have been used extensively in the United States for milk production, there are many nations in which specialized sheep breeds are important as dairy animals. Fur-producing types are also important in many areas and have been tried experimentally in the United States. Historically, in different parts of the world and at various times, economic circumstances have sometimes favored sheep having the ability to produce desired types of wool, and at other times lamb or mutton with desired qualities have been products of primary importance.

Breeders responded to the demands with the development of strains and breeds of sheep especially suited to the particular type of product desired and adapted to the geographic area. Many of the local strains and breeds developed during the long course of history have now disappeared, but we still have a great variety of types and breeds of sheep. The process of breed formation is not necessarily over, several breeds having been developed in the United States alone during this century.

Sheep are raised under many and varied environmental conditions both in the United States and in other areas of the world. The type of sheep production likely to be most profitable depends on the social and economic development of an area. A frontier will support a different type of sheep enterprise than will thickly settled areas. Wars and tariff policies have exerted profound influences on the American sheep industry.

In colonial America, sheep were grown almost entirely for the purpose of producing wool.[1] Original importations were largely of British breed-

---

[1] See Wentworth, E. N., "America's Sheep Trails," The Iowa State University Press, Ames, Iowa, 1948, for details of the fascinating history of sheep in the United States.

ing of the same medium- or long-wool types which later evolved into the modern British long-wool and Down breeds. The wool clip of the early colonies was only 2 or 3 lb per animal per year, but through selection and improved feeding the management it reached 8.57 lb in 1955 and has since remained near this level.

Several local strains or breeds were developed in the early days of our nation's history. Of particular interest is the Arlington Long-Wool breed founded by George Washington and perfected by his adopted son, George Washington Park Custis. This was the first strictly American strain of livestock to receive recognition and was apparently an excellent breed. It was smothered beneath the Merino onslaught of the early nineteenth century.

TYPES OF SHEEP

Fine-wool. Sheep of fine-wool types now popular throughout the world all descend from Spanish foundations. Until near the end of the eighteenth century, exportation of Spanish Merinos was forbidden by law, and Spain controlled the fine-wool trade of the world. Shortly after 1800, extensive importations were made into this country. From these the American Merino was developed, with Vermont exerting the greatest influence for many years. Largely as a result of wars and tariff policies, the Merino sheep business went through fantastic booms and depressions, but it remained the dominant influence in the American sheep business during most of the nineteenth century.

From the early Vermont Merinos, there were later developed several Delaine Merinos in efforts to produce fine wool of sufficient length for combing. Several Merino breed associations were established. The American and Delaine Merino Record Association was founded by the consolidation of several of them, but three other associations still exist. Merino type varies from a small, thin, heavily wrinkled animal of poor carcass quality but producing a fine, dense, heavy fleece of short wool (Type A), to a larger, fairly heavy, and low-set animal of improved carcass quality but generally lacking wrinkles and with a heavier but less fine and dense fleece (Type C or Delaine). There is a third intermediate form (Type B).

From Spanish Merino foundations, the French developed a fine-wool strain of sheep, the Rambouillet, of much greater size than the American Merino. Many of these animals were brought to America about the middle of the nineteenth century, and for a time they were a very popular breed. Interest then declined, to be revived about 1890. The Rambouillet is a large animal producing a heavy, dense fleece. As with

Fig. 14-1. A modern-type grand champion Rambouillet ram. (*Courtesy of American Rambouillet Sheep Breeders' Association.*)

the American Merino, there are variations in amount of skin wrinkles, and at some shows the animals are divided into B and C types. They were brought to America to serve primarily as wool producers. In recent years greater attention has been paid to fleshing qualities, more by some breeders than by others, but the carcasses are not of so high a grade generally as those from the Down breeds. This breed has been called the most typical dual-purpose breed in America, yielding something in density and fineness of fleece to the American Merino and something to the Down breeds in quality of carcass. Foundation flocks for the great Western range sheep industry came partially from the Eastern fine-wool areas but largely from descendants of the early importations of Spanish settlers. The sheep were apparently largely of the unimproved *churro* type rather than Merinos, as is sometimes assumed. American Merinos were widely used to "grade up" these flocks. After 1900 the Rambouillet became the dominant breed in the West, and it has been estimated that 98 per cent of the ewes west of the Mississippi River now carry 50 to 100 per cent Rambouillet blood. Most Rambouillets and Merinos raised today are of the C type.

Long-wool. Robert Bakewell, the noted English livestock breeder of the eighteenth century, greatly improved one of the long-wool breeds, the Leicester. He devoted his major attention to fleshing or fattening qualities and quite neglected wool. The result was a fairly large breed of sheep with rather long and coarse wool and a carcass which, when finished, was overlaid with several inches of fat. To a greater or less extent these qualities are found also in the other long-wool breeds— Cotswolds, Lincolns, and Romneys. The long-wool breeds were created, therefore, primarily to serve as meat animals, supplying the large fat

Fig. 14-2. A representative of the Lincoln breed of sheep— one of the breeds of the long-wool type. (*Courtesy of National Lincoln Sheep Breeders' Association.*)

legs and roasts then in demand. In recent years more attention has been paid to wool, and attempts have continually been made to improve the quality of carcass in these breeds. The fine-wool breeders generally pay particular attention to fleece and less attention to carcass qualities. The long-wool breeders, on the other hand, pay more attention to carcass, less to fleece.

Medium-wool. The various Down, or medium-wool, breeds of sheep— Southdown, Shropshire, Hampshire, Suffolk, etc.—were developed in Great Britain, and in many ways fill the void between the fine- and long-wool types. As a general rule they produce carcasses which come nearest to meeting today's requirements. The Down breeds have been bred and kept relatively pure in commercial flocks in the East and Midwest, whereas in the Western range country they have been used largely for crossing.

As pointed out in an earlier chapter, Hammond and his associates have developed the idea that improved meat type depends upon selection for increased rates of development of later-developing body tissues. Fig. 14-5 gives a visual appraisal of the differences in body proportions of improved and unimproved sheep of various ages.

Fig. 14-3. A Hampshire ram. The Hampshire is one of the largest medium-wool or "Down" breeds of sheep. (*Courtesy of American Hampshire Sheep Association.*)

Fig. 14-4. A Southdown ewe. This breed is one of the smaller medium-wool or "Down" breeds and has long been noted for carcass excellence. (*Courtesy of American Southdown Breeders Association.*)

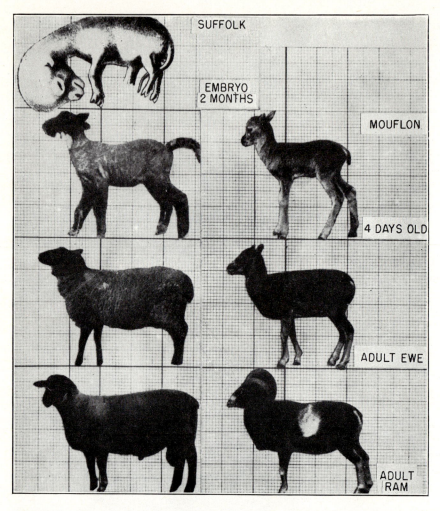

Fig. 14-5. Changes in body proportions of sheep with advancing age. The Suffolk, an improved mutton breed of the "Down" type, is compared with an unimproved breed, the Mouflon. (*Courtesy of the late Dr. John Hammond, Cambridge University, Cambridge, England.*)

FADS IN SHEEP BREEDING

As if all the necessarily diverse factors essential in sheep production were not enough to worry about in a breeding program, the purebred-sheep industry is plagued with a diversity of breed trademarks, some of which are actually harmful from the standpoint of productivity. Two fads that developed in certain sheep breeds are among our best examples of the dangers in allowing selection to be directed toward mere trademarks.

It was long assumed that sheep with wrinkled skin would have more area upon which to grow wool and would therefore produce heavier fleeces than smooth-bodied types. Another apparently equally logical assumption was that a covering of wool extending well down over the face should be related to a strong wool-growing tendency and therefore be indicative of fleece production. The first assumption served as the basis upon which some breeds of fine-wool sheep were selected for an extremely wrinkled condition. Fine-wool and medium-wool breeds were, in some cases, selected for extensive face covering, extending in extreme cases clear to the muzzle.

Systematic studies over a period of years, especially by the U.S. Department of Agriculture and the Texas Agricultural Experiment Station,[1] showed that the wrinkly types of sheep do produce more pounds of grease wool than the smooth types. The differences practically disappear when the dirt and grease are removed in scouring, however, so that the weights of clean wool produced by the two types are almost equal. In addition, in the smooth types staple length is longer and fiber diameter is more uniform. Observation indicates that the smooth types are easier to shear and less susceptible to fly-strike. Thus the evidence seems to be clear that the wrinkly types do not have the wool-producing superiority claimed for them, at least under present-day range conditions, and that they are undesirable in other respects.

Extensive studies have likewise shown that extreme face covering is undesirable. The excessive growth of wool covers the eyes and causes wool blindness, which interferes with the ability of the animal to eat. Other factors may also be important in reducing productivity. In New Zealand woolly faced ewes were less productive even when clipped regularly to prevent wool blindness.[2] Figures in Table 14-1, taken from a report of the Western Sheep Breeding Laboratory,[3] indicate that under range conditions Rambouillet ewes with covered faces are seriously deficient in lamb production. The last column in the table, indicating a difference of 11.1 lb in lamb produced per ewe bred, is especially striking.

Not only were the covered-faced ewes inferior in lamb production, but they were not importantly superior in wool characters. A few statements from the report give a concise summary of the work.

[1] See Shelton, M., et al., *Jour. Anim. Sci.*, **12**:440–450, 1953, for latest paper and references to earlier work.
[2] Cockrem, F., and Rae, A. L., *Austral. Jour. Agric. Res.*, **17**:967–974, 1966.
[3] Terrill, C. E., et al., *Jour. Anim. Sci.*, **8**:353–361, 1949.

Fig. 14-6. *Above,* a champion B-type Rambouillet ram show-ing heavy neck folds and a moderate degree of wrinkling on the body. *Below,* a range Rambouillet ram almost free from wrinkles. Note the difference in degree of face covering in these two rams. (*Courtesy of Dr. C. E. Terrill, Animal Hus-bandry Research Division, U.S. Department of Agriculture.*)

Table 14-1. Relation of Face Covering to Lamb Production*

Face covering	No. of ewe years	Per cent ewes lamb-ing of ewes bred	Per cent lambs born of ewes lamb-ing	Per cent live lambs of lambs born	Per cent lambs weaned of live lambs born	Av. wean-ing wt., lb	Per cent lambs weaned per ewe bred	Lb of lamb weaned per ewe bred
Open	286	95.5	126.7	91.9	89.0	76.3	99.0	75.5
Partially covered	845	95.4	142.2	92.3	88.1	74.8	96.3	72.1
Covered	1,557	91.9	119.7	90.0	88.6	73.4	87.7	64.4

* *Jour. Anim. Sci.*, August, 1949, p. 356.

Ewes with open faces produced 11.3 per cent more lambs and 11.1 more pounds of lamb per ewe bred than those with covered faces. Ewes with partially covered faces weaned 8.6 per cent more lambs and 7.7 more pounds of lamb per ewe bred than those with a covered face. Differences in face covering within these groups were associated with corresponding differences in lamb production. These advantages for ewes with open faces occurred in spite of three periodic clippings around the eyes of all ewes subject to wool blindness.

About 46 per cent of the advantage of open-faced ewes was due to a greater number of lambs born per ewe lambing; 26 per cent was due to higher weaning weights; 19 per cent was attributed to a higher proportion of the ewes becoming pregnant; and 9 per cent was due to greater viability to weaning of offspring.

Open-faced ewes excelled covered-faced ewes in lamb production at each year of age. The greatest advantage of open-faced ewes in pounds of lamb per ewe bred was found at 3 years of age followed in order by 2, 4, 6, and 5 years.

The yearling grease and clean fleece weights and staple lengths of 2,499 Rambouillet ewes and the lifetime grease fleece weights of 798 Rambouillet ewes were slightly greater for covered-faced ewes than for those with open faces. The differences were not significant except for staple length and were not large enough to be economically important.

The great economic importance and high heritability of face covering indicate that it should receive as much or more attention in selection than any other trait in sheep if wool blindness is a problem.[1]

The results of the United States studies have been confirmed in Australia, New Zealand, and other important sheep-growing nations.

[1] *Jour. Anim. Sci.*, August, 1949, pp. 360–361.

Fig. 14-7. Variation in degree of face covering in sheep. The animals shown in G, H, and I are definitely "wool-blind." (*Courtesy of Dr. C. E. Terrill, Animal Husbandry Research Division, U.S. Department of Agriculture.*)

The moral to be drawn from such studies is obvious, namely, that present-day breeders should critically examine existing breed standards, and, if necessary, experiments should be carried out to determine whether they are compatible with the production of the most efficient and profitable animals. Obviously such critical tests should be applied to fads, which are certain to be started in the future, before, not after, they have become so widespread as seriously to hurt a breed.

IMPROVING SHEEP SEEDSTOCK FLOCKS

There are a great many types of sheep in existence. They are raised in the United States under a greater range of environmental conditions and management systems than is true of any other class of livestock. The first job of the sheep seedstock breeder is, therefore, to select the type and breed of sheep best adapted to efficient production under the conditions where operations will be carried on and where breeding stock will be sold.

After the type or types of sheep to be raised have been determined, the second task of the breeder is to maximize the inherent productivity of the type of sheep selected. Although formerly many sheep were kept almost exclusively for wool production, with mature wethers often being used for this purpose, the sheep industry is today almost universally on a dual-purpose basis; that is, both wool and meat animals (usually lambs) are major products, the relative importance of the two varying considerably in different areas. For example, in the semiarid regions of the Southwest, where ranges are so sparse as to make the production of milk-fat lambs impractical, lambs usually sell as feeders at lighter weights and lower prices than those at which milk-fat lambs sell in areas where their production is possible. However, high-quality fine wool can be produced under these conditions. Thus wool is relatively a much more important product than in better range areas or under farming conditions such as prevail in the Middle West.

Regardless of the relative emphasis given to wool and lamb production, the general objectives of sheep breeders are very similar, namely, to produce as many pounds of lamb and as much high-quality wool as possible per 100 lb of ewe.

Measuring Performance. As with other classes of stock, intelligent breeders with clear goals in mind appear to have accomplished much by visual selection, including estimates of weight for age. All evidence, however, points toward the fact that a better job of selection can be done if a careful set of performance records are kept and used.

The keeping of production records on sheep and the intelligent evaluation of these records in making selections necessarily entail a considerable expenditure of time and effort. Because of the variety of types of sheep it is difficult to suggest procedures for making and evaluating records that will be universally applicable. However, the following items are believed to be of greatest importance under most conditions where sheep are raised in the United States:

1. *Prolificacy of ewes* (regularity of lambing and twin production). In some areas where grazing conditions are poor, twin lambs may be a disadvantage, but generally, twin-producing ewes are more profitable than single-producing ewes.
2. *Birth weights of lambs.* This item is of no value in itself but is related to vigor at birth and to rate of gain after birth.
3. *Earliness of lambing.* Most domestic breeds of sheep are seasonal

breeders, lambs being produced during the late winter and early spring months. For this reason the supply of lambs is highly seasonal, and price fluctuations are often extreme from month to month. In many areas late-fall-to-midwinter lambs can be raised satisfactorily and are ready for market at a time of year when prices are usually highest. The limiting factor in such areas is often the inability to get satisfactory breeding behavior at the necessary season.

4. *Weaning weights of lambs.* This is taken individually but it is also used to measure total lamb production of a ewe. This latter may be best expressed as pounds of lamb weaned per ewe (in areas where costs are largely on a per head basis) or as pounds of lamb weaned per 100 lb of ewe (best in areas where costs are related more closely to actual amounts of feed eaten than to total numbers).

   Under most conditions, rapid growth rate of lambs is important. If lambs are all weaned at one time, as is usually the case under range conditions, and often with farm flocks as well, growth rate largely determines weaning weight and therefore is highly related to gross production per ewe. Under farm conditions it is often most profitable to wean lambs at intervals as they reach the most desired or popular market weight. This somewhat reduces the importance of growth rate, but it is still important since prices in most years tend to decline as the marketing season advances. Also, in many areas the growth rate of lambs is reduced by hot summer weather. In these areas it is best to have lambs which grow fast enough to reach market weight prior to this season.

5. *Type and finish of lambs at weaning.* These important factors reflect market value to a large extent. Attempts to evaluate these items objectively by means of measurements have as yet met with little success. Although unsatisfactory at best, the most successful means of evaluating these characters has been through the use of scores based on subjective appraisals of merit.

6. *Fleece weight and quality.* There are specialized uses and thus market demand for many kinds of wool. In general, however, fleeces with the best combination of high clean weight per fleece, long staple length, and uniformly fine fibers will be most valuable. Freedom from dark fibers and coarse hairy fibers (kemp) is necessary for greatest fleece value. In an Australian study,[1] sheep selected for heavy fleece weights had greater ability to consume feed and higher efficiencies of feed use for wool production than was true of a group selected for low fleece production.

7. *Regular annual or semiannual weights of all breeding stock.* Basically, sheep will be most efficient if productivity is evaluated in relation to weight. That is, lamb weight should be high in relation to ewe weight and fleece weight high in relation to size of the animal from

[1] Ahmed, Waheed et al., *Austral. Jour. Expt. Agr. and Anim. Husb.*, 3:269–275, 1963.

which it was shorn. Because of variation in weight with changing feed conditions, it is difficult to know what the "true" weight of an adult animal is. This, together with the fact that many costs in sheep production are on a per-head basis, has discouraged attempts to select for productivity as percentages of live weight. However, it is believed that periodic weights will be useful for evaluating management and feeding regimes, and that some attention can be given to them in selection.

The evaluation of performance records is greatly complicated by environmental factors. Some of these, such as the effects of sex, age of dam, type of birth (single or multiple), and age at weaning, can be measured and definitely taken into account. If these things are not systematically evaluated, selection may be biased in favor of lambs having the most favorable conditions for growth. A study of lambs born during a 14-year period (1921–1934) in the Hampshire, Shropshire, and Southdown flocks at the Agricultural Research Center, Beltsville, Maryland, showed that environmental factors had apparently influenced selection to an undue degree when selection of lambs to be retained for breeding purposes began at three months of age. The summary of the study included the following statement:

The results showed that type of birth and birth weight have influenced the selection of breeding animals. Single lambs have been preferred to twins, early lambs to late ones, and lambs that were heavy at birth to light lambs. With the exception of a portion of the effect of birth weight, the effects of these factors on selection are considered to be environmental in nature, and therefore reduce the chances of selecting the genetically superior animals for breeding purposes.[1]

Extensive studies at the Western Sheep Breeding Laboratory[2] have shown that under range conditions the sex, type of birth and rearing, and age of dam have important effects on body weight at both weaning and yearling ages, on fleece characters, and on type and condition as evaluated by subjective scores. Age of lamb at weaning also had a significant effect upon weaning weight. The environmental factors were relatively unimportant in affecting scores for face covering or neck folds.

Table 14-2, prepared from data in the studies just mentioned, shows the magnitude of some of the differences observed. In general, the differences tend to become smaller with advancing age. The reader should

[1] *U.S. Dept. Agr. Cir.* 538, 1940.
[2] See Terrill, C. E., et al., *Jour. Anim. Sci.*, 7:311–319, 1948, for one report and references to others.

Table 14-2. Differences in Performance of Range Sheep Due to Some Environmental Conditions

	Weaning characters		Yearling ewe characters		
	Body wt., lb	Staple length, cm	Body wt., lb	Clean fleece wt., lb	Staple length, cm
Rambouillet:					
Advantage of:					
Rams over ewes	8.3*	−0.48*			
Mature dam over two-year-old dam	6.1*	0.19	2.6	0.17	0.12
Singles over twins	9.2*	0.05	6.0*	0.23*	0.30*
Columbia, Targhee, and Corriedale:					
Advantage of:					
Rams over ewes	10.8*	−0.54*			
Mature dam over two-year-old dam	8.7*	0.52*	4.16*	1.04†	0.46
Singles over twins	11.7*	0.02	7.12*	1.03†	0.12

* Signifies statistic is large enough to account for 2 per cent or more of the total variation in that trait.
† Grease-fleece weight.

recognize that these figures will not necessarily be applicable to other flocks under different environmental conditions.

In a large operation, the necessary data to arrive at figures such as the above can be collected and the records all adjusted to a single standard, preferably by adding the appropriate number of pounds to the weights of lambs in the lower groups to adjust them to the level of the heavier group.

In actual practice, results similar to selecting on the basis of adjusted records may be obtained by sorting lambs into groups according to sex, age of dam, type of birth, and age of lamb and selecting the same proportion from the various groups.

The weaning weight should be taken as near the age at which lambs are customarily weaned in the area as is possible. Factors based on regression of weight on age for adjusting lamb weights to a standard age (such as 120 or 140 days) have been worked out for certain breeds and flocks, but they apparently are not applicable to lambs in general, since lambs in different flocks of different breeds grow at such different rates.

If lambs are weighed fairly close (probably within 10 days or two weeks) to the standard age, adjustments to a standard weight can be made in the manner suggested for calves. First determine the weight per day of age, multiply this by the day's deviation from the standard age, then make the necessary addition or subtraction. The evaluation of type, degree of finish, amount of face covering, and other traits can best be expressed in terms of subjective scores given by persons with experience.

No one system of scores can be said to be better than another, but the one given in Table 14-3 has been used successfully at the Western Sheep Breeding Laboratory and at various experiment stations.

Plus and minus designations can be used with each numerical score, so that scores of 1+, 1, or 1−, etc., are possible, depending upon the degree of excellence. A total of 15 scores is thus possible.

Fleece evaluation under practical farm conditions will have to be limited to

1. Getting grease weight of each fleece as it is shorn
2. Measuring the length of staple at a given spot, preferably on the side or in the shoulder region
3. Making visual estimates of the fineness at probably three points on the side to estimate average fineness and degree of uniformity
4. Making a careful examination for undesirable characters, such as black fibers or coarse hairy fibers known as *kemp*

Since grease weight represents not only wool but also dirt and grease, the amount of which varies from animal to animal, it would be desirable to have clean-fleece weights. These are seldom obtainable except under experimental conditions. Likewise, laboratory measurements of actual fiber diameter would be better than the visual estimates of fineness. Here again the technique and equipment needed make the determination impractical for most breeders.

Table 14-3. A Widely Used Sheep-scoring System

Numerical score	Definitions as regards type, finish, etc.	Definitions as regards face covering
1	Excellent	Not covered beyond poll
2	Good	Covered to eyes
3	Medium	Covered slightly below eyes, but open-faced
4	Fair	Covered partially below eyes, but not subject to wool blindness
5	Poor	Face covered and subject to wool blindness

A rather simply operated wool *squeeze machine*[1] has been designed
to aid in estimating clean-fleece weights from the grease fleece. It measures the volume of a fleece under a constant pressure. Grease-fleece
weight, squeeze-machine reading, and staple length together accounted
for from 60 to 84 per cent of the variance in clean-fleece weight in
fleeces of various fineness classes.[2] The machine was of most use with
finer wool.

The relative emphasis which should be placed on different characters
in a breeding program depends upon the factors of

1. Economic importance
2. Heritability
3. Genetic and phenotypic interrelationships
4. Variability

The relative economic importance of various traits varies so greatly
with differences in environmental conditions and available markets that
it is doubtful if specific general recommendations can be made. Several
indexes have been developed for specific purposes.[3] In general, for situations in the United States, lamb weight appears to justify more emphasis
than fleece weights.

Heritability of Economic Characters.   A summary of heritability estimates from many sources for characters of economic importance is given
in Table 14-4. Sheep breeds vary so greatly in economically important
traits as compared to other farm animals that it would seem likely that
heritability might differ from one breed or type to another. The heritability of skin folds is an example of this. Sufficient data are not available
for other traits to permit us to more than guess whether real differences
exist or whether the observed variability can be due to chance variation.
Cases where great apparent variability has been evident in different
studies have been indicated by a wider range within which a majority
of studies fall.

The similarity or dissimilarity among estimates from different breeds
and flocks gives us some basis upon which to assume that the estimates
given are (1) applicable to sheep in general or (2) apply only to specific
flocks or breeds.

Heritability estimates for weaning weight have varied greatly. There
is no clear pattern of either high or low estimates being associated with

[1] Neale, P. E., et al., *New Mex. Agr. Expt. Sta. Bul.* 417, 1958.
[2] Price, D. A., et al., *Jour. Anim. Sci.,* 23:350–354, 1964.
[3] See Karam, H. A., et al., *Jour. Anim. Sci.,* 12:148–164, 1953, for one index and
references to others.

Table 14-4. Approximate Averages of Heritability Estimates for Various Characters of Economic Importance in Sheep

Character		Approximate average heritability*
Birth weight	low to medium	10 to 30
Weaning weight	low to medium	10 to 40
Yearling weight	medium	30 to 40
Weaning conformation score	low	10 to 20
Yearling conformation score	low	10 to 20
Weaning condition score	low	5 to 15
Face covering score	high	40 to 60
Skin fold score	high	40 to 60 (fine-wool breeds)
		8† (medium-wool breeds)
Fleece characters:		
Grease-fleece weight	medium	30 to 40
Clean-fleece weight	medium	30 to 40
Per cent yield of clean wool	medium	30 to 40
Weaning staple length	high	40 to 50
Yearling and mature staple length	high	40 to 60
Fineness of fiber	medium to high	30 to 50
Hairiness	high	50 to 60
Reproductive characters:		
Fecundity	low to medium	10 to 30
Earliness of lambing	medium	30 to 40
Number functional nipples	medium	22†

* From many published sources. See references to Terrill (1958), Turner (1963), at end of chapter, and Butcher, R. L., et al., *Jour. Anim. Sci.*, **23**:12–15, 1964, for references to earlier studies.
† Only one known estimate.

breed, management conditions, or method of making estimates. Weaning weight is a complex character depending upon both mothering ability of the ewe and inherent growing ability of the lamb. If milk production were the dominant influence, it might be anticipated that heritability estimates depending upon differences between sire progenies (paternal half-sib estimates) would be low. They have, however, varied from near zero to over 50 per cent. If mothering ability were of great importance, we would expect heritability estimates from dam-offspring regressions to be higher than those from paternal half-sibs since both direct and transmitted influences of the dam would be included. The dam-offspring regression estimates have also been extremely variable; hence no definitive statement can be made regarding the relative importance of mothering ability and inherent growing ability.

In spite of the variability, most estimates are positive, indicating that within-breed selection for improved weaning weight should be effective in most cases.

Heritability of type and condition scores is relatively low in all cases, and it is evident that marked improvement due to within-breed selection is unlikely in most breeds.

Fleece characters, including grease and clean weight, and the factors of staple length, fineness, degree of crimping, and purity—all factors influencing usefulness and value—are for the most part rather high in heritability.

Skin and neck folds have a relatively high heritability in flocks with a high incidence of wrinkling.

In spite of the marked differences between breeds in fecundity (measured by production of single, twin, and triplet, or in rare cases larger, litters of lambs), most within-breed estimates of heritability have been low. However, in what is probably the most thorough study of the problem to date,[1] a heritability of over 30 per cent was found for fertility in three-year-old ewes in a flock of Peppin Merinos in Australia. A considerable part of the variation measured in this study was between having no lamb vs. one or more. Heritability of fertility in two-year-old ewes was low. In both this study and a New Zealand experiment,[2] progress in selection for improved fertility has been demonstrated. Thus, increased attention to this character may be justified. By all means, breeders should make proper allowance for the effect of twinning or individual performance so that they do not fall into the trap of inadvertantly favoring singles for breeding purposes and thus unintentionally exercise selection against high fecundity.

Three factors limit the time at which ewes will settle for the production of late-fall or early-winter lambs. Ewes of most breeds have a limited breeding season and do not exhibit estrus during the spring and summer months. There is evidence that the conception rate of ewes that do experience estrus and ovulation during these times may be low. Rams often produce semen of low quality during the summer months, presumably as a direct result of high summer temperatures. Breed differences are known to exist for at least the first and third of these factors, but little is known about their heritability or even their relative importance in limiting the production of early lambs. Using records of the Kentucky Hampshire and Southdown flocks accumulated over a 22-year period,[3] heritability of date of lambing was estimated at 46 per cent

[1] Young, S. S. Y., et al., *Austral. Jour. Agr. Res.*, 14:460–482, 1963.
[2] Wallace, L. R., *New Zeal. Jour. Agr.*, 97:545–550, 1958.
[3] *Ky. Agr. Expt. Sta. Ann. Rpt.*, p. 38, 1953.

by the paternal half-sib correlation method and 30 per cent from daughter-dam regressions. Repeatability of date of lambing was 0.43. On the basis of these results it was stated: "The repeatability figure and heritability estimates together indicate that effective progress could be made for earlier seasonal lambing by selective breeding."

Selection experiments aimed at increasing the ability to produce early-season lambs are under way in at least three state experiment stations.

In a breeding program, intense selection should first be practiced for traits which are economically important and highly heritable. Progress should be rapid in this phase of the breeding program; and after desirable levels of performance have been attained for these characters, attention can be turned to other, less-heritable traits in which progress can be expected to be slower but nevertheless important once it is attained.

Genetic Correlations. The possibility of the existence of genetic antagonisms which would limit or slow progress in breeding programs in which selection must be made concurrently for several factors has been explored in a limited number of flocks.[1] Genetic correlations between fiber diameter and both clean-fleece weight and staple length are large enough to slow progress if selection is concurrently made for smaller fiber diameter (fineness) and higher clean-fleece weight and longer staple. Possibly important negative genetic relationships exist between staple length and body weight. Genetic antagonisms between staple length and weaning weight and between weaning weight and freedom from folds or wrinkles have been observed in Rambouillet, Columbia, and Targhee flocks.

It is believed that none of these relationships are intense enough to prevent progress. Rather, they will reduce progress as compared with what could be accomplished if each trait were being selected for separately.

Effectiveness of Selection. As with other classes of livestock, observation indicates that selection has been effective over the years in modifying type and productivity in sheep. Numerous examples can be cited to show the possibilities. The work of Neale[2] in New Mexico is unique in that he selected only for staple length—a very important trait in fine-wool sheep under New Mexico conditions—and made considerable progress without observing deterioration in other traits.

In Australia, improvements in fleece weight have been attained

[1] Rae, A. L., *New Zeal. Jour. Agr. Res.*, 1:104–123, 1958; and Morley, F. H. W., *Austral. Jour. Agr. Res.*, 6:77–90, 1955.
[2] Neale, P. E., *New Mex. Agr. Exp. Sta. Bul.* 334, 1948.

through selection, and indications are that improvement of 2 per cent per year is an attainable goal.[1]

Terrill, in a study of data from the Western Sheep Breeding Laboratory, compared the amount of progress actually made for several traits in Rambouillet sheep during the period from 1940 to 1948 with the genetic progress expected. Yearling fleece weight, weaning body weights, and weaning body scores for condition all trended slightly downward in average merit in spite of positive selection for improvement. Improvement was shown in staple length, type scores, face-covering scores, and neck-fold scores. An index based on the above items and thought to reflect all-round merit, was used as the basis for selecting breeding animals in this flock. The average index increased during the period of this study almost as much as would be expected from selection differentials and estimated heritability and it was concluded that average merit of the sheep in the flock had improved.

Breeding Systems for Seedstock Flocks

Individual performance and family averages. These are the basic items upon which selection in sheep must be based. This follows from the facts (1) that although some culling of breeding animals may be practiced on the basis of the first lamb crop or two, they must initially be selected for use in a flock upon the basis of their own and relatives' performance; and (2) that even if a progeny-testing program is followed with prospective stud rams before they are used in the main flock, only a small percentage of the most promising rams can be progeny-tested. These must be selected on the basis of individual and family performance.

Progeny testing. A program in which prospective stud rams are progeny-tested in an auxiliary flock prior to being used in the main flock can be useful under certain conditions. In one study[2] it was concluded that a program of progeny testing yearling rams and using the best ones in the main flock as two-year-olds or three-year-olds should increase progress by about 20 per cent and 5 per cent for weaning and yearling traits, respectively, with characters having heritabilities as low as 10 per cent. For characters with heritabilities of 30 per cent, progeny testing would be expected to increase progress slightly for traits which could be measured at weaning age but would be expected to decrease progress for those traits which could not be measured until yearling age. This

[1] Turner, H. N., *Proc. World Conf. Anim. Prod.*, 2:175–188, 1963, European Association for Animal Production, Rome, Italy.
[2] Dickerson, G. E., and Hazel, L. N., *Jour. Agr. Res.*, 69:459–476, 1944.

latter situation occurs because of the additional year which it takes to evaluate yearling characters of the progeny.

In Australian Merinos, it was concluded[1] that little gain was likely from a progeny-testing program for selecting rams as compared with optimum use of individual and family selection with heritabilities of 0.3. With heritabilities of 0.6, progeny testing might decrease progress. It was assumed that rams would be 18 months old when first used.

Progeny testing would have its greatest potential usefulness in selecting for carcass characters and for traits with low heritability. It will be most effective under conditions in which rams can be progeny-tested as lambs and the best-performing ones used in the main flock as yearlings.

Most of what has been said about mass selection, linebreeding, and inbreeding for seedstock herds of beef cattle will apply equally to sheep.

For all except very large flocks, breeding entirely from home-raised rams (i.e., maintaining a closed flock) will result in some degree of inbreeding. As with other classes of animals, most experimental results indicate deleterious effects of inbreeding.

Inbreeding. In extensive studies at the Western Sheep Breeding Laboratory,[2] rather general decreases were observed in both weaning and yearling weights of inbred Rambouillet, Columbia, and Targhee sheep. The decreases varied from group to group but averaged 0.34 lb for weanling lambs, 0.26 lb for yearling ewes, and 0.49 lb for yearling rams per 1 per cent inbreeding. Decreases in size of Australian Merinos have occurred with inbreeding.[3] For fleece characters, the Western Sheep Breeding Laboratory studies as well as studies in Australia on Merinos have shown a trend toward decreased staple length and fleece weight with increases in inbreeding. The decreases were large enough in some cases to be statistically significant, but of less practical importance than several other characters. The maximum figure reported indicated a decrease of 0.6 lb in grease-fleece weight per 10 per cent increase in inbreeding. Reduced fleece weight with inbreeding may be a consequence of reduced body size and not a specific effect of inbreeding on wool growth.

Inbreeding has rather severe effects on fertility and survival of lambs.[4] The trends are thus consistent with those of other species in showing greater reductions in less heritable traits.

Reductions in productivity with inbreeding are not so high as to pre-

[1] Morley, F. H. W., *Austral. Jour. Agr. Res.*, 3:409–418, 1952.
[2] See Terrill, C. E., et al., *Jour. Anim. Sci.*, 7:181–190, 1948, for one report and reference to others.
[3] Morley, F. H. W., *Austral. Jour. Agr. Res.*, 5:305–316, 1954.
[4] Doney, J. M., *Austral. Jour. Agr. Res.*, 8:299–311, 1957.

vent the development and maintenance of highly inbred lines. Further, in flocks large enough to use three or more rams per year, which would normally mean 5 to 8 per generation, mild linebreeding programs can be followed for many years with increases in inbreeding of only 2 to 4 per cent per generation. At the worst it would be many years before deleterious effects of inbreeding would be important. Except possibly for per cent lambs weaned of ewes bred, consistent selection could well overcome reductions in productivity entirely.

The principal reason for acquiring sires from outside a large flock is that if available animals are truly superior to those produced within the flock, more rapid progress can be made by going outside. If this is known to be the case, the practice should be followed. It is likely to be most effective for highly hereditary characters.

If, however, a flock is already well above average in genetic merit, advantages of using home-raised rams are

1. Better information about their productivity is likely to be available than on purchased animals.
2. It is known how home-raised animals react to a breeder's own environment.
3. Following a mild linebreeding program will tend to slightly increase the consistency of breeding performance of animals sold from the flock.

The formation of highly inbred lines for future use commercially in topcrosses or rotation crosses or for combination for breed improvement is being explored, but results accumulate slowly because of the slow rate at which inbreeding progresses in a species with fairly low reproductive rates. These techniques are still experimental, and further exploration should generally be left to experiment stations or private breeders with inquiring minds and adequate financial resources.

MAXIMIZING PRODUCTIVITY OF COMMERCIAL FLOCKS

The breed or breeds should be selected for any commercial operation using the best available information on suitability for the particular purpose or purposes and for adaptability to the environment. In a broad sense, genotype-environmental interactions are very important in sheep since some breeds are obviously unsuited to some environments. For example, the short-legged, compact, meaty Southdown with its light fleece production and poor flocking instinct is entirely unsuited to most range situations. Genotype-environmental interactions have not been studied intensively within breeds, but available studies suggest that they are relatively unimportant.[1] Thus, selection of foundation stock within a breed from the exact environment in which a commercial operation will be carried on should not be of overwhelming importance as compared to obtaining animals with high inherent productivity.

As a generalization, commercial producers should nearly always follow an outbreeding program to avoid deleterious effects of even low levels of inbreeding. This may or may not involve a cross-breeding program.

Under some rigorous range environments, found particularly in Australia and a few other countries and to a degree in the Southwestern United States, high-quality fine-wool can be produced whereas high-quality lamb and mutton production is impossible or nearly so. For these conditions, fine-wool breeds—Rambouillet in the United States and Merino in Australia—seem to have a combination of adaptability to the environment and ability to produce a quality product which has not yet been duplicated by combinations of other breeds in crosses. Thus, under these conditions, straight fine-wool breeding is recommended with as intense selection as possible for desired wool characters. Rapid visual appraisals giving a fairly accurate estimate of wool production potential have been developed which permit effective selection of replacement ewes. Ram selection should be based on more detailed fleece records. In some large commercial flocks, breeding rams are saved from the flock when available rams from seedstock herds do not have the qualities needed. No studies are known to have been carried out to determine whether or not appreciable heterosis in reproductive performance, survival ability of lambs, and longevity would be obtained in crosses between the different fine-wool breeds. Studies of this kind would appear to have a potential usefulness for production situations of this kind.

Crossbreeding. Crossbreeding has long had an important place in sheep production, especially that of fat lambs. The systems used in Great Britain are perhaps the most widely known. In England and Scotland considerable stress has been placed upon the development of strains adapted to particular areas. These adapted strains are used to furnish the basic stocks for commercial production but are usually used in various crossbred combinations. As an example of this scheme, consider the place of the Black-faced breed of sheep used on the rugged hill land of Scotland. This breed is noted for exceptional vigor and for ability to survive and reproduce under adverse conditions. However, even when

---

[1] Dunlop, A. A., *Austral. Jour. Agr. Res.*, 13:503–531, 1962; and *ibid.*, 14:690–730, 1963.

the environment is favorable and optimum nutrition is provided, carcass quality is mediocre. When ewes of this breed are mated with mutton-type rams, such as the Border Leicester, the resulting lambs, if properly fed, yield satisfactory carcasses. The male offspring of this cross are sold for mutton, and the females are taken to good pasture and bred to mutton-type rams of the Down breeds, with all the offspring going for slaughter.

Systems of sheep breeding similar to this also exist in Australia and New Zealand. Nichols[1] has termed this *stratification* of an industry and has pointed out that it may occur in time as well as space. Such systems develop when ewes suitable for commercial fat-lamb production in areas with high-quality pastures can be raised more cheaply in other areas than ewe replacements can be raised in the fat-lamb areas.

A system of this general nature long served as the basis for much of the sheep industry of the Pacific Northwest. Lincoln × Rambouillet crosses were made, and the ewe lambs were raised economically to yearling age on rather sparse ranges, usually in Montana or Idaho. These ewes were sold as yearlings to sheepmen in better range areas, principally Oregon and Washington, where they were bred to Hampshire or Suffolk rams and all the lambs were marketed. Such sheepmen bought all their replacement ewes. Their own operations were comparatively expensive, making it cheaper to buy replacements than to produce them.

Similarly, Western Ewes serve as the basis for many commercial flocks in the early-spring-lamb area of Kentucky, West Virginia, Virginia, and Tennessee. Western ewes ordinarily available include Black-faced North-westerns (usually Hampshire × Rambouillet crosses), Southwestern ewes (usually Corriedale × Rambouillet crosses but sometimes Suffolk × Rambouillet crosses), and Rambouillets from various Western areas.

Many tests have been conducted under Western range conditions to determine the value of rams of various breeds for improving size and carcass quality of lambs from Rambouillet ewe flocks. The data in Table 14-5, prepared from California results, are typical of the results of such crosses when lambs are raised under *good feed conditions.* Hampshire and Suffolk rams sired faster-gaining lambs than the Shrop-shires, Southdowns, or Rambouillets, whereas all the lambs sired by Down rams graded higher than those sired by Rambouillets, both on foot and in the carcass. Among the Down breeds, Southdown rams sired lambs that were outstanding in live and carcass grades.

Under sparse range conditions, experiments in Montana and New

[1] See Nichols, J. E., "Livestock Improvement in Relation to Heredity and Environment," Oliver & Boyd Ltd., Edinburg and London, 1945, for summary and references to earlier papers.

Table 14-5. *Growth and Quality of Lambs from Rambouillet Ewes by Sires of Various Breeds**

(Six-year average, 1928–1933)

Breed of sire	No. lambs	Av. final wt.	Live grade index†	Per cent lambs marketed‡	Carcass grade index
Hampshire	144	77	3.76	86	3.99
Suffolk	138	78	3.71	88	3.96
Shropshire	135	73	3.81	89	4.16
Southdown	141	70	4.01	93	4.34
Rambouillet	146	71	3.08	70	3.45

* Adapted from *Calif. Agr. Expt. Sta. Bul.* 598.
† The live grade index and carcass grade index are measures of quality determined by assigning an arbitrary value to each grade (Choice 5, Good 4, Medium 3, and Common 2) and calculating the average.
‡ Only the lambs grading Choice, Good, or Medium were marketed.

Mexico have shown no appreciable advantage in either weight or grade for Down × Rambouillet crossbred lambs as compared to straight Rambouillets. Under these conditions, the producer would incur all the disadvantages of crossbreeding with no increase in value of his product.

Under good range conditions, however, the practice of crossbreeding is to be encouraged if market demands are such as to make the practice a profitable one. This will sometimes be the case only when range conditions are good enough for the production of a high percentage of lambs fat enough for slaughter at weaning. This is not necessarily the case, however, since there is a good demand for black-faced crossbred lambs as feeders. Crossbred ewes are usually in demand for replacement ewes in areas such as those mentioned earlier, where conditions are good enough so that virtually all lambs can be sold for slaughter at weaning.

In areas with better-than-average feed conditions where crossbred ewes are in demand as replacements, sheepmen have a feeling that the crossbreds are better lamb producers when bred to Down rams than are straight Rambouillets. There is some experimental evidence to substantiate this.

In the Southern spring-lamb area, the early rapid-growing ewe lambs sell to advantage for slaughter. The sheepman generally has three possible sources of replacement ewes:

1. Retaining his top ewe lambs in spite of their high market value
2. Retaining the late or slow-growing and lower-valued cull ewe lambs from his own flock or purchasing such ewes from neighboring breeders
3. Purchasing replacement ewes from outside his area

Ewes available for purchase from outside the area are predominantly Western ewes whose breeding is Hampshire or Suffolk × Rambouillet (black-faced crossbreds), Corriedale or Columbia × Rambouillet (white-faced crossbreds), or grade Rambouillet. These are usually bred to Down-type rams. A number of experiments in this area[1] have shown that the Western crossbreds are approximately equal in productivity to selected grade or crossbred natives and are distinctly superior to native ewes developed from cull or late ewe lambs and to native ewes of non-descript breeding. Western crossbreds are usually preferred to grade Rambouillets in this area.

Only a few experiments have been reported on crosses between different mutton or Down breeds of sheep ordinarily used in farm sheep production. Several of these breeds are basically of the same type. In Minnesota tests,[2] purebred Shropshire, Columbia, and Hampshire ewes averaged 19 per cent greater weight of lamb weaned per 100 lb of ewe when producing crossbred lambs as compared to purebreds. Part of the advantage came from greater survival of the crossbred lambs and part from faster growth.

A U.S. Department of Agriculture study[3] with the Hampshire, Shropshire, Southdown, and Merino breeds and a Columbia-Southdale strain involving 4,331 lambs born over a 10-year period at Beltsville, Maryland, is the most comprehensive study known to have been reported. These studies included 2-breed, 3-breed, and 4-breed crosses. Most of the 3- and 4-breed crosses were produced by breeding purebred rams to 2-breed and 3-breed cross ewes.

Table 14-6, taken from the report, gives a summary of the results.

Average increases in per cent of lambs weaned of ewes bred were 2.1, 14.9, and 27.3 for 2-, 3-, and 4-breed crosses, respectively, over the comparable averages of the purebred parents. Average gains in weaning weight over the purebred averages were 5.2, 9.5, and 10.4 lb, respectively, for 2-, 3-, and 4-breed crosses. The gains from improved fertility, lamb survival, and weaning weight were additive, with 4-breed crosses producing about 26 lb or 57 per cent more pounds of lamb per ewe bred.

The above results represent one of the most striking examples of heterosis reported in farm animals. It should be emphasized that the results are of all purebreds vs. all crosses made (not all possible crosses were made). Some of the pure breeds were more productive than others. An examination of the detailed data given in the reports indicates that

[1] Kincaid, C. M., and Carter, R. C., *Virginia Agr. Expt. Sta. Bul.* 459, 1953.
[2] Miller, K. P., and Dailey, D. L., *Jour. Anim. Sci.,* **10**:462–468, 1951.
[3] Sidwell, G. M., et al., *Jour. Anim. Sci.,* **21**:875–879, 1962; and *ibid.,* **23**:105–110, 1964.

Table 14-6. Expressions of Hybrid Vigor Due to (1) Per Cent Lambs Weaned, (2) Weaning Weight Per Lamb, and (3) Their Combined Effects[a]

| Type of mating | Hybrid vigor from fertility, prolificacy, and livability as affecting lb lamb weaned per ewe bred | | | Hybrid vigor from weaning wt. as affecting lb lamb weaned per ewe bred | | | Hybrid vigor from combination of % lambs weaned and weaning wt. | |
	% lambs weaned of ewes bred[b]	Lb lambs weaned per ewe bred[c]	Increase, lb	Weaning wt. lb[d]	Lb lambs weaned per ewe bred[e]	Increase, lb	Lb lambs weaned per ewe bred[f]	Increase, lb
Purebred	89.5	45.8	....	51.2	45.8	...	45.8	....
2-breed cross	91.6	46.9	1.1	56.4	50.5	4.7	51.7	5.9
3-breed cross	104.4	53.5	6.6	60.7	54.3	3.8	63.4	11.7
4-breed cross	116.6	59.7	6.2	61.6	55.1	0.8	71.8	8.4
Total			13.9			9.3		26.0

[a] From Sidwell, George M., et al., *Jour. Anim. Sci.*, **23**:105–110, 1964.
[b] Deviations of crossbreds over the respective averages of their purebred parents added to base value for purebreds.
[c] The per cents in column 1 were multiplied by the average weaning weight of purebred lambs.
[d] Deviations of crossbreds over the respective averages of their purebred parents added to base value for purebreds.
[e] The weights in column 4 were multiplied by the per cent lambs weaned for purebred ewes.
[f] The weights in column 7 were obtained by multiplying the per cents in column 1 by the weights in column 4.

the average of *all* 4-breed crosses exceeded the *best* purebred by about 4 per cent in pounds of lamb weaned per ewe bred and that specific crosses exceeded the best purebred by considerably more than this.

When the comparisons are limited to crosses among the three Down breeds—Hampshire, Shropshire, and Southdown—results are somewhat less striking, leading to the conclusion that much of the heterosis was due to crosses between the genetically diverse fine- and medium-wool types. Even within the Down breeds, however, the average of all the possible 3-breed crosses exceeded the best purebred, and the best 3-breed cross exceeded the best purebred by 17 per cent in pounds of lamb weaned per ewe bred.

These results show the potential usefulness of crossbreeding in farm flocks and among breeds of somewhat similar medium-wool types. The study also emphasizes the need for better characterization of the various breeds for their potential usefulness in crossbreeding programs.

Farm flocks tend to be small, and where owners do not purchase replacements, maintenance of systematic crossbreeding systems is difficult. However, the evidence available strongly indicates the desirability of utilizing the heterotic responses of crossbreeding if feasible from the standpoint of flock management.

NEW BREEDS OF SHEEP

The sheep industry of the Western range area is built upon a foundation of fine-wool breeding. Fine-wools have the vigor, hardiness, and flocking instinct necessary for survival on the range, but they are often lacking in size and carcass quality. Long-wool and Down breeds of sheep supply the size and carcass quality lacking in the fine-wool, and for many years crossbreeding has been used extensively for market-lamb production.

In efforts to combine the good qualities of both types and eliminate the need for continual crossbreeding, various new breeds have been developed or are in the process of formation. Their rather great public acceptance thus far is an indication that the objectives have been at least partially realized.

Crosses made originally in 1912 between Lincoln rams and Rambouillet ewes produced foundation stock from which U.S. Department of Agriculture specialists, working mostly at the experiment station at Dubois, Idaho, developed the Columbia breed. The Targhee was also developed by the Department at Dubois. It is based on a ¾ Rambouillet, ¼ long-wool foundation. Both the Columbia and Targhee are now established breeds with their own registry associations.

The Panama has been developed privately by James Laidlow, of Idaho. It is based on stock produced by crossing Rambouillet rams on Lincoln ewes. The Corriedale, a New Zealand breed of sheep now widely raised in the United States, resulted from the use of Leicester and Lincoln rams on Merino ewes.

The Romnelet was developed by the Canadian Department of Agriculture starting in 1935 with crosses of Romney Marsh rams on Rambouillet ewes. In the early years of the new breed, a Targhee ram and six Romeldale rams from a strain developed by A. T. Spencer of Gerber, California, were used to a limited extent. Otherwise the flock was closed from the time *inter se* matings of the first crosses were made.

The selection methods used and trends in productivity during development of this breed have been studied,[1] and the results are of considerable interest both in regard to this breed and for their implications regarding new breed development. Birth weight, weaning weight, and 18-month weight of ewes, and yearling clean-fleece weight declined sharply from the first-cross generation to F_2. There was a further slight decline in birth weight and clean-fleece weight from F_2 to F_3. These early trends are indicative of the difficulty, and probably the impossibility, of maintaining average productivity with *inter se* mating in a population in which productivity depends in part on heterosis.

[1] Peters, H. F., et al., *Canad. Jour. Anim. Sci.*, **41**:102–108, 126–133, and 205–211, 1961.

Fig. 14-8. Representative animals of two new breeds of sheep developed from crossbred foundations. *Above,* the Columbia breed was developed from crosses of the Lincoln and Rambouillet breeds. *Below,* the Targhee descends from foundation stock with ¾ Rambouillet and ¼ long-wool breeding. (*Courtesy of Dr. C. E. Terrill, Animal Husbandry Research Division, U.S. Department of Agriculture.*)

No significant change in any of the traits occurred from F_3 to F_5. From F_5 to F_7 there was a slight further decline in birth weight, and a slight, nonsignificant increase in 18-month weight of ewes and in yearling clean-fleece weight. There was a substantial increase in weaning weight. However, no trait had regained the F_1 level after 20 years of selection.

There was a tendency for selection differentials for the above traits, and especially weaning weight, to increase in the later years of the study. This was possibly due to the fact that horns, wrinkling, excessive face covering, and objectionable body conformation and fleece characters had been largely eliminated through selection, thereby permitting greater emphasis on the measurable performance items discussed above.

The documentation of the formation of this breed permits the objective appraisal of many of the difficulties of breed formation—especially the loss of heterosis and the necessity of selecting for many characters, which automatically reduces the intensity of selection for any one. It is hoped the development of the breed can continue so that future success in improving productivity can be objectively assessed.

WESTERN SHEEP BREEDING LABORATORY

The largest project devoted to sheep-breeding research in the United States in the Western Sheep Breeding Laboratory established at Dubois, Idaho, in 1937. It has operated since that time under the guidance of Directors J. E. Nordby, C. E. Terrill, R. L. Blackwell, and D. A. Price. It is utilizing the facilities developed by the Bureau of Animal Industry Sheep Experiment Station, which was organized at Dubois in 1917, and in addition the experiment stations of the eleven Western states and Texas are cooperating.

The main objective of the program is the improvement of sheep through breeding. The improvement is measured by the application of utility standards as they concern commercial production under ranching conditions. Production is measured on the basis of wool yield, and in pounds of lamb of the desirable type and finish. Obviously, the staple length and the quality and uniformity of fibers, as well as the shrinkage, are important factors that are studied with the yield of wool. All useful type characteristics that influence lamb production, ranch adaptability, and general serviceability are closely integrated in the improvement program. Under ranching conditions at relatively high altitudes, where winter storms are frequently severe, the objectives in sheep improvement must also include a consideration of those characteristics in a fleece that protect the sheep against weather extremes. The fleece must be relatively long and lofty as well as dense and must hold together along the topline and cover well the underline.[1]

Attention in selection is being given to both meat and wool produc-

[1] From mimeographed statement on the Western Sheep Breeding Laboratory and the U.S. Sheep Experiment Station by Julius E. Nordby.

tion but with meat qualities emphasized somewhat more than those relating to wool.

Besides we are interested in the mature form of mutton sheep on the basis of its value for producing the right kind of market and feeder lambs, and replacement ewes. While obviously the mature size, form, and fleshing quality of a sheep are important, the ranchman is perhaps fully as concerned with weight, form, and fleshing qualities at the time of weaning the lamb, because it is at this time that weight and form, but principally fleshing, will determine if the lamb is a market lamb or a feeder lamb. The milking quality in ewes is observed just after lambing, and this is further checked in their ability to produce vigorous and fleshy lambs. In the selection program therefore, much emphasis is placed upon the finish in lambs at the time of weaning. Due consideration is given to twins in this connection. Stud ram prospects are scored carefully for market qualities at weaning, because a ram lamb that has the acceptable growth, ruggedness, form and finish at weaning time is probably more apt to be a potential sire of lambs of the same kind than is a more slowly maturing ram that does not develop these qualities before he is more mature. Vigor is an essential quality in the selection of ranch sheep. Some ewes, and rams also, "wear" uniformly well under heavy service until they are seven or even eight years of age. Others begin to decline early in life and are on the "cull" list at five or before.[1]

Selection is being practiced against too-heavy face covering that might lead to wool blindness and also against heavy skin folds.

Work is being done with the Rambouillet, Columbia, and Targhee breeds, with a few other breeds being used in certain studies and in test crosses. Inbreeding is being done with about 30 Rambouillet lines, and with 10 to 20 lines of each of the other two breeds which are in process of formation. Non-inbred control stocks in which the same type of selection is practiced as in the inbreds are also being carried to see whether the greatest progress can be made with or without inbreeding. In addition, stocks of each breed are maintained without selection other than that which occurs naturally. To maintain these stocks, breeding animals are picked entirely at random and mated in a random manner. Stocks such as these are valuable for estimating genetic parameters (heritabilities and genetic correlations) and for estimating effects of environmental trends on performance. Knowledge of the effects of environmental trends permits more accurate determinations of what fractions of observed changes in selected groups are due to selection and truly represent genetic improvement.

[1] From mimeographed statement on the Western Sheep Breeding Laboratory and the U.S. Sheep Experiment Station by Julius E. Nordby.

Many publications on factors associated with productivity in sheep, methods of evaluating breeding animals, heritability of various traits, and other things about which information is needed in properly carrying out a research program in sheep breeding have already come from the Laboratory. Many of these have been referred to in this chapter.

Since 1946 the Laboratory has been selling surplus rams at auction, with performance records being available to buyers. Analysis of sale records[1] indicates that buyers attach real importance to production traits such as body weight, fleece weight, and mutton conformation in determining the price they will pay for a given ram.

SUMMARY

Sheep breeds and types vary tremendously in size, conformation, fleece characters, and behavior patterns. This complicates sheep improvement but also provides intriguing opportunities. Sheep breeding has been plagued by many fads. Wool blindness and extreme degrees of wrinkling are two of these for which research has shown the production limitations. Fleece characters are medium to high in heritability, and wool length, fineness, and total yield all respond readily to selection. There is a genetic relationship between fiber length and fiber diameter which slows progress in selecting concurrently for both longer fiber length and smaller fiber diameter (fineness). Growth rate is moderately heritable. Conformation and condition scores of weaning lambs are low in heritability as is fertility. Inbreeding depresses the expression of most economically important traits. Crossbreeding to combine characters of different breeds is useful under range conditions and appears to have considerable potential for increasing efficiency of farm sheep production.

REFERENCES

Fraser, A. S., and Short, B. F. 1960. The Biology of the Fleece, *Anim. Res. Lab. Tech. Paper No. 3, Commonwealth Scientific and Industrial Research Organization, Australia.*

Rae, A. L. 1956. The Genetics of the Sheep, *Advances in Genet.*, 8:190–266.

Terrill, C. E. 1958. Fifty Years of Progress in Sheep Breeding, *Jour. Anim. Sci.*, 17:944–959.

Turner, H. Newton. 1963. Influence of Breeding on Production Efficiency (Sheep), *Proc. World Conf. Anim. Prod.*, 2:175–188. (Published by European Association for Animal Production, Rome, Italy.)

[1] Terrill, C. E., *Jour. Anim. Sci.*, 12:419–430, 1953.

CHAPTER FIFTEEN

Retrospect and Prospect

From the modest beginnings of a colonial subsistence agriculture, America's livestock industry has become a veritable giant. The colonists and the animals which they brought with them from Europe and the British Isles represented an unusually meager beginning. Distinct breeds were not available in their homeland, although there undoubtedly were many local varieties available. A number of breeding artisians began to express themselves during the 1700s. Bakewell's methods and results emerged during the last half of that century in Britain. However, from all that can be verified, no constructive leadership for animal improvement was active in America. The colonists were busy carving homes and farms from the wilderness, and a subsistence agriculture dominated the era.

After independence had been won energies could be directed to more specialized undertakings. The influx of immigrants during this period brought new ideas and concepts which could be tested and assimilated into the growing social and economic patterns.

During the late 1700s and early 1800s, stockmen in the United States began giving attention to quality of their livestock and many local races or types were developed. Some of these went on to attain breed status and are still important. The American breeds of swine and several breeds of light horses such as the Morgan and the American Saddle arose in this fashion. With cattle, sheep, draft horses, and some types of light horses, breeds became well-established in Europe before the local types had reached breed status in the United States. As a result they were imported extensively into the United States and displaced the local races.

445

PUREBRED ERA

The Coates herdbook for recording pedigrees of Shorthorns was initiated in England in 1822. Although individual breeders had kept pedigrees prior to this, its founding may be said to have marked the beginning of the purebred era. This pedigree approach to breeding gained many followers in Europe and became popular in the United States at a rather rapid pace.

Numerous European breed importations, especially of sheep, had been made early in the nineteenth century. Importations following the Civil War were much more numerous. Breed societies were developed in the United States both for European breeds and for those American types which gained breed status. Usually these societies recorded ancestry and limited registration in each breed to descendants of imported animals or those identified as foundation animals of the American breeds. Thus, with the closed herdbook all purebred animals had to be offspring of individuals already recorded in the herdbook or of imported animals with recorded ancestry in the British Isles or Europe. The purebreds gained much popularity, as economic and social patterns, during this period, led to the beginning of specialization in livestock production. Standards for perfection were developed by many breed associations. During the late 1800s, some emphasis on recording of performance in dairy animals began.

Purebreds, especially in the hands of interested and specialized husbandmen, appeared superior to the common stock. Rapid expansion of their numbers brought the livestock industry to the point where purebred sires were available for extensive use to grade up common stock. Breed associations were enthusiastic promoters, and their members believed wholeheartedly in the excellence of their stock. During this early purebred era, reliable performance records were not available; hence, there was no objective basis for judging the merit of the animals. In retrospect, the purebreds appear to have made an exceptional contribution. However, over-zealousness in the sale of purebreds too frequently led to the use of animals that had little or nothing to contribute to the commercial stocks.

The purebred boom continued until the early twenties, when the economic depression had a major impact on agriculture. Even today most seedstock herds represent one of the recognized pure breeds, and pedigree systems of breeding, not greatly different from those of the pioneer breeders, are followed. Thus, the purebred era continues. However, emphasis is changing. In the early days it was largely on the expansion in numbers of purebreds. Now it is increasingly on improvement of purebreds, so that they can contribute to the improvement of commercial stock. The need for objective approaches to this task has only gradually been recognized and accepted as a challenge by purebred breeders. Yet with attention to many fancy points it has been difficult for the purebreds to maintain true genetic superiority over grade and commercial stocks. This has been particularly true since the advent of commercial artificial insemination, especially in dairy cattle, in which the gap between the best breeder or seedstock herds and the leading commercial herds has continued to narrow.

REQUIREMENTS FOR IMPROVEMENT

The general requisites for the genetic improvement of our livestock continue to be the same. First, we must assess what we have genetically in our present animals. This requires accurate records of performance on a large number of animals of known ancestry. Second, we must discover how we can increase the number of offspring from those individuals which have the desirable genes at the expense of the individuals with the less desirable genes.

Performance Recording. Tremendous strides have been made in unraveling the heretofore mysterious ways of hereditary. Today the nature of the gene is on the threshold of description. Despite the many possibilities which may be on the horizon, livestock improvement must for the forseeable future begin first with accurate records of performance on a large number of animals, so that their breeding worth can be ascertained. Records of parentage and performance are absolute necessities.

A nationwide system of record keeping for dairy cattle has been in progress since 1906. This program has served as an effective base to provide information for the improvement programs in dairy cattle. Currently about 3 million dairy animals are tested in standard recording schemes. Despite the fact that this is only a small percentage of the total lactating animals, the numbers continue to increase in proportion to the total cattle population. It appears that the increase in the percentage of animals recorded will continue at a slow but steady rate. Much of this is necessary for commercial herds, from an economic standpoint to direct culling for maintaining a high level of production. A valuable byproduct of this testing has been the information for the progeny testing of males in artificial insemination.

Performance-testing programs for beef cattle are growing rapidly. From its beginning in 1953, it has grown to over 500,000 cows in organized performance testing programs in 1964. Most of these programs are based on measurement of the weaning weight and grade of calves. Provisions are also being made for testing postweaning gain in many states. Actually, testing of males offers the greatest promise for genetic improvement. Improvement of rate of gain in beef cattle should be much easier to accomplish than the improvement of milk yield in dairy cattle. A high selection intensity is justified among beef bulls on the basis of their own performance. Genetic improvement in weaning weights may be more difficult to attain.

Records are needed in our swine enterprises. Lack of any uniform recording system is a major deterrent to the location of superior breeding stock. In some states boar testing and performance testing at organized stations is serving to provide breeding stock for a limited segment of the population. Yet in practically all areas an expansion of this activity is needed before sufficient stock from improved sources would be available for commercial use.

Extending the Use of Outstanding Animals.   As the testing and recording programs develop and mature so that truly superior individuals can be identified, the mediums to extend the usefulness of these individuals in seedstock and commercial herds must be found. Artificial insemination can be most helpful in this respect for beef and dairy cattle.[1] Storage of semen in liquid nitrogen can make the best sires available almost anywhere in the world. Presently, over 40 per cent of the total dairy-cattle population in the United States is inseminated artificially and in some countries in excess of 95 per cent of the cattle population is bred in this manner. Increases in the percentage of the total population bred artificially will continue. Economic pressures for efficient production will mean that marginal producers, now using backward managerial procedures and inferior purebred or nondescript scrub bulls, will be forced out of business. More highly specialized managemental programs and managers will allow a continuing slow but steady increase in the percentage of cattle bred artificially. Larger individual herds but fewer herds will also further this trend.

There are many deterrents to the general use of artificial insemination in beef cattle. Nonetheless, the trend toward acceptance in commercial herds is increasing. Presently the percentage of beef cows inseminated is slightly higher than it was for dairy cattle as recently as 1945. Testing programs are now in operation and the quality of sires available could be increased measurably during the next few years. Close attention to management conveniences during the breeding season also must be provided if the artificial insemination programs are to grow in commercial

[1] Recently a dairy sire was credited with 50,690 services in one year.

herds. Limitations are being placed on artificial insemination by many beef breed associations. Yet, it would appear that the demonstrated merit of the technique and is potential for improving genetic merit eventually will result in its widespread use. If this occurs, both the beef and dairy seedstock industries may be faced with the necessity of taking positive steps to maintain broad genetic bases in individual breeds. Further, steps may be necessary to maintain several breeds of each class of livestock. Efficient commercial crossbreeding programs require high quality sires from several breeds.

There is no doubt that the increase in artificial insemination will produce changes in the breeding structure of cattle populations. Purebred breeders have been against such programs because of the greatly reduced market for purebred bulls. Currently, this is a critical point in the beef industry. Although attempts to maintain the status quo may be well founded in the short run, eventually top commercial herds will become genetically equal or superior to the average purebreed herd in economically important traits. Many commercial herds are demanding performance-tested sires. If the purebred industry will not provide them, progressive individuals will turn to artificial insemination.

In dairy-cattle breeding, a distinctive trend towards programs of progeny testing young sires by a limited number of matings in many herds is already evident. The number of artificial-breeding organizations is rapidly decreasing with mergers and consolidations. It would appear that only a few large organizations will continue to service the nation as a whole. These studs have shown that they can handle the proving of males much more accurately than any individual breeder can ever hope to. Nevertheless the females, which are to be the dams of pedigree-selected bulls for progeny testing, will have to come from the clientele which are using the studs. The important contributors to breed improvement in the future will not necessarily be the large independent purebred breeder, but the individual breeder cooperating with his artificial-breeding stud.

Sheep and swine producers have not yet been able to utilize programs of artificial insemination on an extensive practical basis in the United States. Technical problems still prevent the use of frozen semen with these species. Continuing efforts to solve these technical handicaps should bear fruit before too many years. Obviously, artificial insemination alone will not solve all swine-breeding problems. With natural service, more intensity of selection of sires is possible in swine than in the case with cattle, although artificial insemination would permit a further enhancement of selection intensity. A primary concern is to provide a sufficient number of tested males of high merit that are worthy of being used widely. A major advance in production-testing programs in swine must be forthcoming before artificial insemination can be used responsibly on an extensive scale.

GOALS

In developing future goals, animal breeders must continue to seek out the potential requirements of the consuming public. Insensitivity to consumer needs will be tolerated less and less as competition from substitutes for animal products become keener and keener. The expanding United States population, with predictions of 245 million by 1980 and 330 million by 2000,[1] suggest an increase in the demand of food products. If per capita incomes continue to rise, the demand for animal products should continue to be strong. Trends in Europe where incomes have risen point out the strong demand for animal protein.

If our future citizens continue to require annually 90 lb of beef, 60 lb of pork, and the equivalent of 600 lb of milk per capita, by 1980 a 25 per cent increase in production will be necessary. The major portion of the additional meat needs from beef and swine would be expected to come from increased numbers of animals. Efficiencies in production per head would make a measurable but less dramatic contribution to the meat supply. Current dairy-cattle production trends suggest that most of the increased milk needs could be obtained from increasing the production per animal. Secondary increases would come from increasing animal numbers, but present milk-production levels per cow cannot be maintained unless heavy concentrate feeding is continued.

The surplus of cereal grains we have enjoyed for the past two decades has meant that we have been selecting animals under conditions in which the nutritional regimes have included liberal feeding of concentrates. If our future livestock must serve only as harvesters of roughage from nontillable grass lands or as gleaners of by-products which are not satisfactory for human consumption, serious difficulties may result. Would the animals which have been selected under conditions of liberal concentrate feeding respond so poorly under heavy roughage feeding that we would lose much of their productive efficiency? Attention is being given to this point, and in most areas of the world, livestock must continue to perform on roughage and by-product rations. In limited experimentation with dairy cattle and beef cattle, it would appear that the animals which do best on high roughage rations are also those that appear to have the best genotype for utilizing a ration consisting of

[1] Lundsburg, Hans H., "Natural Resources for U.S. Growth: A Look Ahead to the Year 2000," The Johns Hopkins Press, Baltimore, 1964.

a high proportion of concentrate. It is true that the levels of performance vary greatly depending upon whether additional concentrated energy is available or whether roughage is the main source of nutrients. However, there is relatively little evidence to suggest that these interactions are of practical importance. Large quantities of cereal grains will not always be available for livestock production. Perhaps the change will be reasonably gradual so that available genetic variability for utilization of these feedstuffs may be captured in the selection process during the nutritional transition. Actually the entire evolution of the bovine has been based primarily on roughage feeding. The artificial situation of high concentrate feeding is of recent origin. Furthermore, ruminants appear to have a wide range of adaptability to different rations without a great sacrifice in efficiency of utilization.

In shaping our goals in animal breeding, what emphasis in the future must be given to quality? The livestock industry must produce quality products to stimulate consumption. On the other hand, markets must recompense the producer for the extra quality in order to justify the special effort in breeding and production costs. The contrast of protein vs. fat in meat and even in milk is now in open discussion. Increased use of processed and ground meats rather than the complete cuts tends to obscure the marbling and other special features in much of the meat that is marketed. Homogenization of milk destroyed the creamline many years ago, although fat is still in surplus. Both lard and milk fat have been given stiff competition during the past two decades by vegetable oils and fats. There is little evidence that competitive pressures will diminish. Protein from animal sources continues to be in demand when incomes are strong. Substitute products should not have a major depressing effect on the animal protein demand by the 1980s. But as a new generation of consumers emerges, the stigma against substitute products may become less and less.

Breeding Plans and Programs. Since the 1930s considerable advances and sophistication in the breeding plans for animal improvement have emerged. Much of the early work of Fisher and Wright has been translated for usage with populations of farm animals. Considerable effort has been expended in research areas to determine the genetic variability in numerous economic traits through heritability analyses. Undoubtedly there are justifications for continuing certain aspects of these types of studies. They must be more intensive and thorough. We have reached the point where the validity of these estimates needs to be tested experimentally. Much can be gained from studies with laboratory mammals, and more use of such experiments will be forthcoming. Laboratory studies have suggested that for growth rate selection can be effective for many generations. Nevertheless there are many questions peculiar to dairy cattle, beef cattle, sheep, or swine which must be studied with these classes of livestock. Increasing pressures for use of funds make it difficult to maintain the necessary research herds and flocks for these operations. These trends must be revised.

Coupled with information that must be gained from experimental herds, progressive breeders undoubtedly will be willing to cooperate to an increasing degree in some research endeavors. With accurate records to guide the analyses of field results, researchers should be able to detect trends in the larger commercial population to evaluate the effects of recommended breeding practices. This will not alleviate the need for experimental herds, but it will aid in the assessment of recommended practices.

The possibility that we may be reaching genetic limits in our farm populations during the next twenty years is rather remote. Nevertheless we should be on the lookout for signs of diminishing rates of genetic improvement. Along with this, we will need to find more concrete evidence about genetic correlations, especially those negative ones which influence our economic traits. The concept of the selection index has provided a logical framework for planning selection programs, but we still find only limited use of these in practice.

Previous mention was made of the potential importance of genotype-environmental interactions. This topic is receiving considerable emphasis experimentally, and a broader picture of its importance should soon emerge. Although there is no doubt that genotype-environmental interactions can be detected, there is little evidence that they are of major concern under practical operating conditions.

The exploitation of hybrid vigor will surely be of more concern in the years ahead. In most of our species we should first capitalize on the available additive genetic variance by the most effective selection programs we can devise. Nevertheless, in commercial herds some advantage can be gained at all stages of the selection program through the use of crossbreeding. In swine as well as cattle the exploitation of hybrid vigor during the next twenty years will largely result from the use of crossbreeding rather than from the development of inbred lines for crosses. Maternal performance is an important consideration which may justify production of specific two-breed crosses for commercial beef and swine production.

Reproductive fitness rapidly becomes a limiting factor when inbreeding is undertaken. This is particularly true in cattle. If the potentials for genetic gain and economic returns are rewarding enough, we usually can find ways of circumventing limitations. If we reach the stage where inbreeding is the proper alternative for continued improvement and the above conditions are met, the use of embryo transfer in cattle may be a real help in overcoming the fertility barrier. This might permit developing an effective linecrossing program. Embryo transfer also can extend the usefulness of outstanding females, but the advantage would not approach that available from artificial insemination in the male.

As physiological techniques are improved and further understanding of embryo culture is gained, further exploitation of inbred lines may be possible. Perhaps an unlimited number of offspring of the most favorable F_1 combination might be obtained if individual cells could be taken from a culture of the early developing embryo. We would then find ourselves in the position of the corn breeders, trying to locate lines for even more superior crosses.

As our techniques become more refined, additional information about the physiology of the various traits should assist us in our attempts at genetic improvement. The real question is the level at which our inquiry should be directed. It has been proposed that assays of those hormone and enzyme systems which influence the expression of a trait should allow for effective selection. Some attempts in this area have been undertaken. The hormone levels concerned with growth, for example, will be difficult to determine. In addition, we will expect the activity of the various glands to vary according to age, general health, environmental stress, and stage of development of the individuals. Environmental as well as inherent factors will surely influence the level of these hormones. Just when, at what age and under what circumstances, can meaningful assay results be expected?

With a process as complicated as growth or lactation, the interaction of an inumerable complex of hormones and enzymes as well as energy sources must be anticipated. When the number of enzymes involved approaches the number of loci affecting the trait, we can see the improbability of being able to put the individual genes in their proper perspective by the individual-gene approach. Presently we have difficulty putting the individual traits together, even with the mathematical logic of the selection index.

Attempts to guide genetic change at the chromosome and gene level in the past have given results that generally have failed to match expectations. Mutagenic agents have appeared to produce random changes in the genome. Even in plant species, where apparently useful artificial mutants have been incorporated into the genome, extensive screening and discarding of untold numbers of undesirable mutants has been necessary. Ultraviolet and low levels of radiation have recently been employed to produce or release genetic variability that could be seized upon by selection. Results to demonstrate the success of these ventures are as yet inconclusive.

A deep and abiding conviction exists among researchers that "genetic engineering" may be more directive in the future.[1] DNA introductions to implement the formation of gene replacements are freely suggested. Once the enzymatic role in crossing over is more fully understood, it may be possible to increase the recombination of DNA fragments which include desired genes. Viewpoints are also expressed that DNA fractionation and identification will obviate the need for the laborious breeding tests now required for gene identification. Knowledge to control the turning on and off of genetic function and replication may seem to be many years away.

More intensive inquiry will be needed in the future to shore up the logic of our breeding plans. Instrumentation and computers will play an increasing role in breeding research and improvement in commercial herds. These as well as many unforeseen developments will be available to assist us in tackling practical problems we formerly were unable to consider. With all of the available gadgetry, there will be an increasing temptation to substitute the massive approach in lieu of the critical approach. Our problems will be revealed as more complex and difficult than they were conceived previously, but massive busy work will not serve as effective substitutes for inquisitiveness, imagination, and critical interpretation.

[1] Hotchkiss, R. D., *Jour. Hered.*, **56**:197–202, 1965.

Purebred Livestock Registry and

Performance Registry Associations

The following listing was compiled from lists courteously furnished by the Animal Husbandry Research Division, U.S. Department of Agriculture, and by Mr. Allan C. Atlason, Secretary, National Society of Live Stock Record Associations, 3964 Grand Avenue, Gurnee, Illinois.

BEEF AND DUAL-PURPOSE CATTLE

American Angus Association, Glen Bratcher, Secretary, 3201 Frederick Blvd., St. Joseph, Missouri 64500

American Belted Galloway Cattle Breeders Association, Charles C. Wells, Secretary, South Fork Rural Station, West Plains, Missouri

American Brahman Breeders Association, Harry P. Gayden, Exec. Secretary, 4815 Gulf Freeway, Houston, Texas 77023

American Devon Cattle Club, Inc., Kenneth Hinshaw, Secretary, Agawam, Massachusetts 01001

American Dexter Cattle Association, Mrs. Daisy Moore, Secretary, 707 W. Water Street, Decorah, Iowa 52101

American Galloway Breeders Association, Roy S. McDonald, Exec. Secretary, Box 1011, Denver, Colorado 80201

American Hereford Association, Paul Swaffar, Secretary, 715 Hereford Drive, Kansas City, Missouri 64105

American-International Charolais Association, J. Scott Henderson, Exec. Secretary, 923 Lincoln Liberty Life Bldg., Houston, Texas 77002

American Milking Shorthorn Society, Ray R. Schooley, Exec. Secretary, 313 S. Glenstone Avenue, Springfield, Missouri 65804

American Polled Hereford Association, Orville K. Sweet, Exec. Secretary, 4700 E. 63rd Street, Kansas City, Missouri 64130

American Polled Shorthorn Society, Steve Treadway, Secretary, 8288 Hascall, Omaha, Nebraska 68124

American Red Brangus Association, Mike Levi, Secretary, 614 Colorado, Austin, Texas 78700

American Scotch Highland Breeders' Association, Mrs. Margaret Manke, Secretary, Edgemont, South Dakota 57735

American Shorthorn Association, C. D. Swaffar, Secretary, 8288 Hascall, Omaha, Nebraska 68124

Beefmaster Breeders Universal, 348E Gunter Hotel, San Antonio, Texas 78206

International Brangus Breeders Association, Roy Lilley, Exec. Secretary, 908 Livestock Exchange Bldg., Kansas City, Missouri 64102

Pan American Zebu Association, Roy G. Martin, Secretary, P. O. Box 268, Cotulla, Texas 78041

Performance Registry International, Glenn Butts, Secretary, 1140 Delaware, Denver, Colorado 80204

Red Angus Association of America, Mrs. Sybil Parker, Exec. Secretary, Box 391, Ballinger, Texas 76821

Red Poll Cattle Club of America, Wendell H. Severin, Secretary, 3275 Holdrege Street, Lincoln, Nebraska 68503

DAIRY CATTLE

American Dairy Cattle Club, Robert W. Hitchcock, Secretary, Interlaken, New York 14847

American Guernsey Cattle Club, Francis Chapman, Secretary-Treasurer, 70 Main Street, Peterborough, New Hampshire 03458

American Jersey Cattle Club, James F. Cavanaugh, Exec. Secretary, 1521 East Broad Street, Columbus, Ohio 43205

American Red Danish Cattle Association, Marlette, Michigan 48453

Ayrshire Breeders' Association, David Gibson, Jr., Secretary-Treasurer, Brandon, Vermont 05733

Brown Swiss Cattle Breeders' Association, Marvin L. Kruse, Secretary-Treasurer, 800 Pleasant Street, Beloit, Wisconsin 53511

Dutch Belted Cattle Association of America, James H. Hendrie, Secretary, 6000 NW 32nd Ave., Miami, Florida 33142

Holstein-Friesian Association of America, Robert H. Rumler, Exec. Secretary, South Main Street, Brattleboro, Vermont 05301

Purebred Dairy Cattle Club, Karl Musser, Exec. Secretary, 70 Main Street, Columbus, Ohio 43215

DRAFT HORSES

American Suffolk-Shire Horse Association, Edwin R. Henken, Secretary, 300 E. Grover Street, Lynden, Washington   98264

Belgian Draft Horse Corporation of America, Miss Blanche A. Schmalzried, Secretary, P. O. Box 335, Wabash, Indiana   46992

Clydesdale Breeders Association of the U.S., Charles W. Willhoit, Secretary, Batavia, Iowa   52533

Percheron Horse Association of America, Mrs. Anne Brown, Secretary, Rt. 1, Box 118, Fair Oaks, Indiana   47943

LIGHT HORSES[1]

American Quarter Horse Association, Howard K. Linger, Exec. Secretary, P. O. Box 200, 2736 West Tenth, Amarillo, Texas   79105

American Hackney Horse Society, Mrs. J. Macy Willets, Secretary-Treasurer, 527 Madison Avenue, Room 725, New York, New York   10021

American Saddle Horse Breeders Association, Charles J. Cronon, Jr., Secretary, 929 S. 4th Street, Louisville, Kentucky   40203

Appaloosa Horse Club, Inc., George B. Hatley, Exec. Secretary, Box 640, Moscow, Idaho   83843

Arabian Horse Club Registry of America, Inc., Nellie D. Bayley, Asst. Secretary-Treasurer, 332 South Michigan Avenue, Chicago, Illinois   60604

Cleveland Bay Society of America, A. Mackay Smit, Secretary, White Post, Virginia   22663

International Arabian Horse Association, Ralph E. Goodall, Jr., Exec. Secretary, 224 East Olive Avenue, Burbank, California   91503

Jockey Club (Thoroughbred Registry), John F. Kennedy, Exec. Secretary, 300 Park Avenue, New York, New York   10022

Morgan Horse Club, Inc., Seth B. Holcombe, Secretary, P. O. Box 2157, Bishop's Corner Branch, West Hartford, Connecticut   06117

Palomino Horse Breeders of America, Inc., Melba Lee Spivey, Secretary, P. O. Box 249, Mineral Wells, Texas   76067

Pinto Horse Association of America, Inc., Aaron G. Olmsted, Chairman, Somers Road, Ellington, Connecticut   06029

United States Trotting Association, Edward F. Hackett, Secretary, 750 Michigan Avenue, Columbus, Ohio   43215

[1] Partial list. A more complete list is available from the Animal Husbandry Research Division, U.S. Department of Agriculture, Beltsville, Maryland.

PONIES

American Shetland Pony Club, William R. Burns, Exec. Secretary-Treasurer, P. O. Box 1250, Highway 52 North, Lafayette, Indiana   47902

Icelandic Pony Club and Registry, Inc., Mrs. Judith Hassed, Secretary, 56 Ales Acres, Greeley, Colorado   80631

Pony of the Americas Club, Inc., L. L. Boomhower, Secretary, 31 First Street, Mason City, Iowa   50401

Welsh Pony Society of America, Inc., Mrs. Sidney S. Swett, Secretary-Treasurer, Unionville, Pennsylvania   19375

JACKS AND JENNETS

Standard Jack and Jennet Registry of America, Mrs. F. Gerard Johns, Secretary, Rt. 7, Todds Road, Lexington, Kentucky   40502

SHEEP

American Cheviot Sheep Society, Inc., Stan Gates, Secretary, 5051 Flourtown Road, Lafayette Hill, Pennsylvania   19444

American Corriedale Association, Inc., Rollo E. Singleton, Secretary, 108 Parkhill Avenue, Columbia, Missouri   65201

American Cotswold Record Association, C. P. Harding, Secretary, Siegel, Illinois   62462

American and Delaine-Merino Record Association, Charles M. Swart, Secretary, 4000 Water Street, Wheeling, West Virginia   26003

American Hampshire Sheep Association, Roy A. Gilman, Secretary, Stuart, Iowa   50250

American Oxford Down Record Association, C. E. Puffenberger, Secretary, Eaton Rapids, Michigan   48827

American Panama Registry Association, W. G. Priest, Secretary, Jerome, Idaho   83338

American Rambouillet Sheep Breeders' Association, Mrs. A. D. Harvey, Secretary, 2709 Sherwood Way, San Angelo, Texas   76901

American Shropshire Registry Association, Mrs. Jessie F. Ritenour, Box 678, Lafayette, Indiana   47901

American Southdown Breeders' Association, W. L. Henning, Secretary, 212 South Allen Street, State College, Pennsylvania   16801

American Suffolk Sheep Society, Allan Jenkins, Secretary, Union Stock Yard, Ogden, Utah   84400

Black-Top Delaine-Merino Sheep Breeders' Association, Emerson Richards, Secretary, Rt. 4, Howell, Michigan 48843

Columbia Sheep Breeders' Association of America, Mrs. R. Lorraine Vigen, Secretary, Box 802, Fort Collins, Colorado 80521

Continental Dorset Club, J. R. Henderson, Secretary, Hickory, Pennsylvania 15340

Debouillet Sheep Breeders Association, Mrs. A. D. Jones, Secretary, 300 South Kentucky Avenue, Roswell, New Mexico 88201

Karakul Fur Sheep Registry, Mrs. Annette S. Harris, Secretary, Fabius, New York 13063

National Lincoln Sheep Breeders' Association, Ralph O. Shaffer, Secretary, Rt. 1, West Milton, Ohio 45383

Montadale Sheep Breeders' Association, E. H. Mattingly, Secretary, 61 Angelica Street, St. Louis, Missouri 63100

National Suffolk Sheep Association, Mrs. Betty Biellier, Secretary, Box 324, Columbia, Missouri 65201

National Tunis Sheep Registry, Inc., Mrs. Eloise S. Spraker, Secretary, 106 Liberty St., Bath, New York 14810

Romeldale Sheep Breeders' Association, A. T. Spencer, Secretary, Sacramento Drive, Wilton, California 95693

Texas Delaine-Merino Record Association, Mrs. G. A. Glimp, Secretary, Rt. 1, Burnet, Texas 78611

United Karakul Registry, Mrs. Olive May Cook, Secretary, 408 E. Maple Street, Glendale, California 91200

U.S. Targhee Sheep Association, Gene Coombs, Secretary, Box 2513, Billings, Montana 59101

GOATS

American Goat Society, Inc., Carl W. Romer, Secretary, 7900 E. 66th Street, Kansas City, Missouri 64100

American Milk Goat Record Association, Donald Wilson, Box 186, Spindale, North Carolina 28160

American Angora Goat Breeders' Association, Mrs. Thomas L. Taylor, Secretary, Box 409, Rocksprings, Texas 78880

SWINE

American Berkshire Association, Gene Mason, Secretary, 601 W. Monroe Street, Springfield, Illinois 62701

American Yorkshire Club, Wilbur L. Plager, Secretary, 1001 South Street, Lafayette, Indiana 47901

Chester White Swine Record Association, J. Marvin Garner, Secretary, 116 E. 8th Street, Rochester, Indiana 46975

Hampshire Swine Registry, Harold Boucher, Secretary, 1111 Main Street, Peoria, Illinois

Inbred Livestock Registry Association, Helen Mathis, Secretary, Main Street, Box 312, Augusta, Illinois 62311

American Landrace Association, Ray Delaney, Secretary, 313 South Glenstone Avenue, Springfield, Missouri 65802

National Hereford Hog Record Association, Miss Sylvia Schulte, Secretary, Norway, Iowa 52318

O.I.C. Swine Breeders, Inc., Tho. R. Hendricks, Secretary, Greencastle, Indiana 46135

Poland China Record Association, C. W. Mitchell, Secretary, 501 East Losey Street, Galesburg, Illinois 61401

Spotted Swine Record, Gerald Nevins, Secretary, Bainbridge, Indiana 46105

Tamworth Swine Association, Erwin R. Mahrenholz, Secretary, R. R. 1, Box 88, Evansville, Indiana 47712

United Duroc Swine Registry, Bruce Henderson, Secretary, Duroc Bldg., Peoria, Illinois 61602